Government Policy
and Program Impacts
on Technology Development,
Transfer and Commercialization:
International Perspectives

Government Policy and Program Impacts on Technology Development, Transfer and Commercialization: International Perspectives has been co-published simultaneously as *Journal of Nonprofit & Public Sector Marketing*, Volume 13, Numbers 1/2 2005.

Government Policy and Program Impacts on Technology Development, Transfer and Commercialization: International Perspectives

Kimball P. Marshall, PhD
William S. Piper, DBA
Walter W. Wymer, Jr., DBA
Editors

Government Policy and Program Impacts on Technology Development, Transfer and Commercialization: International Perspectives has been co-published simultaneously as *Journal of Nonprofit & Public Sector Marketing*, Volume 13, Numbers 1/2 2005.

Routledge
Taylor & Francis Group
New York   London

First published by

Best Business Books®, 10 Alice Street, Binghamton, NY 13904-1580 USA

This edition published 2013 by Routledge

Routledge
Taylor & Francis Group
711 Third Avenue
New York, NY 10017

Routledge
Taylor & Francis Group
2 Park Square, Milton Park
Abingdon, Oxon OX14 4RN

● ● ●● ●●●● ●●● ●●● ●● ● ●●● ● ●●● ●● ●●●● ●●● ● ●●● ●●● ●● ●●● ●●●● ●● ●●● ● ●●● ●●● ● ●●● ●●●● ● ●● ● ● ●●● ● ●● ● ● ●●●

Government Policy and Program Impacts on Technology Development, Transfer and Commercialization: International Perspectives has been co-published simultaneously as *Journal of Nonprofit & Public Sector Marketing*, Volume 13, Numbers 1/2 2005.

The development, preparation, and publication of this work has been undertaken with great care. However, the publisher, employees, editors, and agents of The Haworth Press and all imprints of The Haworth Press, Inc., including The Haworth Medical Press® and Pharmaceutical Products Press®, are not responsible for any errors contained herein or for consequences that may ensue from use of materials or information contained in this work. Opinions expressed by the author(s) are not necessarily those of The Haworth Press, Inc.

Cover design by Lora Wiggins

Library of Congress Cataloging-in-Publication Data

Government policy and program impacts on technology development, transfer and commercialization : international perspectives / Kimball P. Marshall, William S. Piper, Walter W. Wymer, Jr., editors.
   p. cm.
   "Co-published simultaneously as Journal of nonprofit & public sector marketing, volume 13, numbers 1/2, 2005."
   Includes bibliographical references and index.
   ISBN-13: 978-0-7890-2605-7 (hard cover : alk. paper)
   ISBN-10: 0-7890-2605-8 (hard cover : alk. paper)
   ISBN-13: 978-0-7890-2606-4 (soft cover : alk. paper)
   ISBN-10: 0-7890-2606-6 (soft cover : alk. paper)
   1. Technology transfer–Government policy. 2. Technological innovations–Government policy. I. Marshall, Kimball P. II. Piper, William Sanford III. Wymer, Walter W. IV. Journal of nonprofit & public sector marketing.

HC79.T4G673 2005
338'.064–dc22

                                                                            2004016379

Dedicated
to John V. Gully,
a pioneer in technology marketing.

Government Policy and Program Impacts on Technology Development, Transfer and Commercialization: International Perspectives

CONTENTS

ABOUT THE EDITORS

Kimball P. Marshall, PhD, is Professor of Marketing at the School of Business, Alcorn State University (Lorman and Natchez, MS, USA). He has published professional research articles in such journals as *Journal of Public Policy and Marketing*, *Journal of Business Ethics*, *Journal of Marketing Education*, *Journal of Nonprofit & Public Sector Marketing*, *Journal of Global Business*, and the *International Business Schools Computing Quarterly* among others. Dr. Marshall serves on the editorial boards of *Health Marketing Quarterly* and *Journal of Nonprofit & Public Sector Marketing*. His practitioner and consulting technology experiences include new product development, product management, pricing policies, and public policy technology programs. He has twice served as a NASA ASEE Fellow at the Stennis Space Center where he studied government technology commercialization programs. Dr. Marshall is a member of the Board of Directors and a Past-President of the Atlantic Marketing Association, and a continuing member of the American Marketing Association, the Strategic Marketing Association, and the Federation of Business Disciplines.

William S. Piper, DBA, is Associate Professor of Marketing and Associate Dean of the School of Business at Alcorn State University (Lorman and Natchez, MS, USA). He has presented numerous conference papers on commercial impacts of government technology programs and has published on this topic in the *Journal of Nonprofit & Public Sector Marketing*, *Simulation and Gaming*, and the *International Journal of Technology Management*. Dr. Piper has a Masters Certificate as a focus group director from the Greeley Institute. He is a member of the American Marketing Association, the Atlantic Marketing Association and the Society of Business, Industry and Economics. As a NASA Fellow and Senior Consultant on technology transfer and commercialization with the NASA Stennis Space Center's Remote Sensing Program, Dr. Piper gained valuable insight regarding government technology programs and policies.

Walter W. Wymer, Jr., DBA, is Associate Professor of Marketing at Christopher Newport University in Newport News, Virginia. Dr. Wymer's research has focused on marketing in nonprofit organizations. His work has been published in several academic journals and presented at numerous academic conferences. Dr. Wymer is the editor of the *Journal of Nonprofit & Public Sector Marketing*. He also serves on the editorial boards of the *International Journal of Nonprofit & Voluntary Sector Marketing*, *Health Marketing Quarterly*, and the *Journal of Ministry Marketing & Management*.

Introduction

It is with pride that we present to you this special volume devoted to the topic of government program and policy impacts on technology development, transfer and commercialization. As evidenced by the papers herein, this field encompasses a wide range of macromarketing issues. Accordingly, the articles presented here are diverse, addressing government sponsorship of technology research, impacts of government regulation on technology marketing, economic development, effects of government policies on business practices, implications of government infrastructure development programs on technology diffusion, protection of intellectual property, and societal marketing and ethical issues of dominant social paradigms and distributive justice in cross-national and cross-cultural perspective. Even with so broad a range of topics represented, we can only hope to have provided a sampler of this important macromarketing field. We do, however, hope that the articles we have selected (all of which have survived a rigorous, triple-blind, peer review process) will stimulate further consideration of the implications of government policies for technology development, transfer and commercialization. Hopefully, the perspectives contained here will lead to further research to guide government policymakers and business leaders in efforts to provide fertile environments for innovations that will improve the quality of life of all peoples.

We, the editors of this special edition, express our appreciation to our colleagues who served as reviewers for this special edition. The reviewers often

---

[Haworth co-indexing entry note]: "Introduction." Marshall, Kimball P., William S. Piper, and Walter W. Wymer, Jr. Co-published simultaneously in *Journal of Nonprofit & Public Sector Marketing* (Best Business Books, an imprint of The Haworth Press, Inc.) Vol. 13, No. 1/2, 2005, pp. 1-2; and: *Government Policy and Program Impacts on Technology Development, Transfer and Commercialization: International Perspectives* (ed: Kimball P. Marshall, William S. Piper, and Walter W. Wymer, Jr.) Best Business Books, an imprint of The Haworth Press, Inc., 2005, pp. 1-2. Single or multiple copies of this article are available for a fee from The Haworth Document Delivery Service [1-800-HAWORTH, 9:00 a.m. - 5:00 p.m. (EST). E-mail address: getinfo@haworthpressinc.com].

Available online at http://www.haworthpress.com/web/JNPSM
Digital Object Identifier: 10.1300/J054v13n01_01

provided extensive comments and provocative observations that enhanced the papers included here. We also express our gratitude to Loyola University New Orleans, Alcorn State University and Christopher Newport University for the support that made this special issue possible.

Kimball P. Marshall
William S. Piper
Walter W. Wymer, Jr.

An Overview
of Potential Government Impacts
on Technology Transfer and Commercialization

Kimball P. Marshall

SUMMARY. Government technology transfer is movement of technology across political boundaries, across social sectors, or from one society to another. Technology commercialization involves applying technology in usable products for economic gain. In "free-market" societies one thinks of technology as developed by the private sector, but governments influence development, transfer and commercialization via sponsorship of, and direct involvement in, research activities, and via programs that affect market demand and transfers across political and social boundaries. This paper seeks to stimulate research into the macromarketing implications of government technology programs by developing taxonomies of objectives and actions and by reviewing these types to suggest macromarketing implications and empirical research needs. *[Article copies available for a fee from The Haworth Document Delivery Service: 1-800-HAWORTH.*

Kimball P. Marshall, PhD, is Professor of Marketing, School of Business, Alcorn State University, 1000 ASU Drive, #90, Alcorn State, MS 39096-7500 (E-mail: kimball.p.marshall@netzero.net). He has served as a NASA ASEE Fellow and a consultant to Environmental Protection Agency and Small Business Administration SBIR programs. His research and teaching include broad areas of technology commercialization, new product development, and not-for-profit marketing.

[Haworth co-indexing entry note]: "An Overview of Potential Government Impacts on Technology Transfer and Commercialization." Marshall, Kimball P. Co-published simultaneously in *Journal of Non-profit & Public Sector Marketing* (Best Business Books, an imprint of The Haworth Press, Inc.) Vol. 13, No. 1/2, 2005, pp. 3-34; and: *Government Policy and Program Impacts on Technology Development, Transfer and Commercialization: International Perspectives* (ed: Kimball P. Marshall, William S. Piper, and Walter W. Wymer, Jr.) Best Business Books, an imprint of The Haworth Press, Inc., 2005, pp. 3-34. Single or multiple copies of this article are available for a fee from The Haworth Document Delivery Service [1-800-HAWORTH, 9:00 a.m. - 5:00 p.m. (EST). E-mail address: getinfo@haworthpressinc.com].

Digital Object Identifier: 10.1300/J054v13n01_02

KEYWORDS. Government technology programs, regulation, technology transfer, technology commercialization, macromarketing

We, the People of the United States, in order to form a more perfect union, establish justice, insure domestic tranquility, provide for the common defense, promote the general welfare, and secure the blessings of liberty to ourselves and our posterity, do ordain and establish this Constitution for the United States of America. (Preamble to the Constitution of the United States of America)

INTRODUCTION

For centuries, governments have engaged in the development of new technologies for reasons ranging from desire for military superiority to concerns for public health and the quality of life of citizens. Governments directly employ scientists and engineers to develop inventions, improve upon new technologies or carry out "basic" research. Governments also fund private sector enterprises via grants or contracts to carry out similar activities. Often the resulting technological advances have private sector commercial potential and the potential to enhance the quality of life of all members of the society.

In some cases government-funded technologies have had dramatic commercial effects as private sector industrial and consumer products (Port 2002a,b; Franza and Widmann 1996, Mraz 1996, Piper and Naghshpur 1996, Zeller 1996, *Aviation Week and Space Technology* 1995, 1994, Radosevich 1995, Scott 1995, Kandebo 1994, Proctor 1994, United States Government Accounting Office 1992). Recent examples include radar, microwave ovens, modern cell phone systems, and the Internet. Other examples include the development of new materials for commercial aircraft and personal vehicles, and medical instrumentation. Therefore, it is not surprising that governments have recognized and sought to exploit the military, economic, and political potentials of technology development, transfer and commercialization. In the United States, the result has been the development of myriad programs at the national and local levels.

Government technology development transfer and commercialization programs may be of special interest to academicians and marketing practitioners for at least two reasons. From a "macromarketing" (Fisk 2001, 1999, 1982) perspective, government technology programs and policies affect influence commercial opportunities by influencing which technologies are emphasized, by stimulating markets for products through government purchases, and by constraining or facilitating intellectual property rights. Unfortunately, marketers have not systematically addressed government technology programs, their design or their marketplace impact.

Macromarketing studies of government technology programs might offer two benefits. First, studies could provide new insight into marketplace impacts of government involvement in technology. Second, macromarketing insights might lead to more effective business partner recruitment, as required by many government programs (Piper and Marshall 2000). Program success often requires that private firms act as business partners with government agencies. However, government programs often struggle to recruit effective commercialization partners despite widespread promotion efforts (Small Business Innovation Research Center 1997; Roberts and Malaone 1996; Unisphere 1996; Zeller 1996). A macromarketing perspective might yield more effective policies and procedures for attracting committed commercialization partners.

This paper develops an exploratory framework for classifying government technology objectives and actions. Macromarketing implications of government involvement in technology are considered throughout the discussion, and research questions are identified that, if addressed, could provide policy guidance. No assumption is made that government technology programs are "good" or "bad," or efficient or inefficient vehicles for bringing the benefits of technology to the people of the world. The diverse literature on the benefits of free markets for innovation (Smith 1981, Novak 1997) is recognized. The present goal is only taxonomy. An easily understood, parsimonious framework that could be applied cross-nationally is sought. As the effort is exploratory, exhaustive, mutually exclusive categories are not required. Some readers may wish to break the categories developed here into sub-types, to add categories, or to collapse categories together. Such criticisms are welcomed as important efforts in developing a paradigm to guide research into policies to bring the benefits of technologies to all peoples and societies.

A MACROMARKETING PERSPECTIVE

"Macromarketing" will refer herein to the study and development of social structures, institutions and processes that influence exchange among businesses and consumers in a society (Fisk 2001, 1982). We offer this definition as consistent with Hunt's definition of macromarketing as cited by Meade and Nason (1991) "... the study of (1) marketing systems, (2) the impact and consequence of marketing systems on society, and (3) the impact and consequences of society on marketing systems" (Hunt 1977, p. 6). Economic systems in general, and marketing exchange systems in particular, derive from the institutional structures and cultures of social systems, within nations and cross-nationally. Institutional structures include government policies and programs and are of concern from a macromarketing perspective in that actions of governments influence opportunities and conditions under which exchange occurs. Macromarketing elements develop from both government actions and private-sector initiatives, but, at some point, government involvement is to be expected in economic transactions, at least in the form of regulation of existing institutions and processes, or the stimulation of new institutions or processes. For example, early real estate transactions might have operated with a "handshake" agreement, but in short order government involvement came about to validate deeds, tax property, and restrict land usage. As transactions became more complex, involving additional parties, mortgages, regulated interest rates, and contract enforcement mechanisms guaranteed by courts, social structures and processes developed to create and enforce regulations and procedures. These institutions are infrastructure elements that influence market transactions. Where some people see government processes, taxation and regulation as encroachments on liberty and free markets, others see as benefits that contracts are enforceable in court and land values are protected. Presumably, these benefits allow "free" markets to operate with greater trust, stability and efficiency (Smith 1981). But, this is itself an empirical question to be asked of any government action. Even well-intended government actions might have unintended, undesirable consequences. Government abuses of regulatory power can occur, unjustifiable taxes may be imposed, unfair restrictions might be placed on property, and government buying power international trade regulations might distort free market efficiencies and diminish competition and competitive innovation. In international perspective, government trade treaties and policies might open doors to free trade, or create market imbalances and inefficiencies (Tobin 1991).

Government programs and policies manifest in law, regulation and sponsorship activities (West Publishing Company 1991) are important macromarketing infrastructure components that influence technology markets. These components influence which technologies will be developed, opportunities to stimulate and fill market demand for a new technology, the social impacts of new technologies, and opportunities for and constraints upon marketplace exchanges involving new technologies at the business and consumer levels. It is important for marketing academicians and government policy makers to have a paradigm to guide planning and assessment of programs and policies that influence technology development, transfer, and commercialization. A step toward such a paradigm may begin with consideration of two dimensions of government programs, program objectives and program activities. As a starting point, we define below typologies for each dimension and consider possible market impacts.

OBJECTIVES OF GOVERNMENT TECHNOLOGY PROGRAMS

While a variety of government objectives for involvement in technology programs may be suggested, in the interest of parsimony and a systematic approach we propose nine broad categories of objectives of government involvement in technology programs. These are listed in Table 1. Four categories stand alone: military defense, public safety, improvements in quality of life, and national diplomacy. Five are sub-categories of economic development: economic protectionism, jobs creation, new business formation, export stimulus, and foreign business investment.

Military Defense

Although the proper roles of government are debated, there is consensus for military defense and public safety as legitimate government functions (Smith 1981). All societies must have the ability to protect their geographical boundaries from encroachment by outsiders, secessionist efforts, and in some cases, expansionist efforts by non-government groups. Military defense serves as one important aspect of boundary maintenance. In the modern world, boundary maintenance has expanded to encompass "national security," and military

TABLE 1. Objectives of Government Technology Programs

Military Defense
Public Safety
Quality of Life
Economic Development
Economic Protectionism
Jobs Creation
New Business Formation
Export Stimulation
Attraction of Foreign Firms
International Diplomacy

systems are often used outside of physical borders to protect what are deemed by government leaders to be the "strategic interests" of the country. Ostensibly, violation of a country's strategic interests if unchecked would eventually endanger the recognized physical boundaries of a nation.

Today, with nuclear weapons, intercontinental missiles, instantaneous worldwide communications, and a global economy of multi-national mutual dependence, almost all developed and developing countries are involved in military technology. Military technology programs are carried out in government laboratories and by external organizations under contract and grants. Beyond debates over the morality of war and "banned" weapons, little debate exists regarding the legitimacy of government involvement in technology development for military defense. Debate does occur over the extent to which governments should develop technology in-house or use outside firms, whether commercial market technologies should be sought first, ownership of developed technologies, under what arrangements military technologies should be transferred to private firms for commercialization (Wessner 2003). Such debates involve two fundamental macromarketing questions. First, are there military risks associated with private sector involvement in the development or commercialization of the technologies in question? Second, does government involvement in military technology development enhance or impede free-market processes in regard to technology development, transfer and commercialization?

The first question would seem to be easy to resolve. If national security is at risk, the technology should not be transferred. Still, determinations are not easily made, and government agencies may take cautious positions constraining transfer of technologies. The second question is more difficult. Given the size of defense budgets in some developed nations such as the United States, government involvement in technology development might direct interest in emerging technologies in ways that are not marketplace efficient. Where large defense budgets and grants are available, disproportionate resources can be directed to technologies with little commercial benefit outside of future sales to governments. This problem has confronted the Small Business Innovative Research Program (SBIR) sponsored by the Federal government of the United States even though Federal law mandated that government funded technologies meet agency needs and have private sector commercial potential (Federal Laboratory Consortium for Technology Transfer 1991; Wessner 1999). Thus, key macromarketing issues worthy of future research include:

- Does military spending on technology development lead to the inefficient use of social resources for technology development?

- Under what circumstances, if any, do military involvement in technology transfer impede the transfer and commercialization of technologies for non-military applications?
- Can policies and programs be developed that provide for military technology needs while allowing "free-market" forces to optimally prioritize technology development activities in a modern society?
- Do government expenditures for military technology development programs encourage development of the most appropriate technologies for agency needs, are do such programs inefficiently focus research on agency favored approaches?

In varying degree and with minor editing, the same questions might be asked in regard to government expenditures by any agency whether the agency is concerned with military defense, public safety or the enhancement of quality of life of citizens.

Public Safety

Despite consensus that a key function of government is public safety, debate quickly develops over the meaning of the term and the limits that should be imposed on government involvement (Becker 2002). Public safety includes such diverse areas as police work, security monitoring, guns and explosives safety, transportation safety, and food and drug safety. As with the military, public safety agencies often engage in programs to develop technologies for tools to increase effectiveness and efficiency, but public safety agencies worldwide also address regulations regarding the use of technologies. An example is the ongoing controversies over food irradiation in Japan and in the European Union (Pickett and Suzuki 2000, Emrig 2002).

Budgets of public safety agencies are typically smaller than military budgets, but public safety budgets are substantial and raise similar questions. *To what extent do government technology programs devoted to public safety divert social resources toward agency favored technologies and away from the development of technologies that might be more favored by the commercial marketplace and that might be superior even for agency applications?* Regarding commercialization, it might be expected that public safety agency technologies would have private sector applications and be well received by commercial markets. However, the extent to which technologies developed for a specific agency purpose, even in the area of public safety, are transferable to private sector applications, or to uses by other government agencies, remains an empirical question. Is it realistic to assume that technologies developed for a specific government function would be readily adaptable to commercial purposes that were not anticipated in the development process (Eldred and McGrath 1997a, b)? As with military programs, an important macromarketing question is whether government sponsorship better directs technology development to meet public and government needs as compared to free market forces directing technology development.

The objective of protecting public safety is also applicable to regulatory activities. Debates regarding product regulation abound (Becker 2002, Pickett and Suzuki 2000). Automobile and food and drug safety, are common examples. Even where technology is developed in the private sector, regulatory mandates may restrict or even mandate product designs. One example is Food and Drug Administration monographs that specify acceptable formulations for over the counter (OTC) medications. These FDA regulations emphasize two concerns, physical safety and product efficacy (CDER 1999). However, as Becker argues (2002), market forces in the private sector and civil tort actions drive companies toward safe products, and market forces themselves lead to the failure of products that do not perform as expected. In Becker's view, product safety is relatively easy to determine, but current FDA efficacy demonstration requirements for drugs not covered by current monographs are so expensive that they may prevent the rapid transfer of new technologies into affordable commercial products. Recent changes in FDA processes have begun to address this issue, but the controversy persists (Carey 2003, Weintraub 2003). Thus, another area for macromarketing research would be the risk of latent dysfunctions of government product regulations in the interest of public safety.

Quality of Life

Quality of life and public safety are closely related, but, as used here, quality of life refers to broad improvements in living conditions that go beyond public safety. Quality of life indicators often include such issues as public health, medical care, adequate housing, and a general sense of well being. While such concerns are "justified" under the "promote the general welfare" clause of the United States Constitution (Cheeseman 1997), debates continue as to whether enhancing quality of life is a legitimate role of government. Still, the many government agencies engage in R&D in search of healthy food products, safer automobiles and highways, safer commercial and personal aircraft, safer guns, fuel-efficient homes and automobiles, medical devices, pollution avoidance and correction technologies, and so forth.

Quality of life technology topics reflect the broad scope of agency mandates. Legislation specifies topics, but agencies also set agendas that might be only broadly indicated in enabling legislation. This raises several macromarketing research questions. Do government or market forces more efficiently direct technology development and commercialization to enhance quality of life? Under what circumstances are government technology expenditures to enhance quality of life a necessary aspect of promoting the general welfare and efficient mechanisms for achieving desired benefits? Are there situations in which private sector initiatives are not viable and the social need is so great as to justify government action? What is the risk that government programs might promote one technology in contrast to what would occur in free markets? If quality of life were judged from the perspectives of individuals in society, would commercial markets or government programs more closely reflect those perspectives? The work of Fisk (1998) is an example of the application of macromarketing perspectives to ecological and physical environment quality of life issues by describing and proposing public policies based on "mutual coercion" and "reward and reinforcement" strategies.

Economic Development

Economic development is a major objective of government technology programs. Technology development, transfer and commercialization have been linked by policy makers and researchers to national and firm competitiveness, jobs creation, and the revitalization of economically disadvantaged geographic areas and peoples (Peterson and Sharp 2002, Steensma, Marino and Weaver 2002, Barnholt 1997, Zadoks 1997, Pegels and Thirumurthy 1996, NASA 1996a, b, c, Marshall 1995, Proctor 1994, McKee and Biswas 1993, Mullen 1993, Morbey and Reithner 1990, Brenner and Rushton 1989, Franko 1989, Ayal, Peer and Zif 1987, Goeke 1987, Nason, Nikhilesh and McLeavey 1987, Cundif 1982). However, from a macromarketing perspective, does government involvement risk creating market inefficiencies? Do government technology programs achieve economic development objectives? What latent dysfunctions may result? To provide a framework for considering such questions, Table 1 suggests five broad economic development objectives of government technology programs: economic protectionism, jobs creation, new business formation, export stimulation, and attraction of foreign firms.

Economic Protectionism

As used here, economic protectionism refers to creating trade barriers to buffer domestic firms from non-domestic competitive pressures. Tariffs, import quotas and domestic content requirements are examples of barriers. Critics claim protectionism impinges on free markets by creating artificial conditions that affect costs, prices and supplies in order to limit competition and favor domestic firms. Proponents argue that protectionist policies level the competitive field vis-à-vis firms that benefit from anti-free market policies of other countries. Arguments favoring government economic protectionism are that such policies defend property rights (as in the case of international product counterfeiting), prevent unfair competition (as in the case of anti-dumping laws), and provide time for a domestic industry to adapt to new technologies and gain sufficient strength to compete or adapt to new conditions. Such policies have been used in Japan (Goto 2000, Yu 2000, Odagri and Goto 1999), India (Kumar and Jain 2003), and the European Union (Peterson and Sharp 2002). A classic example is the United States case of Harley Davidson Motorcycles during the Reagan Administration the early 1980s. Besieged by the success of Japanese motorcycles imported into the United States and threatened with bankruptcy, Harley Davidson obtained government protection in the form of import quotas and tariffs, and domestic origin requirements for police motorcycles purchased with Federal funds, while restructuring product lines and production processes and facilities. Although today Harley Davidson Motorcycles is a highly successful company, the benefits and risks of economic protectionist policies as opposed to free-market processes call for further research.

In India, until recently, non-domestic firms seeking to open facilities were required to establish domestic companies with over fifty-percent Indian ownership. After years of relative economic isolation, China today encourages foreign trade and investment, but requires that this be tied to significant technology transfers. The objective is to assure that access to Chinese markets and labor supplies are tied to the development of technology infrastructure needed if domestic industries are to compete beyond cheap labor (Einhorn et al., 2002). EU policies also include protectionist objectives. To escape substantial tariffs, products sold within EU countries must have fifty-one percent EU produced components or labor. It is clear that protectionism is common public policy throughout the world. From a macro-marketing perspective, empirical research is needed to assess the impact of economic protectionist policies on trade opportunities, and the development of competitive strengths by domestic firms. While protectionist policies might provide time for domestic industries and firms to adapt, there is the risk that firms will instead become dependent on economic isolation and captive domestic consumers.

Jobs Creation

Jobs creation is a key economic objective of government technology policies and programs. Theoretically, by creating opportunities for new and improved products, government programs stimulate market demand for new products and enhance competitive advantage of domestic firms and products, and thus stimulating job growth. Consider the following statement from the 1998 Technology Transfer Report of NASA's Goddard Space Flight Center.

> The products and processes developed by transferring technology to U. S. industry increase the nation's competitiveness, create jobs, improve the balance of trade, and enrich the lives of its citizens. (NASA 1999, p. 1)

Despite such statements, little evidence exists to demonstrate real net increases in job opportunities from transferred technologies in developed countries. Logically, net increases in jobs would result from new technologies only if highly demanded new products are created, and if these new products do not lead to the obsolescence and loss of old jobs, or lead to new jobs that more than replace eliminated old jobs. One must distinguish between new job creation and enhancement of current jobs. The author recalls interviews carried out for a NASA satellite imagery program in which executives and workers in traditional aerial photography firms were interviewed regarding the potential benefits of the new technology. An often-mentioned concern was that the government was "competing" with current firms and that this would lead to the loss of business and jobs. Therefore, empirical research is needed to demonstrate the true effect of government technology programs on labor markets overall.

New Business Development

New business development, stimulating formation of new competitive firms, is a common goal of government technology programs that also requires empirical study. New technologies might create business opportunities leading to new business start-ups. This was the case with the Internet, aeronautics, and transportation, microwaves, and biotechnologies (Weintraub 2003). But new business start-ups often fail in risky, new technology situations. Government technology programs might cause new R&D and consulting businesses to emerge to service government grants and contracts, but there is a lack of evidence (Scott 1994, United States General Accounting Office 1992) that government technology programs generate new businesses in great number, or that such businesses survive over many years as going concerns independent of government funding. More likely, but also speculative, the market potential of new technologies may be exploited by larger existing firms that integrate government-sponsored technologies into their own R&D agendas and product lines. Therefore, research needed to assess the new potential for new business start-ups, and the macromarketing environments and government policies under which viable firms are most likely to develop. If government programs encourage development of new firms in technology areas that are not market driven, such firms are likely to fail without continuing subsidies.

Export Stimulation

Export stimulation refers to government activities to increase foreign demand for domestic products. When domestic firms in one nation have technology capabilities that provide superior products over firms of other nations, government efforts (through trade missions, databases to identify foreign business opportunities, and international agreements to reduce protectionist trade restrictions) would benefit domestic firms (Magnusson 2003) and foreign trade partners. However, export stimulation might also take other forms, such as tying foreign aid programs to purchases of products produced by the aid-providing country. Such activities impinge on free market activities and create artificial demand for the favored firms' products. Moreover, if free market forces are not allowed to select the most competitive opportunities for the countries to which aid is targeted, the expenditures of aid monies might inefficient and ineffective in achieving humanitarian or diplomatic goals. This raises empirical questions regarding commercial effects of programs to stimulate technology exports. For future research, it might be hypothesized that the greatest benefits occur when export stimulation programs seek to open free markets, and that market inefficiencies result where export stimulation programs impinge on free market competition by providing favored status to certain firms.

Foreign Business Investment

Foreign business investment, also referred to as foreign direct investment, or FDI, is a common goal of government technology programs. In the United States, Federal, state and local governments actively seek to recruit firms from outside their geographic territories to locate facilities in their own regions, as do governments in many other nations. China (Zang and Taylor 2001), India (Sikka 1996), and Mexico (Smith 2003), for example, have active programs to attract foreign business investment in the form of production facilities and other forms of technology transfers. The intention is to create jobs and stimulate wealth enhancing economic activities. By recruiting technology firms, government agencies hope to generate high paying jobs that will benefit tax revenues and the quality of life of peoples in the targeted region. Examples of FDI success stories in the United States include efforts to recruit vehicle factories in Alabama (Mercedes), Mississippi (Nissan) and South Carolina (BMW).

One danger of efforts to recruit outside firms is that inducements to encourage firms to locate in a nation or region often focus on tax exemptions and physical infrastructure, without addressing the broader needs of targeted industries. Geographic areas that fail to develop advantages matched to needs of targeted industries, advantages beyond low taxes, labor and construction costs (McKee and Biswas 1993), may find that firms leave in search of infrastructure benefits such as educational and research systems and collaborating businesses that are critical to technology activities (Geisler 2003, Jones and Teegen 2003, Yu 2000, Marshall 1995, McKee and Biswas 1993). To develop such advantages, governments, particularly at the local and state level in the United States, have adopted the "industry cluster" model (see for example recent Louisiana economic development through technology plans–Louisiana Economic Development 2003, Louisiana Economic Development Council 2003, Louisiana Business Inc. 2002). The cluster model recognizes the interdependence and common needs of firms in an industry, and develops programs that meet these needs in areas of physical infrastructure, education and human capital development, cultural adaptations, and business collaboration (Geisler 2003, Marino et al. 2002, Steensma et al. 2000, Tornatzky et al. 1996, Marshall 1995, Nason, Dholakia and McLeavey 1987, Goeke 1987). Cluster programs seek to attract technology firms, to stimulate new business formations, and to create better paying, more secure jobs and a better quality of life for people in the region. Austin, Texas is an example of a geographic area that has successfully emphasized technology clusters and education and business networks in economic development marketing (Engelking 1992).

International Diplomacy

The last objective for Government Technology Programs in Table 1 is "international diplomacy." To develop international strategic alliances or to encourage other countries to adopt desired policies, governments might encourage (or discourage) technology transfers through approving (or disapproving) authorizations for trade, joint research programs, or sharing military technologies. Diplomatic objectives may be to attain desired resources, to influence internal political processes, strengthen a friendly government, or gain or enhance a military alliance (Magnnuson 2003). Through technology transfers the target country gains knowledge to create or enhance domestic industries. However, while both countries can benefit, there seldom are guarantees that the technologies transferred will be used only as intended, and the transferring country or its government might feel political repercussions. The

1990s controversy regarding missile technology transferred from the United States to China is an example of a program in which diplomatic objectives created a political backlash. Although the United States government denied a problem, political opponents expressed fears that technologies transferred for civilian economic programs had military applications that could be threatening to the United States (Pomper 1998). Diplomatic policies that affect trade represent a dimension of macromarketing infrastructure and are a worthy, but underdeveloped, area of macromarketing research.

GOVERNMENT TECHNOLOGY PROGRAMS BY ACTIVITY TYPES

Throughout the previous discussion of government technology program objectives examples have been cited of technology related actions taken by governments. In the following sections, technology related actions are presented in a more systematic manner. A typology is presented of actions that are often used in government technology programs. As before, the objective is not to create an exhaustive or mutually exclusive taxonomy. The intention is only to begin the process of organizing a complex field so as to facilitate empirical research. To this end, Table 2 cites six major types of actions that are often used in government technology programs. Each of these is then classified into sub-categories and briefly discussed.

Intellectual Property Protection Laws and Enforcement

The main types of actions are typically used to protect intellectual property: patents, copyrights, trademarks, and service marks. In all three cases, only persons or organizations that have received permission from the holder of the protection may use the resources, and then only in the manner for which permission was given. Patent rights may take several forms, such as a product patent, a process patent or a design patent, and may be granted as a general patent or may be limited to a specific application. In the United States the developer of a technology must formally apply for a patent and demonstrate its utility before protection is granted. Copyright protection may extend to written materials, computer software, and works of art. Copyright protection is granted when the work is first developed even if the developer does not formally register the material. However, without formal registration, developers may find it difficult to demonstrate primacy. Trademarks are spoken and graphically styled words, phrases or images that are used to uniquely identify a product or firm. Trademarks do not have to be registered, although service marks (trademarks associated with service businesses) do have to be regis-

TABLE 2. Types of Government Technology Programs Activity Area

Intellectual Property Protection
    Patents
    Trademarks
    Copyrights

Direct and Indirect Technology Development
    Government Labs
    College and University Grants and Contracts
    Grants and Contracts to Private Sector Firms

Direct Commercial Technology Transfers
    Licensing
        Exclusive
        Non-Exclusive
    Assignment of Patents and Copyrights

Infrastructure Development
    Physical Infrastructure Incentives
    Social Infrastructure Initiatives
        Education Program Development
            Grants
            Contracts
        Set Aside Programs
    Import Quotas
    Foreign Content Restrictions
    Import Tariffs
    Duty Free Zones

Market Stimulation Activities
    Government Purchasing Contracts
    Government Trade Missions
    Free-Trade Agreements
    Regulatory Mandates for Technology and Design Standards
    Public Education Social Marketing Programs

Demarketing Programs
    Banned Products
    Domestic Taxes
    Regulatory Restrictions
    Export Restrictions
    Public Education Social Demarketing Programs

tered. As with copyrights, registration of trademarks provides greater protection in the United States.

Patent protection may be especially problematic in international trade where countries do not always recognize one another's patent assignments or attribute the same levels of protection. Recent controversies regarding Chinese and United States' computer manufacturers demonstrate the problem and raise questions of when reverse engineering, and look-alike products cross the line into patent, copyright and trademark infringement (Einhorn, Rogers and Crockett 2003, Einhorn, Elgin and Reinhardt 2002, Einhorn Burrows and Magnusson 2002, Einhorn et al. 2002). Such issues may cause companies to be reluctant to transfer technologies to other nations for fear that critical intellectual property and product differentiators could be lost, perhaps even with support from host countries' government technology policies. This problem is especially difficult in regard to software, where pirating in China is said to be commonplace and the government passive (Meredith 2003). Problems of pirating and counterfeiting are not limited to computer technology or to China. In addition to software, pirating entertainment materials such as music, movies and books (Wildstrom 2002, Green, Glover and Hof 2003) is commonplace worldwide–a result of digital technology itself that has had substantial effects on the legal and distribution systems involved in the marketing of such commodities.

The importance of laws protecting intellectual property cannot be overestimated as a source of economic development from innovation. Novak (1997) traces the background of government-enforced protections to demonstrate that they provide the opportunity for the originator to profit from his work and so motivate research efforts. Moreover, intellectual property protections are source of core marketing concepts such as product differentiation, positioning and branding. "Orphan drugs" and the success of Genzyme, Inc. is an example of the importance of exclusivity of intellectual property to technology commercialization (Watson 2003). In 1983, the U. S. Congress passed legislation to create this category for drugs to treat illnesses suffered by fewer than 200,000 persons in the United States. Most drug developers are reluctant to address the needs of so small a market. The legislation established that after FDA approval of an orphan drug no other company may market a competing drug for seven years without demonstrating superiority. This law has been used by Genzyme, Inc. to build a billion-dollar business serving health care niche markets (Watson 2003).

Generic drugs in the United States provide another example of government influence on technology commercialization. Under the Patent Term Restoration Act, firms list their patents in the Federal Drug Administration's (FDA) "Orange Book." These listings are used to verify that generic drugs are not infringing on patents. Critics claim the process allows originators to extend patents beyond appropriate limits, and call for removing the FDA from patent monitoring, allowing the courts alone to judge infringement (Hollis 2001). Critics believe that this would reduce unfair extensions of patent protection and open many drugs to the generic market. This issue is an opportunity for macromarketing research to assess social and economic effects of government regulation on drug development, production and distribution.

Direct and Indirect Technology Development

Direct technology development refers to R&D activities carried out by a government using its own personnel and resources. Examples are R&D carried out in United States national laboratories, military laboratories, and other government agency laboratories such as NASA. Indirect technology development refers to R&D sponsored by government agencies through grants and contracts with private sector for-profit and not-for-profit organizations. Government research may be directed to almost any field represented by a government agency (see for example U. S. Department of Energy 1999, U. S. Department of Housing and Urban Development 2003, and NASA 1999, 1996a, b, c). When direct government R&D yields new technologies, governments may patent these, register copyrights, use the technologies for internal purposes, and license private sector partners to commercialize the technologies.

Indirect technology development refers to government programs to induce private sector firms or other governments to carry out R&D (Coburn 1995). Two common approaches are grants and contracts. Grants provide funds to carry out research with less specification of the activities than is required by contracts. Contracts typically provide clear specification of the work and deliverables. Intellectual property rights associated with technologies developed by recipients of grants and contracts may reside with the sponsoring government or with the developer depending upon prior agreements.

An example of an indirect technology development program is the Small Business Innovative Research (SBIR) program of the U. S. Government (Gans and Stern 2003). SBIR allows private sector firms to apply for grants to develop technologies applicable to stated needs of Federal agencies. Federal agencies with R&D budgets must put three percent of those budgets into the SBIR program. Sponsored firms retain the rights to developed technologies and are required to make realistic efforts toward commercializing the technologies. In some cases, there may be national security restrictions. The intention of SBIR is economic and technology development. A critical macromarketing question is whether SBIR stimulates small business growth or whether SBIR directs business activities into directions that are unproductive in commercial markets. While some successful commercial products have developed from SBIR grants (National Technology Transfer Center 1996), there is no clear evidence that SBIR is an efficient mechanism for stimulating commercially successful new technologies or products Gans and Stern 2003).

Direct Commercial Technology Transfers

Direct commercial technology transfers refer to situations in which government agencies hold patents or copyrights for technologies and allow private sector firms to use the technologies for commercial purposes. The government agency negotiates a licensing agreement and receives royalties when the technology used. In some cases, the Government agency may elect to transfer the patent or copyright to the private firm. In the United States, such agreements are intended to benefit citizens, but there are no clear guidelines and there is the risk that profitable technologies may benefit the firm at the expense of the citizenry. It is difficult to know the true commercial market value of a technology before introduction, and government officials might not be experienced in assessing market potential.

Licensing might be carried out as exclusive or non-exclusive arrangements. Under exclusive arrangements, the license is granted to only one firm. Under non-exclusive agreements, multiple firms could be given commercialization rights. Licenses may be limited to certain applications or markets. In theory, non-exclusive agreements would create competition and speed commercialization. However, for-profit firms might be not invest in commercialization activities under non-exclusive arrangements because competitive forces could increase risk and limit opportunities to maximize profits. Similar logic applied to the special treatment of "orphan drugs" discussed earlier. An important question for macromarketing researchers is what arrangements optimally benefit, the government, the citizenry, and commercializing firms.

Infrastructure Development

Infrastructure development as here applied to technology development refers to activities involved in creating the physical, social, economic and industrial capabilities to support technology development, transfer and commercialization initiatives. Such activities were discussed earlier under the heading of economic development objectives. Government *physical infrastructure development initiatives* include such activities as developing or enhancing transportation, power, water and other utility systems, as well as construction of buildings needed to accommodate private firms targeted for formation, recruitment, retention or expansion. Governments may carry out construction projects directly, or hire private firms. Through economic development programs such as discussed above, governments frequently provide new roads, water and sewage systems and electrical systems as part of "deals" to attract major firms.

In addition to physical infrastructure development incentives programs to recruit specific companies, governments can invest in physical infrastructure needed for the general populace to adopt a technology. Examples include rural electrification programs that provide electrical power to rural areas, resulting in speeding the adoption of household technologies such as washing machines and television. A more recent example is Spain's extensive investment in broadband computer communications capabilities to make Internet access available to schools and libraries in rural areas (Prada 2003). The result was a twenty-nine percent surge in personal computer sales and a rise of ten percentage points in Internet use over two years.

Social infrastructure development initiatives address educational, health and other types of "human capital" programs. Social infrastructure development is intended to provide quality of life resources required to meet the needs of targeted technology companies and industries. Educational programs development and enhancement might be the most common types of social infrastructure development activities. Public and private schools are given grants or contracts to develop new curricula and enhance current programs that train workers in the skill sets needed by a targeted industry. A key risk is that programs developed to support a technology cluster concept may misread the market and invest in skills that are sophisticated but not demanded.

Set-aside programs are grants and contracts reserved for specified groups in order to involve members of these groups in social infrastructure development. Groups may be defined by race, ethnicity or economic status. Set-aside programs are intended overcome current disparities in economic development through involvement in commercial activities related to new technologies. Unfortunately, the same disparities that set-aside programs are intended to overcome can prevent effective participation by targeted groups. An important macromarketing question is the degree to which set aside programs lead to "distributive justice," or are ineffective ways of addressing social disparities in comparison to free market forces.

Import quotas, foreign content restrictions (domestic content requirements), and *import tariffs* are widely used to encourage foreign firms to locate plants and facilities in host countries, and to hire local workers so as to gain access to local market demand beyond quota and content limits, or to avoid tariff costs. Import quotas restrict the amount of a product that may be brought into a host country for commercial sale. Foreign content restrictions and their counterpart–domestic content requirements–are regulations that require that a specified portion of a product's value added components be produced in the host country. Similarly, regulations could constrain outsourcing service jobs to foreign firms and place immigration restrictions on foreign technology workers. Recent tensions in the U. S. information technology (IT) service industry over the use of Indian firms led the U. S. Congress to consider bills to restrict visas of foreign nationals employed to service U. S. IT customers, and several states began considering laws to ban outsourcing government IT services to firms in low wage countries (Kripalani, Einhorn and Magnusson 2003). Import tariffs are taxes on foreign goods and used or sold within the host country. By adding to the distribution costs of imported goods, import tariffs raise prices of these goods and provide competitive advantage to domestic products.

These types of government activities are often tied to economic protectionism objectives as discussed earlier, but may also be tied to quality of life and public safety objectives as illustrated by United States Government restrictions on the importation of prescription drugs using the claim that quality cannot be assured (Cohen 2003). Such objectives beg macromarketing research to document impacts of such artificial market constraints on quality of life of citizens, product quality and safety, product availability, prices, and economic development (Mullen 1993). In theory, efforts to limit access to markets violate free market principles as does forcing domestic investment in plant, equipment and jobs. Under free market principles, firms would be allowed to carry out production activities wherever best fits their business. Location decisions are complex (Jones and Teegen 2003), and trade constraints might create market inefficiencies. Still, import quotas, foreign content restrictions, and tariffs are widely used.

Duty free zones are geographic areas designated as exempt from the above types of regulations. These areas are often located near an international transportation facility such as a dock or airport, or near an international border as with the Mexico-U.S. border. Firms located in the zone may import goods into the zone for further processing. Such goods are exempt from import quotas and tariffs so long as the goods do not leave the zones. When raw materials and component parts are manufactured into completed products, the completed products may be exported to other nations, or sold within the host country, at which time tariffs would be paid as the goods leave the zone. Duty free zones tie to economic development objectives of jobs creation, new business formation, and foreign business investment as they are set up to attract businesses that will invest in technology, facilities, local labor.

Market Stimulation Activities

Market stimulation activities as part of government technology programs "create" or encourage demand for technology products. Such activities include direct government purchasing contracts, trade missions, free trade agreements, regulatory standards, and public education social marketing. The most direct approach is *government purchasing contracts* by which government agencies purchase products for their own use. Because government purchases are typically large and occur over time, government adoption of a product or technology can ensure a minimum level of demand and so encourage private sector adoption. An illustration is recent endorsement by over 24 countries throughout the world of "open source" software such as Linux as alternatives to the Microsoft Windows operating system for computers (Lohr 2002). If governments follow endorsements with actual procurements, the "guaranteed" software market for Linux will be substantial. As another example, one objective of the U. S. SBIR program noted earlier is to develop products for use by the sponsoring agency, and for sales to other agencies are considered an indicator of commercial success (Gans and Stern 2003). Government purchasing contracts are not here suggested as means of creating artificial demand, but government contracts, if substantial, can help a company achieve production volumes that provide economies of scale needed to make products available at competitive prices.

As used here, *government trade missions* refer to efforts by governments to represent their countries' business interests by identifying and facilitating foreign trade opportunities. This might be done through formal trade mission offices in a target country, or by trade tours in which business people, with members of their own government, travel to other countries to promote products. The latter such missions were made famous by Ron Brown, the United States Secretary of Commerce in the 1990s in the Clinton administration. Foreign trade missions and their government agents can also facilitate foreign trade opportunities by carrying out market research for industries. This area represents fertile ground for macromarketing research into how governments can facilitate foreign trade without violating free market principles and in ways that benefit the general citizenry of both the originating and target countries.

Free-trade agreements are treaties entered into by one or more countries to reduce trade barriers such as import quotas and tariffs. The best known examples are the General Agreement on Tariffs and Trade (GATT) involving many countries through the world, the North American Free Trade Agreement (NAFTA) involving Canada, the United States and Mexico, and the European Union. Countries enter into such agreements with the intention of opening new markets or expanding current markets for products for which they have competitive advantage. While consistent with free market principles, governments entering into such agreements often find opposition from industry associations and labor unions that have benefited from trade restrictions. Macromarketing research provides some support for such concerns and has suggested mixed quality of life results from international trade (Mullen 1993). Although less technologically sophisticated industries may be vulnerable to foreign competition, industries and firms with advanced technologies might benefit from open access to their trade partners' technology markets. However, research is needed into the cultural and physical, social and economic infrastructure conditions that must exist if technological products from developed countries are to be accepted in less technologically developed countries when free trade agreements remove government imposed restrictions on trade.

Regulatory mandates for technology and design standards is another activity that governments can adopt to foster the growth of specific technologies. As used here, the term refers to government-imposed requirements that certain capabilities be built into products designated for certain purposes. These are different from domestic content requirements discussed earlier. A recent example of a mandated technology in the United States is the Federal Communication's Commission's (FCC) adoption of one from among several competing high definition television protocols, and the mandate that all new televisions sold in the United States after a certain date incorporate this standard (Hart 1994, Karr 1996). In so doing the FCC assures a future market for that technology. Another example is the EU's adoption of wireless communication technology standards that align EU industries with policies and market standards in Asian and Pacific Rim countries with rapidly growing telecommunications industries (Lembke 2002). The adoption of such standards can provide EU companies with a competitive advantage in those markets.

Mandated standards might be necessary if industries and firms require consistency before adopting a technology. Historians will recall the battles between Edison and Westinghouse over whether the AC or DC electrical transmission standard would be adopted. The need for standards and the selection process would be interesting areas for macromarketing research. To illustrate, let us again consider high definition television (HDT). HDT has been slow to be adopted in the United States because television broadcasters must invest in expensive new camera and transmission equipment in order to broadcast HDT signals. However, recipients of television signals must have compatible television receivers if they are to receive and process the HDT signal. Broadcasters cannot risk investing in new equipment that would be incompatible with newly manufactured television sets. Therefore, standards must be established so that producers of television cameras and transmitters, firms in the broadcast industry, manufacturers of television receivers, and television viewers can be assured that investments are appropriate and that all equipment will be compatible. Government technology and design mandates can provide a stability that is needed for technology commercialization. From a macro-marketing perspective a key issue is whether government mandates assure that the best technology standards are adopted, or whether and under what conditions the evolution of standards in the free market would allow the best performing technologies to become the standard.

Public education social marketing programs (Kotler and Andreasen 1991) may be carried out by governments to encourage adoption of new technologies, or of behaviors that make use of new technologies. By extolling the benefits of the new technologies or ways of doing things, demand is stimulated for products that incorporate the new technology. For example, public service announcements that caution against obesity and fatty foods might stimulate the adoption of foods using fat free food processing technology.

Whereas traditional marketing approaches embrace the marketing concept of fulfilling customers' current needs and wants, new technology marketing might require a different vision in which marketing programs educate the market to recognize new needs and wants that can be fulfilled by new technologies (Davidow 1986, Shanklin and Ryans 1984). Where new technologies are concerned, product-markets are uncertain, potential buyers resistant, and older products based on past technologies are the standard. Products based on new technologies may be viewed as substitutes that the customer might not yet be willing to adapt. Such may be the case with biotechnology in food production (Emirog 2002). Indeed, the classic consumer adoption process and innovation diffusion models (Kripalani, Einhorn and Magnusson 2003, Kotler 2000 p. 355, Rogers 1983, 1976, 1962) demonstrate the risks, and imply that a long time frame is needed for full adoption of a new technology as individuals and institutions go through processes of awareness building, interest, evaluation, trial and, if the trial is successful, adoption. The speed of this process and the likelihood of adoption of new products has been found to be influenced by five factors: (1) perceived relative advantage, (2) compatibility with prior approaches, (3) complexity and difficulty of adoption and use, (4) divisibility in the sense of trials, and (5) observable and easily communicated benefits (Kotler 2000, Sultan, Farley and Lehmann 1996, Mahajan, Muller and Mass 1995, Hahn, Park and Zoltners 1994, Ram and Jung 1994, Gatignon and Robertson 1985, Rogers 1983, 1962). Where required infrastructures are in place, consideration of these factors may be brought into public education social marketing programs.

Demarketing Programs

Demarketing programs (Kotler 2000, Kotler 1973, Kotler and Levy 1971) refers to efforts to reduce demand for a good or service on a temporary or permanent basis. Private sector marketers might carry out demarketing activities to phase out a product or manage demand beyond the firm's capacity. As used here, "government demarketing programs" refers to efforts by government agencies to reduce demand for a product or to limit or completely prevent market access to a product. Such activities might be carried out with such objectives as public safety or quality of life. Four types of government demarketing activities are suggested here: banning products, domestic taxes, regulatory restrictions on use, public education, and export restrictions.

Banning products refers to declaring the sale, possession or use of a product to be illegal. Civil or criminal penalties might be attached to the enabling law or regulation. Penalties are intended to discourage use of the product by increasing distribution and usage costs to a point at which the product no longer has value. Efforts to ban products are typically unsuccessful, as has been seen with pornography, prostitution, and drugs and weapons. For this reason, and on free market principles, some writers argue that it is more practical to allow market forces to prevail (Block 1991). In regard to public safety, it might be argued, as does Becker (2002) regarding FDA regulations, that threats of lawsuits and market rejection are better constraints.

Domestic taxes imposed at very high levels for sale, possession, or use of a product is sometimes suggested as an alternative to banning as a means of reducing market usage and demand. Similarly, "penalty taxes" may be seen as an alternative to direct regulation, particularly in regard to industry and environmental pollution (Buchanan and Tullock 1975). As with civil and criminal penalties, the intention is to increase the cost of the product to the producer and consumer in order to discourage demand. Such efforts have been made at the consumer level, for example, in regard to cigarette sales (Solow 2001). One potential problem here worthy of research is the question of perceptions of the legitimacy of the tax. If the tax is seen as punitive, the government action may be resented, government prestige may be diminished, and contraband may be encouraged.

Regulatory restrictions on use are an intermediary demarketing step. Rather than ban products, sales, possession and use may be constrained to particular circumstances. Examples include drugs that may be purchased only with the prescription of a licensed physician, and various types of explosives and firearms that may be purchased, possessed or used only by persons holding government certifications. An additional example is the restriction on R12 freon that until the 1990s was commonly used in automobile air conditioning systems and sold over the counter to "do-it-yourself" mechanics. In the early 1990s, in response to ecological concerns, sales of freon were restricted to licensed technicians, and costs increased greatly. In part, these increased costs were due to government taxes, but the increased costs were also due to the constraints place on the market distribution system (Consumer Reports 1997, Freedman 1996). Here also we see an important need for macromarketing research into the impacts of government imposed regulatory restrictions on product availability. Do restriction achieve the desired goals, or do restrictions drive restricted products into hidden, unregulated markets? A related question is the impact on quality of life that results from government demarketing programs in the form of regulatory restrictions.

Export restrictions may be imposed on a technology to prevent commercialization of a technology in foreign markets. Such restrictions may be tied to military defense or to economic development. When technological products are transferred to other countries, the associated technological knowledge is also transferred, as has been observed with computer technology in China (Einhorn, Elgin and Reinhardt 2002). Export restrictions on products are a mechanism by which attempts are made to protect such knowledge. However, it is an open question as to whether export restrictions can long prevent the acquisition of desired technological knowledge and capabilities by determined foreign governments or private sector foreign competitors.

Public education social demarketing programs can discourage continued use of obsolete and undesired technologies and associated behaviors, just as social marketing programs can encourage behaviors associated with adoption of a new technology. Public service announcements and government sponsored educational programs might warn people about the dangers of a drug or seek the rejection of high gasoline consumption automobiles. Social demarketing programs might be especially effective in encouraging rejection of obsolete or dangerous products when coupled with introducing new technologies offering improved personal and societal benefits, and when usage is consistent with prior behaviors.

CONCLUSIONS

This review has developed taxonomies for classifying objectives of government technology programs and the associated activities governments engage in to achieve these objectives. As the objective and activity types have been discussed, macromarketing and free market implications have been considered. These taxonomies have not been represented as exhaustive of all types of objectives and potential technology program activities. Instead, this work should be viewed as an early step in the development of an organized perspective of government technology development, transfer and commercialization programs and issues. Still, it is a step, and it is hoped that the marcromarketing related research questions and issues that have been suggested will stimulate reflective thought and formal research. Such work is needed to further our understanding of the implications of technology for our social, economic and political systems, the impacts of technology transfers into different cultural and structural contexts, and the capabilities and responsibilities of marketers in the improvement of quality of life for all people through technology development and marketing.

REFERENCES

Aviation Week & Space Technology (1995), "Technology Transfer Special Report," Vol. 143, No. 17 (October 23): 57-63.

Aviation Week & Space Technology (1994), "Technology Transfer Special Report," Vol. 141, No. 19 (November 7): 42-50.

Ayal, I., A. Peer, and J. Zif (1987), "Selecting Industries for Export Growth–a Directional Policy Matrix Approach," *Journal of Macromarketing*, Vol. 7, No. 1 (Spring): 22-33.

Barnholt, Edward W. (1997), "Fostering Business Growth with Breakthrough Innovation," *Research-Technology Management*, (March-April): 12-16.

Becker, Gary S. (2002), "Economic Viewpoint–Get the FDA Our of the Way, and Drug Prices will Drop," *Business Week*, (September 16): 16.

Block, Walter (1991), *Defending the Undefendable*, San Francisco, CA: Fox and Wilkes.

Brenner, M. S. and Rushton, B. M. (1989), "Sales Growth and R&D in the Chemical Industry," *Research-Technology Management*, (March-April): 8-15.

Buchanan, James M. and Gordon Bullock (1975), "Polluter's Profits and Political Response: Direct Control versus Taxes," *American Economic Review*, Vol. 65, No. 1 (March): 139-147.

Carey, John 2003), "McClellan's Friendlier, Speedier FDA," *Business Week*, (June 16): 33-34.

CDER (Center for Drug Evaluation and Research (1999), *CDER 1999 Report to the Nation: Improving Public Health Through Human Drugs*, Washington, DC: Food and Drug Administration.

Cheeseman, Henry R. (1997), *The Legal and Regulatory Environment*, Englewood Cliffs, NJ: Prentice-Hall.

Coburn, Christopher, Ed. (1995), *Partnerships, A Compendium of State and Federal Cooperative Technology Programs*, Battelle: Columbus, OH.

Cohen, Harold E. (2003), "It's Time to Stop This Madness," *Drug Topics*, Vol. 147, No. 11: 13.

Consumer Reports (1997), "Psst! Wanna Buy Some Freon?" *Consumer Reports*, Vol. 62, No. 4 (April): 6.

Cundiff, Edward W. (1982), "A Macromarketing Approach to Economic Development," *Journal of Macromarketing*, Vol. 2, No. 1 (Spring): 14-19.

Davidow, William H. (1986), *Marketing High Technology: An Insider's View*, New York, NY: The Free Press.

Einhorn, Bruce, Peter Burrows and Paul Magnusson (2003), "Making a Federal Case Out of It," *Business Week*, February 10: 36, 38.

Einhorn, Bruce, Ben Elgin, Cliff Edwards, and Linda Himelstein (2002), "High Tech in China, Is it a Threat to Silicon Valley?" *Business Week*, October 28: 80-84.

Einhorn, Bruce, Ben Elgin, and Andy Reinhardt (2002), "The Well-Heeled Upstart on Cisco's Tail," *Business Week*, October 28: 91.

Einhorn, Bruce, Dexter Roberts and Roger O. Crockett (2003), "Winning in China," *Business Week*, January 27: 98-100.

Eldred, Emmett W. and Michael E. McGrath (1997a), "Commercializing New Technology–I," *Research-Technology Management*, Vol. 40, No. 1 (January-February): 41-47.

Eldred, Emmett W. and Michael E. McGrath (1997b), "Commercializing New Technology–II," *Research-Technology Management*, Vol. 40, No. 2 (March-April): 29-33.

Emirog, Lu Haluk (2002), "Foods Produced Using Biotechnology: How Does the Law Protect Consumers?" *International Journal of Consumer Studies*, Vol. 26, No. 3: 198-109.

Engelking, Susan (1992), "Brains and Jobs: The Role of Universities in Economic Development and Industrial Recruitment," *Economic Development Review*, Vol. 10, No. 1 (Winter): 36-40.

Federal Laboratory Consortium for Technology Transfer (1991), *Technology Innovation: Chapter 63 United States Code Annotated–Title 15, Commerce and Trade, Sections 3701-3715 (As Amended Through 1990 Public Law and with Annotations)*, St. Paul, MN: West Publishing Company.

Fisk, George (2001), "Reflections of George Fisk: Honorary Chair of the 2001 Macromarketing Conference," *Journal of Macromarketing*, Vol. 21. No. 2: 121-123.

Fisk, George (1999), "Reflection and Retrospection: Searching for Visions in Marketing," *Journal of Marketing*, 63 (January), 115-21.

Fisk, George (1998), "Green Marketing: Multiplier for Appropriate Technology Transfer." *Journal of Marketing Management*, Vol. 14: 657-676.

Fisk, George (1982), "Editor's Working Definition of Macromarketing," *Journal of Macromarketing*, Vol. 2, No. 1: 3-4.

Franko, L. G. (1989), "Global Corporate Competition: Who's Winning, Who's Losing, and the R&D Factor as One Reason Why," *Strategic Management Journal*, Vol. 10: 449-474.

Franza, Richard M. and Roberts S. Widmann (1996), "An Examination of Agreement Type, Firm Size, and Other Factors Affecting the Commercialization of Air Force Technology," *National Contract Management Journal*, Vol. 27, No. 1: 51-71.

Freedman, Eric (1996), "Big Demand, Dwindling Supply," *Automotive News*, Vol. 70, No. 5650 (March 18): 101.

Gans, Joshua and Scott Stern (2003), "When Does Funding by Smaller Firms Bear Fruit?: Evidence from the SBIR Program," *Economics of Innovation and New Technology*, Vol. 12, No. 4: 361-384.

Gatigon, Hubert and Thomas S. Robertson (1985), "A Propositional Inventory for New Diffusion Research," *Journal of Consumer Research*, March: 849-867.

Geisler, Eliezer (2003), "Benchmarking Interorganizational Technology Cooperation: The link between infrastructure and sustained performance," *International Journal of Technology Management*, Vol. 25, No. 8: 675-702.

Goeke, Patricia (1987), "State Economic Development Programs: The Orientation is Macro but the Strategy is Micro," *Journal of Macromarketing*, Vol. 7, No. 1: 8-21.

Goto, Akira (2000), "Japan's National Innovation System: Current Status and Problems," *Oxford Review of Economic Policy*, Vol. 16, No. 2 (Summer): 103-113.

Green, Heather, Ronald Glover and Robert D. Hof (2003), "Music Merchants Rush in Where Labels Have Failed," *Business Week*, February 10: 36.

Hahn, Mini, Sehon Park, and Andris A. Zoltners (1994), "Analysis of New Product Diffusion Using a Four-Segment, Trial-Repeat Model," *Marketing Science*, Vol. 13, No. 3: 224-247.

Hart, Jeffrey A. (1994), "The Politics of HDTV in the United States," *Policy Studies Journal*, Vol. 22, No. 2 (Summer): 213-218.

Hollis, Aidan (2001), "Closing the FDA's Orange Book," *Regulation*, Winter 2001: 14-17.

Hunt, Shelby D. (1977), "The Three Dichotomies Model of Marketing: An Elaboration of Issues," in *Macro-Marketing: Distributive Processes from a Societal Perspective*, edited by Charles C. Slater, Boulder: Boulder Business Research Division, University of Colorado, pp. 52-56.

Jones, Gary K. and Hildy J. Teegen (2003), "Factors Affecting Foreign R&D location Decisions: Management and Host Policy Implications," *International Journal of Technology Management*, Vol. 25, No. 8: 791-813.

Kandebo, Stanley W. (1994), "Technology Transfer Special Report–Firms to Build Car Engine Using Advanced Materials," *Aviation Week & Space Technology*, Vol. 141 (No. 19, November 7): 49-50.

Karr, Albert R. (1996), "FCC is Expected to Advance Proposal to Mandate High Definition TV Rules" *Wall Street Journal–Eastern Edition*, Vol. 227, No. 92 (May 9, 1996): B5.

Kotler, Philip (2000), *Marketing Management, The Millennium Edition*, Prentice Hall: Englewood Cliffs, NJ: 6.

Kotler, Philip (1973), "The Major Tasks of Marketing Management," *Journal of Marketing*, October: 42-49.

Kotler, Philip and Alan Andreasen (1991), *Strategic Marketing for Non-Profit Organizations, Fourth Edition*, Prentice Hall: Englewood Cliffs, NJ: 402-429.

Kotler, Philip and Sidney L. Levy (1971), "Demarketing, Yes, Demarketing," *Harvard Business Review*, November-December: 74-80.

Kripalani, Manjeet, Bruce Einhorn and Paul Magnusson (2003), "India: A Tempest over Tech Outsourcing," *Business Week*, June 16: 55.

Kumar, V. and Jain, P. K. (2003), "Commercialization of New Technologies in India: An Empirical Study of Perceptions of Technology Institutions," *Technovation*, Vol. 23, No. 2: 113-120.

Lembke, Johan (2002), "Global Competition and Strategies in the Information and Communications Technology Industry: A Liberal-Strategic Approach," *Business & Politics*, Vol. 4, No. 1: 41-69.

Lohr, Steve (2002), "An Alternative to Microsoft Gains Support in High Places," *The New York Times*, September 5. (http://www/nytimes.com/2002/09/05/technology/05CODE.html).

Louisiana Business Inc. (2002), *Louisiana Technology Guide*, Baton Rouge, LA: Louisiana Business Inc.

Louisiana Economic Development (2003), *Louisiana Economic Development: Driving Success*, Baton Rouge, LA: Louisiana Economic Development. See also www.led.state.la.us.

Louisiana Economic Development Council (2003), *Louisiana: Vision 2020, 2003 Update, Master Plan for Economic Development*, Baton Rouge: State of Louisiana. See also http://vision2020.louisiana.gov.

Magnusson, Paul (2003), "A Man of Many Missions: Trade Honcho Bob Zoellick has a Strong Diplomatic Agenda," *Business Week*, March 31: 94-95.

Mahajan, Vijay, Eitan Muller, and Frank M. Bass (1995), "Diffusion of New Products: Empirical Generalizations and Managerial Uses," *Marketing Science*, Vol. 14, No. 3, Part 2: G79-G89.

Marino, Louis, Karen Strandholm, H. Kevin Steensma, K. Mark Weaver (2002), "The Moderating Effect of National Culture on the Relationship Between Entrepreneurial Orientation and Strategic Alliance Portfolio Extensiveness," *Entrepreneurship Theory and Practice*, Vol. 26, No. 4 (Summer): 145-160.

Marshall, Kimball P. (1995), "New Directions in United States Competitiveness: Strategic Market Planning for Community and Regional Development," *Delta Business Review*, Vol. 5, No. 1 Summer/Fall: 26-33.

McKee, Darryl and Abhijit Biswas (1993), "Community Resources and Economic Development Marketing Strategy: An Empirical Investigation," *Journal of Macromarketing*, Vol. 13, No. 1 (Spring): 33-47.

Meade, William K. and Robert W. Nason (1991), "Toward a Unified Theory of Macromarketing: A Systems Theoretic Approach," *Journal of Macromarketing*, Vol. 11, No. 2 (Fall): 72-82.

Meredith, Robyn (2003), "Microsoft's Long March," *Forbes*, February 17: (www.forbes.com/free_forbes/2003/0217/078.html).

Morbey, G. K. (1989), "R&D Expenditure and Profit Growth," *Research-Technology Management*, (May-June): 20-23.

Morbey, G. K. and Reithner, R. M.(1990), "How R&D Affects Sales Growth, Productivity and Profitability," *Research-Technology Management*, (May-June): 11-14.

Mraz, Stephen J. (1996), "Sweeping Up With New Technology," *Machine Design*, Vol. 68, No. 19 (October 24): 59-60.

Mullen, Michael R. (1993), "The Effects of Exporting and Importing on Two Dimensions of Economic Development: An Empirical Analysis," *Journal of Macromarketing*, Vol. 13, No. 1 (Spring): 3- 19.

NASA–National Aeronautics and Space Administration (1999), *NASA's Goddard Space Flight Center 1998 Technology Transfer Report*, Greenbelt, MD: NASA's Goddard Space Flight Center (see also http://techtransfer.gsfc.nasa.gov).

NASA–National Aeronautics and Space Administration (1996a), *Review Report: Review of National Aeronautics and Space Administration New Technology Reporting (P&A-96-001)*, (September 30), Washington, DC: NASA–Office of the Inspector General.

NASA–National Aeronautics and Space Administration (1996b), *NASA Commercial Technology–Implementation of the Agenda for Change*, (May), Washington, DC: NASA.

NASA–National Aeronautics and Space Administration (1996c), *NASA Strategic Plan*, (February), Washington, DC: NASA.

Nason, Robert W., Nikhilesh Dholakia, and Dennis W. McLeavey (1987), "A Strategic Perspective on Regional Redevelopment," *Journal of Macromarketing*, Vol. 7, No. 1 (Spring): 34-48.

National Technology Transfer Center (NTTC) (1996), *NASA Solutions* (CD-ROM), Washington, DC: NASA, and Wheeling, WV: NTTC.

Novak, Michael (1997), *The Fire of Invention: Civil Society and the Future of the Corporation*, New York: Brown and Littlefield, Publishers, Inc.

Odagri, Hiroyuki and Akira Goto (1999), *Technology and Industrial Development in Japan: Building Capabilities by Learning, Innovation and Public Policy*, New York, NY: Oxford University Press.

Pegels, C. Carl and Thirumurthy, M. V. (1996), "The Impact of Technology Strategy on Firm Performance," *IEEE Transactions on Engineering Management*, Vol. 43, No. 3 (August): 246-249.

Peterson, John and Margaret Sharp (2002), *Technology Policy in the European Union*, New York, NY: St. Martins Press.

Pickett, Susan E. and Tatsujiro Suzuki (2000), "Regulation of Food Safety Risks: The Case of Food Irradiation in Japan," *Journal of Risk Management*, Vol. 3, No. 2 (April): 95-109.

Piper, William S. and Kimball P. Marshall (2000), "Stimulating Government Technology Commercialization: A Marketing Perspective for Technology Transfer," *Journal of Nonprofit and Public Sector Marketing*, Vol. 8 (3): 51-63.

Piper, William S. and Shahadad Naghshpur (1996), "Government and Technology Transfer: The Effective Use of Both Push and Pull Marketing Strategies," *International Journal of Technology Management*, Vol. 12, No. 1: 85-94.

Pomper, Miles A. (1998), "Probes of Chinese Missile Technology Focus on Satellite Export Rules," *CQ Weekly*, Vol. 56, No. 28: 1886-1890.

Port, Otis (2002a), "A Taxi You Hail and Bail," *Business Week*, December 16: 125.

Port, Otis (2002b), "We all Scream for Thermoacoustics," *Business Week*," December 16: 125.

Proctor, Paul (1994), "Regional Agencies Help Small Firms Get Foothold," *Aviation Week & Space Technology*, Vol. 143, No. 17 (October 23): 62.

Prada, Paulo (2003), "Tech Slump? Not in Spain," *Business Week*, June 16: 14.

Radosevich, Raymond (1995), "A Model of Entrepreneurial Spin-Offs from Public Technology Sources," *International Journal of Technology Management*, Vol. 10, No. 7-8: 879-893.

Ram, S. and Hyung-Shik Jung (1994), "Innovativeness in Product Usage: A Comparison of Early Adopters and Early Majority," *Psychology and Marketing*, January-February: 57-68.

Roberts, Edward B. and Dennis E. Malone (1996), "Policies and Structures for Spinning Off New Companies from Research and Development Organizations," *R & D Management*, Vol. 26, No. 1 (January): 17-48.

Rogers, Everett M. (1983), *Diffusion of Innovations, 3rd Edition*, New York, NY: The Free Press.

Rogers, Everett M. (1976), "New Product Adoption and Diffusion," *Journal of Consumer Research*, Vol. 2 (March): 290-301.

Rogers, Everett M. (1962), *Diffusion of Innovations*, New York, NY: The Free Press.

Scott, William B. (1995), "Satellite Control Concepts Bolster Civil, Defense Systems," *Aviation Week & Space Technology*, Vol. 142, No. 10 (March 6): 43-46.

Scott, William B. (1994), "Tech Transfer Impact Remains Elusive," *Aviation Week & Space Technology*, Vol. 141, No. 19 (November 7): 42-44.

Shanklin, William L. and John K. Ryans, Jr. (1984), *Marketing High Technology*, Lexington, MS: Lexington Books.

Sikka, Pawan (1996), "Indigenous Development and Acquisition of Technology: An Indian Perspective," *Technovation*, Vol. 16, No. 2 (February): 85-90.

Small Business Innovation Research Center (1997), *National Conference: Small Business Innovation Research*, Small Business Innovation Research Conference Center: Sequim, WA.

Smith, Adam (1981), *An Inquiry into the Nature and Causes of the Wealth of Nations, Vol. II*, R. H. Campbell, and A. S. Skinner (ed.), Indianapolis, IN: Liberty Classics.

Smith, Geri (2003), "Wasting Away," *Business Week*, June 2: 42, 44.

Solow, John L. (2001), "Exorcising the Ghost of Cigarette Advertising Past," *Journal of Macromarketing*, Vol. 21, No. 2 (December): 135-145.

Steensma, H. Kevin, Louis Marino, K. Mark Weaver, and Pat H. Dickson (2002), "The Influence of National Culture on the Formation of Technology Alliances by Entrepreneurial Firms," *Academy of Management Journal*, Vol. 43, No. 5: 951-973.

Steensma, H. Kevin, Louis Marino and K. Mark Weaver (2000), "Attitudes Toward Cooperative Strategies: A Cross-Cultural Analysis of Entrepreneurs," *Journal of International Business Studies*, Vol. 31, No. 4 (Fourth Quarter): 591-609.

Sultan, Fareena, John U. Farley, and Donald R. Lehmann (1996), "Reflection on 'A Meta-Analysis of Applications of Diffusion Models,'" *Journal of Marketing Research*, May: 247-249.

Tobin, James (1991), "The Adam Smith Address: On Living and Trading with Japan: United States Commercial and Macroeconomic Policies," *Business Economics*, Vol. 26, No. 1 (January): 5-16.

Tornatzky, Louis G., Yolanda Batts, Nancy E. McCrea, Marsha S. Lewis, and Louisa M. Quittman (1996), *The Art and Craft of Technology Business Incubation*, Athens, OH: National Business Incubation Association.

UNISPHERE–International Ventures Network (1996), *Progress Report on the NASA-UNISPHERE Relationship*, Washington, DC: UNISPHERE.

United States Department of Energy (1999), *From Invention to Innovation*, Washington, DC: United States Department of Energy Inventions and Innovation Program–Office of Renewable Energy. (DOE/GO-10099-810). (See http://www.oit.doe.gov/inventions).

United States General Accounting Office (1992), *Federal Research: Small Business Innovation Research Shows Success but Can Be Strengthened*–GAO/RCED-92-37, Washington, DC: U. S. General Accounting Office.

U. S. Department of Housing and Urban Development (2003), *Getting Building Technology Accepted: Developing and Deploying New Building Technologies*, (PTH11296), Washington, DC: U. S. Department of Housing and Urban Development.

Watson, Noshua (2003), "This Dutchman is Flying," *Fortune*, June 23: 89-90.

Weintraub, Arlene (2003), "Five Hurdles for Biotech," *Business Week*, June 2: 61-65.

Wessner, Charles W. (Editor) (2003), *Government-Industry Partnerships for the Development of New Technologies: Summary Report*, Washington, DC: National Academy Press.

Wessner, Charles W. (Editor) (1999), *Small Business Innovation Research Program: Challenges and Opportunities*, Washington, DC: National Academy Press.

West Publishing Company (1991), *Technology Innovation: Chapter 63 United States Code Annotated–Title 15 Commerce and Trade, Sections 3701 to 3715*, West Publishing Company: St. Paul, MN.

Wildstrom, Stephen H. (2003), "A Firefight Over Burning DVD's," *Business Week*, January 13: 24.

Yu, Tony Fu-Lai (2000), "A New Perspective on the Role of the Government in Economic Development," *International Journal of Social Economics*, Vol. 27, No. 7-10: 994-1013.

Zadoks, Abraham (1997), "Managing Technology at Caterpillar," *Research-Technology Management*, Vol. 40, No. 1 (January-February): 49-51.

Zang, Wei and Robert Taylor (2001), "EU Technology Transfer to China: The Automotive Industry as a Case Study," *Journal of the Asian-Pacific Economy*, Vol. 6, No. 2 (June): 261-274.

Zeller, Martin (1996), "A Technology Transfer Center in Action," *Database*, Vol. 19, No. 5 (October/November): 36.

Perceived Impacts
of Government Regulations
on Technology Transfers

Marina Onken
Caroline Fisher
Jing Li

SUMMARY. This paper examines the effects of government regulation
on the technology transfer process. Technology transfer is an important
component of an economic development effort in communities, states,
and nations. Understanding the process used to transfer technology is
needed to promote policies that develop an effective infrastructure to encourage technology transfer. This paper uses qualitative and quantitative
methodologies to examine managerial perceptions of the effects of government policies on the technology transfer process. The impacts of tax
policies, environmental regulations, health and safety regulations, labor
regulations, international trade regulations, and the differences in regu-

---

Marina Onken, PhD (E-mail: mho@onken.com), is Associate Professor, Touro
University International, 5665 Plaza Drive, Third Floor, Cypress, CA 90630.

Caroline Fisher, PhD (E-mail: fisher@loyno.edu), is Bank One/Francis Doyle Distinguished Professor of Marketing, specializing in consumer analysis and research, and
Jing Li, PhD (E-mail: jingli@ loyno.edu), is Associate Professor of Management, specializing in operations management, both at Loyola University New Orleans, 6363 St.
Charles Avenue, New Orleans, LA 70118.

[Haworth co-indexing entry note]: "Perceived Impacts of Government Regulations on Technology
Transfers." Onken, Marina, Caroline Fisher, and Jing Li. Co-published simultaneously in *Journal of Nonprofit & Public Sector Marketing* (Best Business Books, an imprint of The Haworth Press, Inc.) Vol. 13, No.
1/2, 2005, pp. 35-55; and: *Government Policy and Program Impacts on Technology Development, Transfer
and Commercialization: International Perspectives* (ed: Kimball P. Marshall, William S. Piper, and Walter
W. Wymer, Jr.) Best Business Books, an imprint of The Haworth Press, Inc., 2005, pp. 35-55. Single or multiple copies of this article are available for a fee from The Haworth Document Delivery Service
[1-800-HAWORTH, 9:00 a.m. - 5:00 p.m. (EST). E-mail address: getinfo@haworthpressinc.com].

lations between countries are studied. Items used to measure the success of technology transfer are proposed. *[Article copies available for a fee from The Haworth Document Delivery Service: 1-800-HAWORTH. E-mail address: <docdelivery@haworthpress.com> Website: <http://www.HaworthPress.com> © 2005 by The Haworth Press, Inc. All rights reserved.]*

KEYWORDS. Government regulation, technology transfer, economic development, government policy, tax policies, environment, health, safety, labor, international trade

INTRODUCTION

Technology transfer is an important source of economic development. The global competitiveness of nations depends on the level of technological innovation (Pang and Garvin, 2001; Porter, 1990). Technology transfers from abroad play a large part a country's technological and economic development (Kumar and Marg, 2000). Developing countries often use foreign direct investment (FDI) as a way to acquire foreign advanced technology (Yin, 1999).

The adaptation of technology from developed countries also drives long-term growth in most developing countries (Zattler, 2002). Developing countries experience increased international trade, increased economic growth, and increased levels of productivity from technology transfers (Okabe, 2002). Finally, technology transfers have a positive effect on consumers' welfare in both sending and receiving countries when both are developed markets (Petit and Sanna-Randaccio, 1998). Indeed, technological innovation benefits consumers and producers. The producer experiences lower costs and higher profits. Consumers benefit through lower prices in both the origin of the technology and the foreign country receiving the transfer.

Given the importance of technology transfers in both developing and developed economies, this study examines how public policies enhance technology acquisition. Following a review of potential roles of governments in technology transfer and a review of potential private sector firms' needs and motivators for conducting technology transfers, qualitative and quantitative analyses are presented that address managerial perceptions of factors that impact technology transfers, including government policies and regulations.

GOVERNMENT POLICY AND TECHNOLOGY TRANSFER

Because of the impact of technology transfers on their economies, government agencies encourage such transfers. The role of government in encouraging technology transfer can take several different forms (Bozeman, 2000). The most conservative role is limited to removing barriers to the free market to allow the free transfer of technology.

A second role a government can take is developing well-specified industry and market development goals and supporting research and development to meet those goals. Japan, for example, selects industries in which the country wants to excel, such as telecommunications. The government then puts its resources towards technology development in that area. This role includes providing resources and creating policies to stimulate technology research and transfer.

A third government role is to provide a link between the public and private sectors, either by providing research itself, or by developing policies that affect research development and technological innovation (Bozeman, 2000). This role may include facilitating the transfer of technologies from public research universities to commercial organizations.

A fourth role is also highly influential in stimulating technology. Governments can stimulate certain kinds of research by partnering in the commercialization of technology (Bozeman, 2000). The government may create economic development programs and incubators to stimulate the research process. The link between government and the private sector usually takes one of two different forms: government agencies can produce the technology, and the private sector then receives the technology; or governments can stimulate the development of certain kinds of technology by the private sector through policies and regulations.

Both the first and fourth roles of government include the development or change of government regulations and policies. Some government policies that may affect technology transfers include placing restrictions on foreign equity holdings, licensing arrangements or joint ventures, and screening of foreign investments (Davidson and McFetridge, 1985). In addition, governments influence international technology transfers through trade policies, intellectual property rights protection, or policies that affect the attractiveness and character of foreign investment (Martinot, Sinton and Haddad, 1997). For a country-specific example, China amended its technology transfer rules in 2002 to facilitate its participation in the World Trade Organization (WTO) and increase transfers of technology (Adcock, 2003).

Currently, little is known about how government policies actually affect technology transfers. Muralidharan and Phatak (1999) found no support for the hypotheses that government's requirements for technology transfers from multi-national corporations (MNCs) that enter their countries or that intellectual property protection laws were associated with increased levels of research and development activity. No published studies have adequately documented an empirical relationship between government policies and the success of technology transfers, nor is data readily available on the number or success of technology transfers to conduct such an analysis of the impact of government policies.

Without such data, researchers need to use another approach to assess the effects of government rules and regulations on private-sector technology transfers. One approach is to study how MNCs react to government policies and regulations using executives within these organizations as expert key informants. Technology transfer is a means for an organization to transfer intangible assets, or technological knowledge. The impetus to transfer technology arises out of a need to gain a competitive advantage in the marketplace. Usually, transfers take place only when there is "an awareness of a profitable opportunity" to apply the technology in a different location (Hakanson and Nobel, 2001). Although this is the primary driving force behind international technology transfers in the private sector, government policies developed to protect local industries can lead to decreased foreign direct investments. Corporations have learned that they must establish foreign subsidiaries or enter into joint ventures to reach otherwise protected markets (Martinot, Sinton and Haddad, 1997), and thus avoid dealing directly with governmental policies that inhibit foreign direct investments.

DRIVERS BEHIND INTERNATIONAL TECHNOLOGY TRANSFER

In an increasingly competitive and global business environment, firms are under more pressure to operate efficiently. One of the greatest drivers for any strategic action is to use the firm's assets to generate the greatest economic impact (Osman-Gani, 1996). International technology transfer is often conducted to cut costs, and to make production processes more efficient and to become more competitive in its industry. This arguably is one of the strongest private sector drivers for any firm to conduct any strategic technological action.

Tax implications act as drivers during international technology transfer, affecting how the technology will be transferred between countries (Loewenstein, Klass, Hickey, Leek and Joseph, 1999). For example, accounting for profits that arise out of intangibles, such as a patent, affects the firm's taxes. Even value-added taxes need to be considered when planning for the transfer of technology. Tax implications can affect the flow of technology and act as the driver behind transferring the technology in the first place. Both of these situations affect the strategic decision making to effectively manage technology transfers.

Socially responsible actions can also act as a driver behind the transfer of technology, but more often, government regulation is a stronger driver for a firm to incorporate technology to comply with health, safety, and environment regulations (Videras and Alberini, 2000). Although firms must participate in international technology transfer if it results in compliance with environment, health, or safety regulations, if a firm sees its transfer of technology as having a positive impact upon the environment or its own worker health or safety, the firm should have a greater motivation to participate in technology transfer, even on a voluntary basis. Although altruistic behavior on its own is rarely a motivator or driver for a firm, attention to fiduciary responsibility should be of paramount concern to the general manager or CEO. Firms see voluntary participation in governmental programs as a way to transfer technology and to pay attention to its own environmental issues (Videras and Alberini, 2000).

THE CURRENT STUDY

This study combines exploratory qualitative and quantitative approaches to develop greater understanding of managers' perceptions of the impact of government policies and regulations on technology transfers. The research was supported by a grant from the Louisiana Board of Regents to study international technology transfer in the state. The grant was funded to help the state of Louisiana develop public policy that would enhance inbound technology acquisition, and to inform business leaders, educators, and policy makers of the scope of international technology transfer in Louisiana.

The state of Louisiana developed an economic development plan in 1998 called "Louisiana: Vision 2020." Goal two of this plan actively sought diversification into emerging technology areas. Louisiana was particularly interested in fifteen technology-driven clusters as targets for its economic development efforts. Eight of these clusters were previously existing in the state and seven were emerging. Innovation and technology were identified as driving forces behind the growth and diversification of the state economy. Increasing the competitiveness of firms in Louisiana begins with understanding the flow of technology, including the knowledge to transfer technology. The state of Louisiana recognizes that, for Louisiana firms to be competitive in the global marketplace, they must be able to understand the drivers behind technology transfer, and the best practices for managing the transfer.

This study specifically examined the best practices and processes that create a competitive advantage for firms that practice international technology transfer. These best practices were determined by interviewing managers who had been involved in an international technology transfer during the prior five years. The people who are most aware of the success of technology transfers and the impacts of government policies and regulations on technology transfers are the executives who have participated in them. No published articles detail managers' perceptions of government programs and their impacts on technology transfers. Such knowledge is required to develop effective government programs.

METHODOLOGY

This research had two stages, the qualitative interview stage and the quantitative standardized questionnaire stage. A sensible way to understand a complex process where few constructs are already specified and little existing research exists at the focal level of analysis is the grounded theory approach (Eisenhardt, 1989; Glaser & Strauss, 1967; Strauss & Corbin, 1990). Following this approach, the initial part of this investigation involved depth interviews with managers in firms that had been involved in technology transfers. This qualitative technique was used to determine the relevant variables to include in the quantitative part of the study.

The depth or long interview is "a sharply focused, rapid, highly intensive interview process that seeks to diminish the indeterminacy and redundancy that attends more unstructured research processes" (McCracken, 1988). It is used to gather data related to cultural categories and shared meanings. It uses open-ended interview questions designed to probe specific areas of interest to the researcher. While standard questions were asked, depth interview techniques were used to probe for comprehensive answers as to the practices and processes and the variables that impacted technology transfers.

The qualitative study used the maximum variation sampling technique of purposefully picking a wide range of variation on dimensions of interest (Ragan, 1987). The stages of international technology transfer, and the variables that have an impact upon the process were used as a guide in developing the research protocol. The investigators were able to document diverse variations that have emerged given different conditions during the international technology transfer process. The reason for using this particular sampling technique was to identify important common patterns that cut across variations.

All firms in the sample were U. S. firms that had experienced international technology transfers during the prior five years. Each firm is an international company that has divisions across the world. Table 1 provides a profile of the eight organizations that participated in the qualitative part of the study. Half of these business organizations had between 500 and 600 employees; three were smaller. Most were in the chemical industry, a reflection of the types of inter-

TABLE 1. Organizations in the Qualitative Study

Firm	Size	Corporate Structure	Product or Process
A	300 employees	Business units	Infrared testing materials
B	500 employees	Business units	Intermediate chemical product
C	135 employees	Joint venture	Polymers
D	500+ employees	International units	Synthetic materials
E	N/A	International units	Polymers
F	500 employees	Team-based structure	Abrasives
G	600 employees	Team-based structure	Chemical materials
H	5 employees	Subsidiary of large foreign firm	Waste conversion

national business organizations in the state. The corporate structure varied widely.

Each depth interview took place in the offices of the selected firms with participating managers. The researchers used an interview guide to facilitate comparison of the interviews and to assure that all areas of interest were covered in the interview. The questions were open-ended and the participants were encouraged to go beyond the questions asked by the researcher in describing the technology transfer processes. All interviews were taped and transcribed. All eight firms requested, and were granted, confidentiality in reporting of results.

Data analysis was performed using two different forms of content analysis, (1) conceptual analysis and (2) procedural analysis. Using the two forms of content analysis allows the investigator to identify common themes among the managers' interviews, as well as to develop a general model of the process of technology transfer. Conceptual analysis refers to the traditional technique of determining what words or concepts are present in a text (Carley, 1990). In this case, conceptual analysis identified the concepts most frequently mentioned by the managers in describing the technology transfer processes. Procedural analysis centers on the procedures that the author of a text uses to perform some task described in the text (Carley, 1990); in this case, the processes for accomplishing the technology transfer described by the manager in each interview transcript. Three different researcher evaluators, none of whom were involved in gathering the interview data, developed investigator triangulation by independently coding the data according to the research protocol, and then coding the data together, discussing differences. Each sentence in the interview data was coded independently and then together with all three evaluators.

To facilitate analysis of the data, the researchers used QSR Nud*ist 4© software. This program manages data and documents and allows the researcher to create, manage, and explore ideas and categories within the data. It is designed to allow researchers to discover themes, construct and test theories, generate reports (including both text and coding patterns), and build models by linking with graphical display software. Each of the interviews was loaded into the software package, and each sentence within each interview was coded according to the category or theme that it best represented. Once the interviews were coded, the researchers were able to look at one category across all interviews to determine trends and commonalities among the interviews. Word searches are possible with the software; for instance, the researcher could search for the word, "transfer," and then be able to see among all the interviews the instances in which the interview subjects used that word.

Reproducibility, the extent to which classification produces the same results when more than one researcher codes the text, was the most important reliability issue in this study. The researchers addressed this problem by developing and using a dictionary of words, concepts, categories, and relationships in coding the qualitative data. Each of the three different researchers coded all eight of the interviews to allow inter-coder reliability checks. All the coders used the same computerized qualitative analysis software program. The coders discussed differences in coding and reached full agreement with respect to coding of the data.

This qualitative research design had two major limitations. First, the sample size was small, making generalization to the larger population of firms difficult, if not impossible. Second, the study used retrospective accounts by the managers that have sometimes been associated with errors of memory as its primary source of data. However, for purposes of determining the relevant variables to include in the second part of the study, the methodology is appropriate and highly recommended (McKennell, 1974). This methodology is similar to that used in identifying customer needs, where a sample size of eight was found to identify over 80 percent of the total needs identified from eight focus groups plus nine one-hour interviews (Griffin and Hauser, 1992).

Quantitative Standardized Questionnaire

Once practices, processes, and variables were elicited using depth interviews, following the counsel of McKennell (1974) to conduct interviews with members of the target population, they were used to create a standardized questionnaire. Variables were identified and categorized. Then questions were developed to elicit information about the level and impact of these variables in organizations that had conducted technology transfers. Likert-type response scales were used for these questions.

A questionnaire was developed from the interview information provided in the qualitative part of the study. Items relating to government policies and regulations and the success items were developed from the qualitative data using Nud*ist software. Table 2 shows the questionnaire items that related to government and regulation or to success of the technology transfer. The questionnaire items were rated on a six-point scale, from strongly disagree to strongly agree.

The questionnaire was sent to companies that had completed an international technology transfer within the prior five years. The questionnaire was developed from the stage one qualitative research to cover the areas that emerged as being relevant to technology transfer (Joyner and Onken, 2002). Table 2 provides the questions used to measure government policies and the success of the transfer.

To find out what organizations had completed international technology transfers, a letter was sent to 3,600 organizations from two mailing lists: German companies with subsidiaries in the U.S. and Louisiana Manufacturing Organizations with 50 or more employees (1999). The letter asked if the firm had completed such a transfer and if the manager would be willing to complete a questionnaire. Positive responses from 69 companies indicated willingness to participate in the study. Questionnaires were sent to these companies. Follow-up phone calls were used to increase the response rate, resulting in the receipt of 36 completed questionnaires for a 53.6 percent response rate.

TABLE 2. Questionnaire Items

Government	Tax policies had an impact on the decision to transfer this technology.
	Existing government policies and regulations increased the cost of the technology transfer.
	Government policies and regulations that differed between countries posed a problem to the technology transfer.
	Policies and regulations regarding the environment affected the technology transfer process.
	Policies and regulations regarding health and safety affected the technology transfer process.
	Policies and regulations regarding labor affected the technology transfer process.
	Policies and regulations regarding international trade affected the technology transfer process.
Success	Our customer base has increased because of the technology transfer.
	Operating costs have gone down because of the technology transfer.
	The firm is more competitive since the technology transfer.
	The technology transfer has had a positive impact on the firm with respect to taxes.

The sample data were analyzed using SPSS software. Means and standard deviations were calculated for all questions. Correlations were calculated between both the success scale and the individual questions that measured opinions about the impact of government policies and questions that measured the results of the technology transfers. These are shown in Table 10.

RESULTS OF THE QUANTITATIVE SURVEY

Characteristics of Responding Organizations

Tables 3 to 8 provide information about the organizations that participated in the quantitative part of this study. The organizations that responded varied greatly in their number of employees, as shown in Table 3. Within this wide variation, the majority of the respondents to the survey were from units with less than 500 employees and organizations with between 1,000 and 50,000 employees in total.

Experience with international technology transfers showed a U-shaped distribution, as can be seen from Table 4. A large number of the responding organizations (40 percent) reported that they had been involved in over ten technology transfers in the prior five years. On the other end, 25.7 percent of the organizations had been involved in only one or two technology transfers over that period.

Many of the responding organizations (50 percent) were in either the chemical or manufacturing industry, as shown in Table 5. The only other industry category that was represented by more than one firm was "other." Since the respondents self selected to participate in the survey based on experience with

TABLE 3. Number of Employees in Organization

	Employees at Unit		Employees in Total Corporation	
Range	#	%	#	%
1-100	13	36	8	22
101-500	12	33	1	3
501-1,000	3	8	6	17
1,001-5,000	8	22	7	19
5,001-10,000	0	0	4	11
10,001-50,000	0	0	8	22
Over 50,000	0	0	2	6
Total Responding	36	100	36	100

TABLE 4. Number of Technology Transfers Unit Was Involved in During Prior Five Years

Number	Frequency	Percent
1	5	14.3
2	4	11.4
3	3	8.6
4	2	5.7
5	2	5.7
6-10	5	14.3
11-50	7	20.0
Over 50	7	20.0
Total Responding	35	100

TABLE 5. Organization's Industry Type

Industry	Frequency	Percent
Government/military	1	2.8
Retail	1	2.8
Utility	1	2.8
Computer-related	1	2.8
Medical	3	8.3
Chemical	8	22.2
Manufacturing	10	27.8
Other	11	30.6
Total Responding	36	100.0

technology transfers, these results suggest that technology transfers are more frequent in the chemical and manufacturing industries. This would be an expected characteristic of organizations participating in the study, since many oil and petrochemical firms do business in Louisiana.

The most common form of organizational structure reported by the respondents was centralization of core competencies and decentralization of other competencies, as illustrated in Table 6. Decentralized and nationally self-sufficient organizations were the least common among these respondents that had participated in international technology transfers.

No one role of overseas operations appeared significantly more often than the others for the respondents, as Table 7 shows.

TABLE 6. Organizational Structure

Structure	Frequency	Percent
Decentralized and nationally self-sufficient	6	16.7
Centralized and globally scaled	12	33.3
Sources of core competencies centralized, others decentralized	18	50.0
Total Responding	36	100.0

TABLE 7. Role of Overseas Operations

Role	Frequency	Percent
Sensing and exploiting local opportunities	10	37.0
Implementing parent company strategies	9	33.3
Adapting and leveraging parent company competencies	8	29.6
Total Responding	27	100.0

The most common method of development and diffusion reported by the respondents was no diffusion, as shown in Table 8. More than half reported that the knowledge developed during the technology transfer was developed and retained within each unit.

Government Regulations and Technology Transfer

To examine their perceptions of the effects of government policies on the technology transfer process and the success of the technology transfer, respondents were asked to respond to the questions shown in Table 2. A scale of one to six was used to measure the degree to which they agreed with each statement, where one equaled strongly disagree and six equaled strongly agree.

Table 9 provides the means, standard deviations, and percent of respondents who agreed or strongly agreed with the statement for the government and regulation items and the success items. The means to the government and regulation items were all 3.00 or below, indicating some amount of disagree-

TABLE 8. Method of Development and Diffusion of Knowledge

Method	Frequency	Percent
Knowledge developed and re-tained within each unit.	17	51.5
Knowledge developed and re-tained at headquarters.	3	9.1
Knowledge developed at head-quarters and transferred to overseas.	13	39.4
Total Responding	33	100.0

TABLE 9. Means and Standard Deviations for Questionnaire Items

Item	Mean	Standard Deviation	Percent Agree or Strong Agree
Tax policies had an impact on the decision to transfer this technology.	2.33	1.53	12.9
Existing government policies and regulations increased the cost of the technology transfer.	2.87	1.67	36.7
Government policies and regulations that differed between countries posed a problem to the technology transfer.	3.00	1.77	35.7
Policies and regulations regarding the environment affected the technology transfer process.	2.55	1.48	21.9
Policies and regulations regarding health and safety affected the technology transfer process.	2.72	1.46	25.8
Policies and regulations regarding labor affected the technology transfer process.	2.37	1.19	16.7
Policies and regulations regarding international trade affected the technology transfer process.	2.41	1.36	25.0
Our customer base has increased because of the technology transfer.	4.59	1.24	87.5
Operating costs have gone down because of the technology transfer.	3.72	1.63	65.7
The firm is more competitive since the technology transfer.	4.86	1.07	94.4
The technology transfer has had a positive impact on the firm with respect to taxes.	3.18	1.47	40.9

ment with these items. The lowest mean and lowest percent agreement were for "Tax policies had an impact on the decision to transfer this technology" and for "Policies and regulations regarding international trade affected the technology transfer process." The highest means and percent agreement were for "Government policies and regulations that differed between countries posed a problem to the technology transfer" and "Existing government policies and regulations increased the cost of the technology transfer." More than one-third of the respondents agreed with these two statements.

The means to the success items were all above 3.00, indicating some amount of agreement with these items. The highest mean scores and percent agreement were for "The firm is more competitive since the technology transfer" and "Our customer base has increased because of the technology transfer." The lowest mean and percent agreement were for "The technology transfer has had a positive impact on the firm with respect to taxes."

A success scale was created by summing the responses to each of the questions measuring success. The Cronbach's alpha for this scale was 0.7687, indicating that the items in the scale were measuring the same underlying concept. Table 10 gives the Pearson Product-Moment correlations between the questions measuring government and regulatory factors and the success scale and the individual questions measuring success of the transfer.

The only items showing significant positive correlations with the success scale were "Policies and regulations regarding health and safety affected the technology transfer process" and "Policies and regulations regarding labor affected the technology transfer process." Both of these were significantly positively correlated with the overall success scale. Health and safety seemed to impact the customer base and the firm with respect to taxes, while labor regulations had a significant impact on the organization's taxes. Another item, "Tax policies had an impact on the decision to transfer this technology," was negatively correlated with "Our customer base has increased because of the technology transfer." The results using Spearman's rho and Kendall's tau-b showed the same patterns of significance and are not separately reported.

DISCUSSION

Do managers perceive that government regulations influence their technology transfers? Perhaps the biggest perceived impact upon the firms' success was for the item concerning health and safety policies and regulations. This was positively and significantly correlated with the international technology transfer success scale. The higher the managers rated the impacts of policies and regulations regarding health and safety on the transfer process, the greater

TABLE 10. Pearson Correlation Coefficients Between Government and Success Questions

	Success Scale	Our customer base has increased because of the technology transfer.	Operating costs have gone down because of the technology transfer.	The firm is more competitive since the technology transfer.	The technology transfer has had a positive impact on the firm with respect to taxes.
Tax policies had an impact on the decision to transfer this technology.	.021	−.378**	.017	.038	.197
Existing government policies and regulations increased the cost of the technology transfer.	.213	.075	.003	.127	−.273
Government policies and regulations that differed between countries posed a problem to the technology transfer.	.201	.160	.040	.101	−.196
Policies and regulations regarding the environment affected the technology transfer process.	.328	.145	.096	.183	−.074
Policies and regulations regarding health and safety affected the technology transfer process.	.478**	.518***	.019	.203	.406*
Policies and regulations regarding labor affected the technology transfer process.	.430*	.102	.163	.219	.464**
Policies and regulations regarding international trade affected the technology transfer process.	.365	.027	.269	.157	.142

* Significant at alpha = .10
** Significant at alpha = .05
*** Significant at alpha = .01

was the success of the transfer. This health and safety item was also significantly positively correlated with two of the individual success items: the increase in the firm's customer base and the positive impact of the transfer upon the firm's taxes. Perhaps some managers pay attention to health and safety concerns during the process as a significant issue, and this attention may have a positive effect upon the success of the transfer.

Two other government-regulation items were positively correlated with some of the items that make up the success scale. The managers' ratings of the impact of government tax policies on their technology transfers were negatively correlated with their perceptions of an increase in the firm's customer base from the transfers. Finally, the item that measured managers' perceptions of policies and regulations regarding labor's effect upon the technology transfer process was significantly correlated with positive impacts of the technology transfer on the firm with respect to taxes.

Individual Items

More than one-third of the respondents felt that existing government policies and regulations increased the cost of the technology transfer. Governments are perceived as negatively influencing the cost of technology transfers, which could keep the number of such transfers down.

The policies and regulations thought to have a positive impact by the lowest percentage of respondents were those related to the environment, health and safety, and international trade. Less than 20 percent of respondents thought that tax and labor policies had a positive impact on technology transfers. Governments probably should not consider these types of regulations when they are trying to increase technology transfers.

Respondents felt that differences between government policies and regulations among countries posed a problem for their technology transfers. Decreasing or eliminating these differences should be a goal of all governments, especially those of developing countries. Perhaps these regulations should be part of the discussions between countries on trade and tariffs, if not already so. Further research is needed to determine which government policies and regulations differ and detrimentally affect technology transfers.

As far as the success of the technology transfers, the respondents were overwhelmingly positive. These results could be due to a respondent bias; those that did not feel their transfers were successful may not have responded to the questionnaire. Of those who did respond, most felt that the transfer made the organization more competitive, increased their customer base and decreased costs. Less than half agreed that the transfer had a positive impact on the firm with respect to taxes. The strong perceived impacts of technology transfers on the success of organizations should provide additional impetus for governments to encourage technology transfers.

Success Measure

One of the strongest contributions of this study is the success measure of technology transfer. The Cronbach's alpha indicated that organizations tend to respond similarly to all items measuring the success of their technology transfers. The overall success scale could therefore be used to determine which regulations were significantly related to success. The correlation results showed the highest relationship between overall success and policies regarding health and safety, the same item that had the highest level of agreement and highest mean. This area is the first one that governments should investigate when trying to determine what they might do to increase the number of technology transfers.

The other policy area that was significantly correlated with overall success was labor, an item that had one of the lowest mean scores and lowest agreement percentages. When looking at specific success items, labor was significantly correlated with only one item, positive impacts on taxes. Interestingly, this success item had the lowest mean and percent agreement. This one aspect of technology transfers, tax effects, would seem to be impacted by labor regulations and to be the only aspect impacted by these regulations.

One final significant correlation found was the negative correlation between impacts of tax policies on the decision to transfer technology and increase in customer base as a result of technology transfers. Apparently, if tax policies were perceived to have a major impact on the transfer decision, then the customer base did not increase as much as if the perceived impact was less strong. Perhaps the transfers were undertaken to affect the firm's taxes rather than to increase its competitiveness. This could well be a situation where governments created an unnatural incentive to pursue technology transfers. Organizations may then have conducted the transfers even though they were not particularly beneficial to the organization in other ways.

This study suggests that government regulations and policies can have an impact both on the number of technology transfers and on the success of those transfers. This is certainly a result that is worth further investigation. If technology transfers create more competitive business firms with lower operating costs and larger customer bases, then governments should certainly want to encourage them through effective policies and regulations. On the other hand, artificial tax incentives may encourage businesses to conduct technology transfers that are not particularly beneficial to the firm in other respects.

There are some limitations to the study. First, studying technology transfer is not easy, not only because establishing the boundaries of "technology" is difficult, but because the process of technology transfer is nearly impossible to delineate into distinct steps (Bozeman, 2000). However, this paper examines the relationships between government laws and regulations and the successful transfer of technology transfer by the firm, a narrow part of the entire technology transfer process. Second of all, the study's results are difficult to generalize to a larger population because of the size of the sample. However, by using a combination of qualitative and quantitative techniques, this study is a thorough exploration of some of the key success factors a firm needs to possess to successfully transfer technology across national borders.

Implications

This study reinforces the resource-based view of the firm by emphasizing that the intangible resources, the knowledge and technology, are what makes a firm most competitive. The success of international technology transfer depends upon the firm's ability to leverage government policies when creating its strategies. This study suggests that a firm should be giving due diligence to health, safety, and labor regulations, especially with respect to the impact upon the firm's tax situation. However, if a firm's objective is to increase its customer base, taking into account the government's tax policies should not be done during the firm's strategic analysis. Rather, the managers of the firm should be focusing upon cutting costs and increasing its customer base to increase revenues, the basic tenets of business. Navigating the government's policies and regulations is a necessary part of the business, but it is not what will create a competitive advantage for the firm. Only by leveraging the firm's intangible assets, its knowledge and technology, will the firm be able to create a competitive advantage that is not imitable by its competitors.

Lessons for Organizations

Firms that intend to participate in international technology transfer should first evaluate the policies and regulations by examining health and safety codes that could be involved, as well as labor issues. A formal process that incorporates an evaluation of the laws and regulations should result in the firm to successfully transfer the technology. Firms with no prior experience in technology transfer may not have the managerial experience in place to successfully complete the transfer. Therefore, they need to incorporate a formal process of evaluating the government laws and regulations to ensure a successful transfer.

This is consistent with the resource-based view of the firm in which the firm uses its intangible resources to develop a sustainable competitive advantage (Teece, 1977; Barney, 1991). A firm can develop organizational knowledge that will help it complete international technology transfer through codification. Although some of the laws and regulations are of a technical nature and need a professional, such as a lawyer, to interpret them, managers can gain experience by iteratively completing international technology transfers. This strengthens organizational learning and helps the firm gain an expertise so that it can successfully complete even more complex international technology transfers in the future. The development of managerial experience is one of the most important intangible resources that a firm can develop to create a sustainable competitive advantage that other firms are not able to imitate (Barney, 1991).

In fact, by applying a resource-based view of international technology transfer, the firm can view the entire process as a learning experience to develop its skills and to enhance its overall strategy (Tsang, 1997). Part of the learning curve for a firm is to be able to effectively assess each project's effectiveness, and this needs to be done on a case-by-case basis. Despite a firm's attempts to codify its learning, such as codifying other countries' laws and regulations regarding health and safety, into a database, the overall strategy of the firm needs to be considered. Planning for the effective use of the technology often requires a more detailed and customized plan than an "off the shelf" version that might have already been applied in another international technology transfer case (Contractor and Sagafi-Nejad, 1981). Customizing the firm's actions to the specific needs of the other country that is involved in the transfer will enhance the ability of the firm to successfully transfer technology.

Once the knowledge has been formally codified into a database, a structured training program can take place. Most of the firms in the qualitative sample described in this paper used computer databases as a way to store knowledge, but fewer firms used computer-based training as a method of disseminating the information. However, an e-learning system can enhance a firm's ability to manage the information and to conduct training (Nagle, 2002). Using a learning management system to train the workers about the key components of the technology transfer, the codified knowledge suddenly becomes a key resource that can be applied consistently and added to with each additional international technology transfer project. Again, the leverage of an intangible resource can be used to gain a sustainable competitive advantage, especially if competitors cannot imitate it (Barney, 1991). Experience and knowledge, if codified correctly, is a key factor in successful international technology transfer.

Further research needs to be done that ties the resource-based view of the firm, international technology transfer, and organizational learning. International technology transfer is clearly a process of transferring an organization's learning across firm and national boundaries, whether it is the technological database, process, or even the know-how of the operation of a piece of hardware. Bartlett and Ghoshal (1989) describe an organization's competence in being able to successfully do business globally as its ability to manage knowledge and learning. The knowledge and learning competence translates into a resource that needs to be managed for a competitive advantage. Future research in international technology transfer should examine the process of organizational learning and how to best manage knowledge to gain a competitive advantage.

REFERENCES

Adcock, A. (2003). Getting tech transfer right in China. *Managing Intellectual Property*, March, 2003, 44-46.

Barney, J. B. (1991). Firm resources and sustained competitive advantage. *Journal of Management, 17* (1), 99-120.

Barlett, C.A. and S. Ghoshal (1989). *Managing Across Borders: The Transnational Solution*. Harvard Business School Press.

Bozeman, B. (2000). Technology transfer and public policy: A review of research and theory. *Research Policy, 29*, 627-655.

Carley, K. (1990). Content analysis, originally prepared for *The Encyclopedia of Language and Linguistics*, Asher, R. E. et al. (Eds.), Edinburg, UK: Pergamon Press, July 30, 1990.

Contractor, F. J. and T. Sagafi-Nejad (1981). International technology transfer: Major issues and policy responses. *Journal of International Business Studies*, Fall, 113-135.

Davidson, W. H. and D. G. McFetridge (1985). Key characteristics in the choice of international technology transfer mode. *Journal of International Business Studies, 16* (2), 5-22.

Eisenhardt, K. M. (1989). Building theory from case study research. *Academy of Management Review, 14* (4), 57-74.

Glaser, B. G. and A. L. Strauss (1967). *The Discovery of Grounded Theory: Strategies for Qualitative Research*, Aldine de Gruyter.

Griffin, A. and J. R. Hauser (1992). The voice of the customer. *Marketing Science Institute*, Working Paper, Report Number 92-106.

Harkanson, L. and R. Nobel (2001). Organizational characteristics and reverse technology transfer. *Management International Review, 41* (4), 395-420.

Joyner, B. E. and M. H. Onken (2002). Communication Technology in International Technology Transfer: Breaking Time and Cost Barriers. *American Business Review*, June, 17-26.

Kumar, N. and K. S. K. Marg (2000). Foreign direct investment and technology capabilities in the developing countries: A review. *International Journal of Public Administration, 23* (5-8), 1253-1268.

Loewenstein, U. W., J. Klass, J. B. Hickey, J. Leek, and G. Joseph (1999). Technology transfers: International tax implications. *The International Tax Journal, 25*(2), 1-34.

Martinot, E., J. E. Sinton, and B. M. Haddad (1997). International technology transfer for climate change mitigation and the cases of Russian and China. *Annual Review of Energy Environment, 22,* 357-401.

McKennell, A. (1974). *Surveying Attitude Structures.* Amsterdam: Elsevier.

Muralidharan, R. and A. Phatak (1999). International R&D activity of US NMCS: An empirical study with implications for host government policy. *Multinational Business Review, 7* (2), 97-105.

Nagle, B. (2002). Technology and training, *Canadian HR Reporter,* February 25, 15-16.

Okabe, M. (2002). International R&D spillovers and trade expansion: Evidence from East Asian economies. *ASEAN Economic Bulletin, 19*(2), 141-154.

Osman-Gani, A. A. M. (1996). International technology transfer for competitive advantage: A conceptual analysis of the role of HRD. *Competitiveness Review,* (9), 9-19.

Pang, L. C. and J. Garvin (2001). Technology transfer in Northern Ireland: The development of university policy. *Irish Journal of Management, 22* (1), 193-212.

Petit, M. L. and F. Sanna-Randaccio (1998). Technological innovation and multinational expansion: A two-way link? *Journal of Economics, 68* (1), 1-26.

Porter, M. E. (1990). *The Competitive Advantage of Nations,* New York: The Free Press.

Ragan, C. C. (1987). *The Comparative Method,* Berkeley, CA: University of California Press.

Strauss, A. L. and J. Corbin (1990). *Basics of Qualitative Research.* Newbury Park, CA: Sage.

Teece, D. (1977). Technology transfer by multinational firms: The resource cost of transferring technological know-how. *Economic Journal, 87,* 242-261.

Tsang, E. W. K. (1997). Choice of international technology transfer mode: A resource-based view. *Management International Review, 37* (2), 151-168.

Videras, J. and A. Alberini (2000). The appeal of voluntary environmental programs: Which firms participate and why? *Contemporary Economic Policy, 18*(4), 449-461.

Yin, X. (1999). Foreign direct investment and industry structure. *Journal of Economic Studies, 26* (1), 38-57.

Zattler, J. K. (2002). Growth policies in developing countries: A transaction cost perspective. *Intereconomics, 37* (4), 212-222.

Building Capable Organization
via Technological Innovation:
A Study of a China's Leading Textile
Machinery Company

Jifu Wang
Sharon L. Oswald

SUMMARY. One of the most interesting structural adjustments from a centrally planned government to a market economy has taken place in China. When China's began preparation for entry into the World Trade Organization, the SOEs found themselves with little time to prepare for outside competition. The long-term influences of traditional centralized systems left companies in a state of redundancies, stagnation and operational obsolescence. Most lack flexibility in terms of structural change, were weak in technological innovation, had a large number of surplus employees, as well as inefficient production and business operations. This paper chronicles the transformation of one of China's largest lead-

---

Professor Jifu Wang, PhD, (in Management from Auburn University) is Assistant Professor, University of Houston, Victoria.

Sharon L. Oswald, PhD, is Professor and Department Head of Management, Auburn University, holds the Colonel George Phillips Privett Professorship in Business.

Address correspondence to: Jifu Wang, Department of Management and Marketing, School of Business, University of Houston, Victoria, 3007 N. Ben Wilson, Victoria, TX 77901-5731.

[Haworth co-indexing entry note]: "Building Capable Organization via Technological Innovation: A Study of a China's Leading Textile Machinery Company." Wang, Jifu, and Sharon L. Oswald. Co-published simultaneously in *Journal of Nonprofit & Public Sector Marketing* (Best Business Books, an imprint of The Haworth Press, Inc.) Vol. 13, No. 1/2, 2005, pp. 57-73; and: *Government Policy and Program Impacts on Technology Development, Transfer and Commercialization: International Perspectives* (ed: Kimball P. Marshall, William S. Piper, and Walter W. Wymer, Jr.) Best Business Books, an imprint of The Haworth Press, Inc., 2005, pp. 57-73. Single or multiple copies of this article are available for a fee from The Haworth Document Delivery Service [1-800-HAWORTH, 9:00 a.m. - 5:00 p.m. (EST). E-mail address: getinfo@ haworthpressinc.com].

Digital Object Identifier: 10.1300/J054v13n01_04

ing textile machinery manufacturing companies, Jingwei Textile Machinery Company Ltd. (Jingwei) in its transformation from a relatively inefficient and ineffective SOE to competitive leader in the textile machinery manufacturing industry. In just one year (1998 to 1999), Jingwei realized an astounding 96.2 percent increase in sales. Much of the company's success is attributed to a commitment to technological advancements and the foresight of top management. *[Article copies available for a fee from The Haworth Document Delivery Service: 1-800-HAWORTH. E-mail address: <docdelivery@haworthpress.com> Website: <http://www.HaworthPress. com> © 2005 by The Haworth Press, Inc. All rights reserved.]*

KEYWORDS. Competitiveness, China, government policy, business transformation

INTRODUCTION

The transformation of state-owned enterprises (SOEs) to competitive market entities lies outside the realm of traditional strategy research. The last ten years has seen emerging economies of the East struggling at different stages of development and facing different challenges and market responses. It is expected that the challenges and responses differ between firms, depending upon the nature of their individual competitive environments. As noted by Hoskisson, Eden, Lau and Wright (2000) the implications of the economic and political changes for these SOEs haunt strategy researchers.

The essence of strategic management posits that competitive advantage lies in the speed and accuracy in which a company responds to changes in demand. For the SOEs, the challenge went far beyond demand. The long-term influences of traditional centralized systems left companies in a state of redundancies, stagnation and operational obsolescence. Most lacked flexibility in terms of structural change, were weak in technological innovation, had a large number of surplus employees, as well as inefficient production and business operations (Jefferson & Singh, 1999; Steinfeld; 1998 Peng, 1997; Child and Lu, 1996). While much attention had been focused on the transformation of SOEs in the former countries of the Soviet empire, one of the more intriguing structural readjustments occurred in China. Behind the United States, China was the most attractive market for attaining foreign direct investment. Yet, its most critical challenge was to transition from a centrally controlled economy to a market driven economy. This transition was eminently important to China in order to become a full member of the World Trade Organization.[1] Consequently, the SOEs had little time to prepare for outside competition. Cutting the SOEs loose from state control, and making them competitive with the outside world was one of the government's top priorities (Peng and Luo, 2000).

One such SOE that made this transition was China's, Jingwei Textile Machinery Company Ltd (Jingwei), a major player in textile machinery manufacturing, producing three categories of products: natural fiber textile machinery, chemical fiber textile machinery and components, and special parts for textile machinery. By the mid-1990s, the vigorous development of China's industry resulted in an increasing demand for machine tools. Specifically, the consumption rate of machine tools and the market share of numerical controls and high-efficiency precision machine tools had increased from 6 to 30 percent of all machine tools in a six-year period of time. Domestic production of similar grade machine tools was insufficient to meet market demand resulting in a drastic increase in US imports. When the imports reached $688 million in machine tools during the first six months of 1998, it became evident that something needed to be done on the domestic front.

This paper chronicles Jingwei's transformation from a relatively inefficient and ineffective SOE to a competitive leader in the textile machinery manufacturing industry. In just one year (1998 to 1999), Jingwei realized an astounding 96.2 percent increase in sales. Much of the company's success is attributed to a commitment to technological advancements and the foresight of top management.

SOE: Historical Perspective

SOEs were a big player in China's national economy, accounting for 35-40 percent of the nation's gross national product and 60 percent of all state revenue. Of the 305,000 state enterprises, 118,000 were industrial (World Bank, 1997). The SOE constituted the nation's entire heavy industrial base: steel makers, machine builders, auto and truck manufacturers, and petroleum producers. These enterprises were the firms that provided basic industrial inputs for the economy, employment and social welfare for the vast majority of China's urban workers, as well as the bulk of fiscal revenues for most governmental levels (Steinfeld, 1998).

Most of the big state-owned enterprises, especially those with strong production capability and technical competence, were built in the 1950s. The national economy from the 1950s to 1980s was dominated by Soviet-style centralized planning. Consequently, the national economic goal was to provide employment to everyone with production capability. As was typical of centrally planned economies, the industrial base focused on machinery, electronics, shipbuilding and textiles. Many on the SOEs developed into "small all-inclusive societies" heavily involved in social welfare. This was best illustrated by the fact that most SOEs owned, in addition to their core businesses, anything from kindergartens to hospitals.

However, the long-term influences of traditional centralized systems left companies in a state of redundancies, stagnation and operational obsolescence. Most lacked flexibility in terms of structural change, were weak in technological innovation, had a large number of surplus employees, and were crippled by inefficient production and business operations (Jefferson & Singh, 1999; Steinfeld, 1998; Peng, 1997; Child and Lu, 1996).

Because of the poor state of SOEs, the government's decision to transform the industrial base to a market economy brought with it many challenges. It became clear that businesses would have to experience retrenchment and divestiture in order to gain high levels of efficiency and effectiveness. This was the opportunity that faced Jingwei.

The Move Toward Reform

The new efficiency policies put great pressure on state-owned enterprises. Controlled by the bureaucracy and operated with state subsidies for decades, many lacked the ability to adapt to external changes or innovate enough to survive. But, because China at the time was committed to a rapid admittance into the World Trade Organization, quick transformation was essential. Still, the SOEs were left with little time to prepare for outside competition. Reform brought a greater dependence by the SOEs on public subsidies. The government's desire to contain its own budget deficit resulted in the state banking system providing the necessary financial support.

Beginning in 1980, the government gave many enterprises increasing autonomy over their operations. They were permitted to retain a share of profits for wage bonuses and new investment, they were given greater autonomy over production decisions and wages, and they were allowed to adopt a "management responsibility system" (seen as a counterpart to agriculture's household responsibility system), and in some cases they were permitted to recruit new management. Central and local governments usually negotiated these new freedoms on a firm-by-firm basis. As a result, the operational environment varied enormously between firms, across regions, and across sectors. In 1984 the Enterprise Bill of Rights formalized these changes and created an additional impetus for growth.

The increased autonomy of state enterprises allowed them to benefit from China's dual-track pricing system, which was introduced in the early 1980s. Under this system, government planners typically set commodity prices but allowed all output above the government planned allotment to be sold at market prices. Since the volume of planned output rarely changed, enterprises sold an increasing amount on the open market. Thus all growth and development occurred at market prices, improving resource allocation. Because no enterprise was made worse off, the reforms received enthusiastic support. Today more than 95 percent of industrial output is sold at market prices (Cheng, 2003).

The decentralization of management decisions boosted firm productivity across industries. But relative to the rest of the economy, state enterprises languished, with slow growth and declining profits. In part this was because state enterprises, unlike their non-state competitors, were required to provide to their employees job security and a range of social services including housing, education, and health care. Yet slacking performance also reflected a deeper malaise rooted in the poor investment decisions of the past and in an "iron rice bowl" system that did not penalize low productivity.

More recently, however, lower subsidies, tighter credit, and growing competition unmasked the inefficiency and poor financial condition of many state enterprises. Consequently, reform came in the shape of mergers, lease agreements, management contracts, worker and management buyouts, and bankruptcies. The central government selected 1,000 (from more than 100,000) state-owned industrial firms, including Jingwei, to form the core of China's modern enterprise system.

THE RESEARCH METHODOLOGY

In order to best analyze the transformation of Jingwei from a SOE to competitive leader in the textile machinery manufacturing industry, a research design had to be established. Two fundamental questions exist at the core of strategy research (Bowman, 1995): "What makes some firms more successful than others?" and "How do I make this particular firm successful?" Researchers, such as Chandler, Andrews, Mintzberg and others, relied on case studies and company histories to generate a wealth of theories and insights into corporate practice (Allison, 1971, Bower, 1970; Chandler, 1962; Cyert & March, 1963; Mintzberg, 1979).

Eisenhardt (1989) provided a road map to developing theories from case studies that may be appropriate in an emerging market context. An integral part of this approach is the development of research instruments that can be used in quantitative studies. In addition, the combination of quantitative and qualitative data in emerging market research can be particularly useful in yielding novel, relevant, and reliable insights.

Semi-structured interviews were conducted with key respondents regarding how they successfully initiated strategic change, how they achieved their goals in terms of competitive product development, and how they had to change their policies and procedures to achieve the fit among the structure, strategy and culture. People designated in the interview were managers from departments of technology, HR, production, marketing and planning.

As previously determined, the decision to use case-based research was well documented and seemed most appropriate for this study. However, the actual methodology was something less clearly established. According to Hoskisson, Eden, Lau and Wright (2000), the strategy literature is only recently beginning to come to grips with the economic implications and political changes for SOEs. Consequently, a paucity of research on SOEs has existed. Among the problems challenging case-based research in emerging market economies include:

1. Theories promulgated for developed market economies may not be appropriate for emerging economies;
2. Researchers are troubled by sampling and data collection problems, difficulties in measuring firm performance, and a variety of timing issues; and
3. Emerging economies are not a homogeneous or clearly identifiable and recognizable group. What's more, it is likely that different segments of emerging economies find themselves in different states of development.

Another problem associated with case-based research on emerging economies, and in particular in China, is the absence of a culture involving independent researchers. Under a controlled economy, SOE managers were highly bureaucratic and political. They did not need to provide information to outsiders, and were able to operate behind a wall of secrecy. Thus, Lee and Miller (1999) highly recommended the research method in collaborative projects with local researchers using face-to-face interviews, which can be a key means of gaining access to reliable data sources. Thus, this was the method adopted for the present study.

THE CASE OF JINGWEI

Competitive Situation

The vigorous development of China's industrial base brought increasing demand for machine tools. According to a China Machine Tool & Tool Builders' Association report, during the 8th five-year plan period (1990-1995) the import of machine tools steadily increased to $2.2 billion USD in 1995. In 1996 the import value reached the record breaking level of $2.52 billion USD. This surge was driven by the belief and subsequent occurrence of the Chinese government's cancellation of the tariff exemption on imported capital equipment at the end of the year. The import of machine tools in 1997 dropped to $1.58 billion USD. Starting from January 1, 1998, the Chinese government resumed tariff incentives and preferential treatment on imports of high technology and machine tools for foreign-funded projects. China imported 688 million USD worth of machine tools in the first six months of 1998, up 9.77 percent from the previous year.

Additionally, there was a rapid increase in the demand for NC (numerical controls), and high-efficiency and precision machine tools of medium, medium-high, and high grades. According to a China Machine Tool & Tool Builders' Association report, the consumption rate of machine tools and the market share of NC and high-efficiency precision machine tools increased from 6 to 30 percent of all machine tools during the period of 1990-1996. Domestic production of similar grade machine tools has been insufficient to meet market demand, and, as a result, imports of foreign manufactured machine tools increased dramatically during the previous years.

Strategic Challenges

Many strategic challenges faced Jingwei in its transformation. Probably the greatest challenge for Jingwei in the new market economy was technology innovation. Jingwei traditionally bought the most advanced products or technologies produced by its competition in order to better understand its marketplace challenges. Frequently, Jingwei purchased the blueprint and the patent, and spent several years digesting the new technology, later turning it into commercial products. By the time the products hit the market they bordered on obsolete. Rather than be a leader in the industry, Jingwei was in a constant state of "catch up."

Compensation was another major problem. Centrally planned economies promoted equal pay for all; consequently, employees and managers were not paid competitively. This perpetuated the lack of motivation shown by the staff. While global competition affects products, technology, quality, production costs, and marketing strategies, it is foremost about human talents. And, humans want to be appropriately rewarded for their efforts. As a result, Jingwei could not assume a leadership role without a competitive compensation system.

Still another challenge Jingwei had to overcome was that of high turnover. It was particularly difficult to keep junior technicians. Jingwei had become a training ground for joint venture companies, as well as private and foreign owned companies. Two or three years after graduation, headhunters "stole" these technicians away from Jingwei. Shitong Chen, CEO of Jingwei, illustrated this point with his personal experience. "My son, who graduated from a university, majoring in computers, went to work for a foreign owned company. His monthly salary was higher than mine. I have been working for so long, and managing such a large company, which has been very successful.

However, I was paid less than new college graduates. On the other hand, if I need people such as my son, I have no way to afford to attract them here."

Chen noted that this was a problem shared by all SOEs. "The government knows it and we, the executives of SOEs, know it, but it is just hard for us to solve it under the present conditions, which is an accumulation and legacy of the centrally controlled system of the past 50 years." The needed social infrastructure was lacking.

The fourth major challenge for Jingwei was quality. CEO Chen argued that labor costs were not low in China. While on the surface, individual wages seemed quite low, the inefficiencies of the full employment system under a centrally planned government allowed for several people to be employed to do the work of one. And, the skills of these people were quite low, particularly in comparison to Taiwan, as was noted by Chen. Taiwanese firms focused on mastering the core technology and assembly secrets, and outsourcing other factory jobs, resulting in considerably higher profits than in Mainland China.

Strategic Responses

With the challenges of technology, pay, turnover and quality looming, Jingwei set out on a path toward competitiveness. The first step was to establish a contemporary integrated manufacturing system (JW-CIMS). The goal of JW-CIMS was to improve information integration, implement process re-engineering and optimization, and achieve enterprise integration. This system required huge investments. Jingwei invested 15 million yuan into its research and application over a 10-year period prior to 1996. By 1996, Jingwei had attained the operating scale and company size required by the government to issue public stock. The 10 million yuan of equity was used for the development and modifications of JW-CIMS subsystems.

JW-CIMS consisted of manufacturing automation system (MAS), computer-aided quality system (CAQ), management information system (MIS), as well as computer-aided design (CAD), computer-aided process planning (CAPP), and computer-aided manufacturing (CAM). A team of about 100 worked on the development and application of CIMS. In 2000, Jingwei had three super small machines, 70 CAD work stations, 250 peripheral PCs, and 100 graphic printing machines, which constituted the information network covering all the systems, departments for various products and businesses. With the support of the data system, considerable work was done in enterprise management, product development, technology innovation, product manufacturing, sales and marketing.

Engineering Design System

Since 1990, GT-CAD was used to establish a database of parts for old products, and to provide support for improvement modifications. Two- and three-dimensional technology was used for new product design and development. Optimization technology was used for designing crucial parts and providing for product standardization. By 1996, drafting boards were eliminated and all new products were designed through the engineering design system. As a result, many new application methods were developed for CAPP and CAM systems to effectively improve the engineering design technology.

Manufacturing Automation System

Jingwei invested heavily in Computer Numerical Control (CNC) and Flexible Automation System (FAS) and spent another 90 million yuan on computer numerical machines, production center, and flexible production lines. In 1999, computer numerical control machines accounted for more than 30 percent of all the fixed assets in Jingwei, and CNC technology greatly enhanced the manufacturing capability, providing a solid foundation for new product development and product quality. Two FAS lines were integrated through networks and MIS, using CAD/CAPP/CAM. A large-scale flexible production workshop was established, thus providing Jingwei with the capability to develop and manufacture new, high-technology products.

Quality Control System (QCS)

A Quality Control System (QCS) worked together with MIS to collect and analyze information quality data. The major functions of QCS were the planning of quality, scheduling of quality index, evaluation and assessment. QCS was also used to implement and evaluate the ISO9000 standards. As a result of QCS product quality and quality management at Jingwei was greatly improved.

Enterprise Management System

Enterprise Management System (EMS) was a modified version of Manufacturing Resource Planning II (MRP II). EMS covered all the strategic functional units, such as marketing, material procurement, production planning, manufacturing control, inventory management, accounting and finance, equipment management, human resource management and office automation. It was an advanced and effective management information system, which integrated the material flow, information flow, and value flow. The entire operational process was integrated, resulting in more efficient production scheduling, shorter production cycles for major products, and timely and accurate response to customer demand. It further enhanced the firm's ability to adapt to changing markets as well as improved capital utilization and inventory management.

Key Success Factors

During the previous 10 years, Jingwei paid special attention to training and development of top management, considered the central nervous system of the organization. Having analyzed the real situations, both internally and externally, the top management team made several decisions that proved crucial to Jingwei's long-term success.

Accumulating Capital from the Stock Market

Most Chinese firms financed their operations through bank loans. Many firms got into a malignant cycle of borrowing, and suffered from heavy debt. Jingwei's top management team understood this trap and chose an alternative path. In the late 1980s, Jingwei started its preparation to enter the stock market. In China, few firms attempted to enter the stock market because the government had strict regulations and high requirements for firm performance and size. After 10 years of hard work, Jingwei succeeded in issuing stock to the public in 1996, and, in turn, gathered enough funds for product research, development and innovation. In May 2000, Jingwei issued stocks for the second time to finance its strategic actions.

Proactive Actions in the Market Place

Jingwei was one of the few companies in China with advanced CIMS, thus increasing the firm's ability for quick market response. The integration of information processing and research and development capabilities provided Jingwei with a competitive advantage. As previously noted, each year Jingwei purchased, for research purposes, the most advanced equipment and facilities in the world, to include computer numerical controlled machines, flexible production lines, etc. Based on developments in the global market, Jingwei tried to create next generation products and gain the first mover's advantage. Chen explained that market followers were always playing catch-up and were always attempting to meet the strategic minimum in the market. Without proactive actions, he continued, Jingwei would never win in the market place.

Strategic Asset Restructuring

When asked about the strategic responses to challenges facing Jingwei, Chen commented that strategic asset restructuring was one of the core actions leading to a competent organization. In the past, the lack of scientific scheduling and market demand-supply mechanisms resulted in over-capacity in the market. Today, most firms were restructuring assets by spinning off some of their units along the value chain in attempt to find a core competency. Jingwei, on the other hand, adopted a related diversification strategy. Jingwei's previous machinery could only produce fine cotton yarn, which is the last of five cotton processes. In a strategic move, Jingwei diversified into a full line of high-quality machines for cotton processing. Chen made an analogy: "In the past, we were only one war ship, and other producers were independent ships fighting on their own, whereas now we have grown into a fleet of ships, enjoying economies of scope." In fact, what Jingwei had done is to combine the advantages of different assets and optimize the operation–taking the lead in its industry.

Strategies for Human Talents

In order to attract the best people in the industry, Jingwei moved its headquarters from Shanxi Province to Beijing, leaving the production base in Shanxi. The technology center was set up in Beijing, with research and development responsibilities. Because the production base and technology center were physically located in two different localities, different pay and compensation systems were used, thus alleviating problems with paying technical employees at a higher wage than their production counterparts. Other economic incentives were implemented. For example, the firm purchased an apartment worth 1.4 million yuan. Every month the firm paid the majority of the mortgage. This was used as an incentive to employees to stay with Jingwei.

Jingwei's Performance Evaluation

Sales

Jingwei reported sales of 803.59 million yuan ($97.09 million USD) for the year ending December 1999. This represented an unprecedented increase of 96.2 percent (Table 1) over 1998 when the company's sales were 409.63 million yuan.

During 1999, the company's sales increased at a faster rate than the three major competitors in the Asian market:[2] Shanghai Erfangji Textile Machinery Co., Lakshmi Machine Works Limited of India, and China Textile Machinery Stock Ltd. Shanghai Erfangji Textile Machinery Co. reported sales of 291.04 million Chinese Renmimbi ($35.16 million USD) in 1999. In 2000, Lakshmi Machine Works Limited of India had sales of 4.31 billion Indian Rupees ($92.58 million USD), of which 46 percent was spinning preparatory machinery. China Textile Machinery Stock Ltd. generated sales of 276.89 million yuan (US$33.45 million) in 1999 (Table 1).

TABLE 1. Sales Comparisons

Company	Year Ended	Sales (US$mlns)	Sales Growth	Sales/Emp (US$)	Largest Region
Jingwei Textile Machinery Company Ltd	Dec 1999	97.089	96.2%	11,265	N/A
Shanghai Erfangji Textile Machinery Co.	Dec 1999	35.163	−5.4%	N/A	N/A
Lakshmi Machine Works Limited	Mar 2000	92.581	5.0%	N/A	India (100.0%)
China Textile Machinery Stock Ltd	Dec 1999	33.453	0.7%	5,396	N/A

Stock Performance

Prior to 1998, Jingwei's stock performed poorly. For example, in 1997 the stock traded at 3.28 yuan, versus 1.52 Chinese RMB on February 2001. In 1997, the stock retreated significantly from its high because of its strategic expansion, and by the end of the year was at 0.82 yuan. Since that time the stock price more than doubled. For the 52 weeks ending February 23, 2001, Jingwei's stock increased by 126.4 percent to 1.52 yuan. During the previous 13 weeks, the stock has increased 38.8 percent.

During the 12 months ending December 31, 1999, earnings per share totaled 0.21 yuan. Thus, the Price/Earnings ratio was 7.23. Earnings per share rose 950.0 percent in 1999 from the previous year.

As seen in Table 2, the three major competitors to Jingwei varied greatly in terms of price to sales ratio: trading from 0.23 times all the way up to 4.91 times their annual sales. Jingwei was trading at 0.78 times book value. Since the price to book ratio is less than 1, this means that theoretically, the net value of the assets is greater than the value of a company as a going concern.

The market capitalization was 643.10 million yuan ($77.70 million USD) at the time of this case. Closely held shares (i.e., those held by officers, directors, pension and benefit plans and those shareholders who own more than five percent of the stock) amount to over 50 percent of the total shares outstanding: thus, it was impossible for an outsider to acquire a majority of the shares without the consent of management and other insiders. The capitalization of the floating stock (i.e., that which is not closely held) was 21.32 million *yuan* ($2.58 million USD).

Profitability Analysis

On the 803.59 million Chinese RMB in sales reported by the company in 1999, the cost of goods sold totaled 595.94 million Chinese RMB, or 74.2 per-

TABLE 2. Summary of Company Valuations

Company	Date	P/E	Price/Book	Price/Sales	52 Wk Pr Chg
Jingwei Textile Machinery Company Ltd	2/23/01	7.2	0.78	0.80	126.40%
Shanghai Erfangji Textile Machinery Co.	2/19/01	105.2	1.59	4.91	179.25%
Lakshmi Machine Works Limited	2/23/01	4.2	0.41	0.23	−51.47%
China Textile Machinery Stock Ltd	2/19/01	N/A	5.83	3.21	220.00%

cent of sales. In other words, the gross profit was 25.8 percent of sales. This gross profit margin was slightly lower than that achieved by the company in 1998, when cost of goods sold totaled 73.4 percent of sales (Table 3).

The company's earnings before interest, taxes, depreciation and amortization (EBITDA) were 111.89 million Chinese RMB, or 13.9 percent of sales. This EBITDA margin was better than the company achieved in 1998, when the EBITDA margin was equal to 10.0 percent of sales.

In 1999, earnings before extraordinary items at Jingwei Textile Machinery Company Ltd were 88.60 million Chinese RMB, or 11.0 percent of sales. This profit margin was an improvement over the level the company achieved in 1998, when the profit margin was 2.1 percent of sales. The company's return on equity in 1999 was 12.0 percent. This was significantly better than the 1.2 percent return the company achieved in 1998.

Inventory Analysis

As of December 1999, the value of the company's inventory totaled 447.71 million *yuan*. Since the cost of goods sold was 595.94 million yuan for the year, the company had 274 days of inventory on hand. In terms of inventory turnover, this was a significant improvement over December 1998, when the company's inventory was 312.83 million yuan, equivalent to 380 days in inventory.

DISCUSSION AND CONCLUSIONS

China is an important nation that portends to become the world's largest economy during the new millennium. In purchasing power parity, China is al-

TABLE 3. Profitability Comparison

Company	Year	Gross Profit Margin	EBITDA Margin	Earns Bef. Extra
Jingwei Textile Machinery Company Ltd	1999	25.8%	13.9%	11.0%
Jingwei Textile Machinery Company Ltd	1998	26.6%	10.0%	2.1%
Shanghai Erfangji Textile Machinery Co.	1999	N/A	N/A	2.3%
Lakshmi Machine Works Limited	2000	16.2%	14.6%	5.0%
China Textile Machinery Stock Ltd	1999	N/A	N/A	−63.5%

ready the world's second largest economy behind the United States, touting the world's largest population base of nearly 1.3 billion people. Behind the United States, it is the most appealing market for attracting foreign direct investment. Yet, its most critical challenge is to complete a transition of its centrally controlled economy to a market driven economy. This transformation puts the greatest pressure on the government's state-owned enterprises, which must learn to compete in an increasingly open and competitive world. Cutting them loose from state control, and making them competitive with the outside world, was one of the government's top priorities.

Jingwei Textile Machinery Company Ltd (Jingwei), a major player in textile machinery manufacturing, transformed in little more than a year from a relatively inefficient and ineffective SOE to a competitive leader in the textile machinery manufacturing industry. Jingwei identified technological innovation as its first and foremost challenge. The top management believed that they were facing a monumental challenge of global competition. Management realized that competitive survival meant standing up to both the domestic and foreign opponents. To win in the marketplace, management knew it would have to re-engineer the company–change its practices–rather than continue to operate in a catch-up mode.

Technology underpins China's fastest growing industries. It was believed that this would provide the tools needed to compete in every business, and drive growth in every major market segment. Like Jingwei, advanced technologies has been the vehicle used by SOEs to improve productivity and profitability–allowing them to pay higher wages and increase employment more rapidly than less technical firms. Advances in technology have spurred the growth and creation of core competencies, facilitated the conduct of business worldwide, and accelerated the development of new products, services, and capabilities, allowing the flexibility to respond changing global market demands. Technology continues to be the decisive factor in the speed with which former SOEs like Jingwei have found a position in the international marketplace. Even before joining the WTO, China's approach to globalization and constant economic growth during a time when the rest of Asia was in a financial crisis, caught the eye of much of the world market (Cheng, 2003; Baark, 2001).

Admission into the WTO was an important step for China. Prior to accession, China had the largest economy of those countries not WTO members and was believed to be the world's third largest economy. Some predictions have suggested that China's share of the world trade market could reach 9.8% by 2020. Interestingly, in the first year after WTO admission China's exports increased by 19.4 percent. This rapid growth in trade has been directly attributed to WTO entry (Cheng, 2003; China.org.cn, 2002).

Since accession into the WTO, China's technology market has grown as expected. While the fastest growth has occurred in the information technology, other segments of the technology-based markets have seen similar increases. While it is early to tell, predictions suggest that the textile industry will benefit from WTO membership (Cheng, 2003). Future growth is expected as long as companies are flexible enough to de-emphasize slow growing business and capitalize on new growth areas (Webb, 2003).

While Jingwei is a representative of the most successful firms, other SOEs have made similar strides. Having transformed their "branch plant" operations to vibrant competitive organizations that are beginning to enter the global marketplace, these SOEs have transformed their business activities. To improve their capital structures, they have adopted modern corporate structures and convinced government ministries and banks to convert excessive debt into stock holdings with positions on the board of directors. Some firms have issued stock in Hong Kong and the US capital markets. Others have developed joint venture partners to provide additional capital, as well as needed know-how and market access. Several have improved financial structures by spinning off operations into new joint ventures, merging with similar domestic operations, or selling assets for cash.

As illustrated by Jingwei, survival was, at least in part, predicated on investing in distinctive competencies and in improving technology. Jingwei is only one of the success stories for China. With its emphasis on technological improvements and entry into the WTO, China is a trading force to be reckoned with. Already the influx of foreign investment has increased and more multinationals have put their money in China. In the first nine months of 2002, China's actually utilized foreign investment had increased nearly 23 percent.

Some of the big investors in China have been other ASEAN states, in particular Japan, South Korea, and Singapore (Cheng, 2002). China's emphasis on quality and its investment in technology have helped to place the country on the worldwide market. With the entry into the WTO, China has begun an irreversible trend "to integrate its economy into the global economic framework" (China.org.cn, 2002). And, companies like Jingwei have paved the way. Companies like Jingwei have proven that investment in technology and innovation is an essential component to competitiveness in today's global economy.

AUTHOR NOTE

Jifu Wang served as an executive in top management for several firms in Shenzhen Special Economic Zone, China and has rich management experiences in international business. Dr. Wang has published several research articles on international strategy with emphasis on capable organizations.

Sharon L. Oswald has published more than 50 articles in academic journals and has been an invited lecturer both domestically and internationally. Much of her research interest today lies in strategic international business issues and project management.

NOTES

1. China became a full member of the WTO on December 11, 2001.
2. Note: not all of these companies have the same fiscal year: the most recent data for each company are being used.

REFERENCES

Allison, G. T. (1971), *Essence of Decision: Explaining the Cuban Missile Crisis*, Boston: Little, Brown.

Adler, N. J., Brahm, R., & Graham, J. L. (1992), "Strategy Implementation: A Comparison of Face-to-Face Negotiations in the People's Republic of China and the United States," *Strategic Management Journal*, 13: 449-466.

Adler, N. J., Campbell, N. & Laurent, A. (1988), "In Search of Appropriate Methodology: From Outside the People's Republic of China Looking in," *Journal of International Business Studies*, Spring: 61-74.

Baark, E. (2001), "Technology and Entrepreneurship in China: Commercialization Reforms in the Science and Technology Sector," *Policy Studies Review*, 18(1): 112.

Bower, J. L. (1970), *Managing the Resource Allocation Process*, Boston: Division of Research, Graduate School of Business Administration, Harvard Business School Press.

Bowman, E. H. (1995), "Strategy History: Through Different Mirrors," *Advances in Strategic Management*, 11(A): 25-45.

Chandler, A. D. (1961), *Strategy and Structure: Chapters in the History of American Industrial Enterprise*, Cambridge, MA: The M.I.T. Press.

Chen, Shitong (2000), Personal Interview.

Cheng, J. Y. S. (2003), "Regional Impact of China's WTO Membership," *Asian Affairs*, 217-236.

Child, J. & Lu, Y. (1996), "Institutional Constraints On Economic Reform: The Case of Investment Decisions in China," *Organization Science*. Vol. 7: 60-67.

China.org.cn (2002), China's WTO Updates.

Cyert, R. M. & March, J. G. (1963), *A Behavioral Theory of the Firm*, Englewood Cliffs, NJ: Prentice-Hall.

Eisenhardt, K. M. (1989), "Building Theories From Case Study Research," *Academy of Management Review*, 14(4): 532-550.

Jefferson, G. H., & Singh, I. (1997), "Ownership Reform as a Process of Creative Reduction in Chinese Industry," in A. M. Babkina, Ed., *Domestic economic modernization in China*, Commack, NY: NOVA Science Publishers, Inc.

Hoskisson, R. E., Eden, L., Lau, C. M. & Wright, M. (2000), "Strategies in Emerging Economies," *Academy of Management Journal* 43(3): 249-267.

Lee, J., & Miller, D. (1999), "Strategy, Environment and Performance in Two Technological Contexts: Contingency Theory in Korea," *Organization Studies*, 17: 729-750.

Patton, M. Q. (1987), *How to Use Qualitative Methods in Evaluations*. Newbury Park, CA: Sage.

Peng, M. W. (1997), "Firm Growth in Transitional Economies: Three Longitudinal Cases from China, 1989-96," *Organizational Studies*, 18: 385-413.

Peng, M. W., & Luo, Y. (2000), "Managerial Ties and Firm Performance in a Transition Economy: The Nature of a Micro-Macro Link," *Academy of Management Journal*, 43(3): 486-501.

Riordan, C. M., & Vandenberg, R. J. (1994), "A Central Question in Cross-Cultural Research: Do Employees of Different Cultures Interpret Work-Related Measures in an Equivalent Manner?" *Journal of Management*, 20: 643-671.

Steinfeld, E. S. (1998), *Forging Reform in China: The Fate of State-Owned Industry*, Cambridge University Press.

Wang, J. (2001), *Strategic challenge, strategic responses, and strategies: Study of Chinese state-owned enterprises*. Unpublished doctoral dissertation, Auburn University, Alabama.

Webb, Alysha (2003), "China: Where It Pays to Stay Nimble," *Business Week Online*, March 5, 2003.

World Bank. (1997), *China 2020: Development Challenges in the New Century*.

Yin, R. K. (1994), *Case Study Research: Design and Methods*, 2nd ed., SAGE Publications.

The Effects of Changing Technology
and Government Policy
on the Commercialization of Music

Gregory K. Faulk
Robert P. Lambert
Clyde Philip Rolston

SUMMARY. Government policy in the form of copyright protection laws fostered the commercial development of music. Technological innovation spawned the recording, movie and radio and television broadcast industries, expanding the economic horizon of music. Readily available recording devices, digital recording and Internet peer-to-peer song swapping have led to piracy problems. Although digital electronic music delivery could increase industry revenues, lack of consumer portability and conflicts between copyright protection for the industry and fair use provisions for individuals precluded widespread adoption. Choices and costs for consumers and industry growth and profits have

Gregory K. Faulk, PhD, (E-mail: faulkg@mail.belmont.edu), is Associate Professor of Finance, and Robert P. Lambert (E-mail: lambertr@mail.belmont.edu) is Professor of Marketing, both at College of Business, Belmont University, 1900 Belmont Boulevard, Nashville, TN 37212.

Clyde Philip Rolston is Associate Professor of Music Business, Mike Curb School of Music Business, Belmont University (E-mail: rolstonc@mail.belmont.edu).

[Haworth co-indexing entry note]: "The Effects of Changing Technology and Government Policy on the Commercialization of Music." Faulk, Gregory K., Robert P. Lambert, and Clyde Philip Rolston. Co-published simultaneously in *Journal of Nonprofit & Public Sector Marketing* (Best Business Books, an imprint of The Haworth Press, Inc.) Vol. 13, No. 1/2, 2005, pp. 75-90; and: *Government Policy and Program Impacts on Technology Development, Transfer and Commercialization: International Perspectives* (ed: Kimball P. Marshall, William S. Piper, and Walter W. Wymer, Jr.) Best Business Books, an imprint of The Haworth Press, Inc., 2005, pp. 75-90. Single or multiple copies of this article are available for a fee from The Haworth Document Delivery Service [1-800-HAWORTH, 9:00 a.m. - 5:00 p.m. (EST). E-mail address: getinfo@haworthpressinc.com].

Available online at http://www.haworthpress.com/web/JNPSM
Digital Object Identifier: 10.1300/J054v13n01_05

been affected more by changes in technology and government policy than by variations in industry structure. Deregulation and technological development have changed the radio broadcast and concert promotion industries from monopolistically competitive to oligopolies, which in turn has threatened the growth of the music industry. *[Article copies available for a fee from The Haworth Document Delivery Service: 1-800-HAWORTH. E-mail address: <docdelivery@haworthpress.com> Website: <http://www.HaworthPress. com> © 2005 by The Haworth Press, Inc. All rights reserved.]*

KEYWORDS. Commercialization of music, music copyright protection, technological innovation in music, music industry structure

INTRODUCTION

Government policy creates the economic environment within which a particular industry can develop commercially without violating the rights of citizens. Technological innovation can foster commercial development in that industry as well as create new industries that provide opportunities for the existing industry. As technology develops, government policy adapts the economic environment to incorporate technological advances. The industry economic structure, be it monopolistic competition (many competitors offering slightly differentiated products), oligopoly (only a few firms that dominate an industry), duopoly (an oligopoly with two firms) or monopoly (a single firm in an industry), can be influenced by the convergence of government policy and technology. Consumer choices, product prices and industry profits can be affected by industry economic structure. In general, consumers have more choices, and prices and industry profits are lower in monopolistically competitive industries than in monopolies.

The consumer choice, product price and industry profits tradeoff is not as clear cut in oligopolies or duopolies. An oligopoly with a contestable market is effectively the same in terms of the consumer choice/product price/industry profits tradeoff as monopolistic competition since firms have no market power over consumers. Bertrand oligopolies also resemble monopolistic competition since firms react to each others pricing decisions. Stackelberg oligopolies, characterized by a lead firm that sets output levels and followers that react once the lead firm makes a decision, have consumer/industry tradeoff characteristics of both monopolistic competition (following firms) and monopolies (lead firm). Cournot oligopolies, where firms react to each others output, have consumer/industry tradeoff characteristics closer to monopolies than monopolistic competition. Although illegal in the United States, oligopolies characterized by collusion are effectively monopolies (Baye and Beil, 1994).

Business models also develop as a consequence of industry structure, government policy and technology. Occasionally, a technological advance arises that completely reshapes industry economic structure and business models. This paper analyzes the effect of technology development and government policy on the commercialization of music.

The legal foundation of the music industry in the United States is copyright protection which gives artisans the incentive to create new music. The first group to commercialize copyright protection were music publishers, who profited from the sale of sheet music to the public. The technological innovations of recorded music, talking motion pictures, radio and television broadcasting created new industries that furthered the commercial development of music. The record label in particular became the driving economic force in the commercialization of music. Due to the costs of discovering music and delivering music, the industry structure became an oligopoly.

The technological development of tape recording both stimulated (increased sales) and repressed (piracy) music industry revenues. The ability of individuals to freely copy and distribute a copyrighted product that they purchased created a legal conundrum. Government policy as reflected in copyright law had the duty to protect the artisans and their business agents (copyright protection) as well as the individual users of their creations (copyright fair use). Criminal prosecution by the music industry was its only remedy for the piracy problem, and it was problematical at best.

The development of digitally based Internet song delivery and reproduction exponentially increased the piracy problem within the music industry. Congress revised copyright law to deal with the digital/Internet age and gave the music industry stronger protection against copyright infringement (piracy). However, the piracy problem persists. An industry duopoly was formed when record labels allied with Internet portals (web sites) to electronically deliver music on a subscription fee basis. This business model can substantially reduce music delivery costs and has the potential to increase revenues through "on demand" delivery of music. However, it has met with consumer resistance due to lack of portability as well as the availability of free songs through peer-to-peer Internet portals. Artist to consumer electronic delivery systems have developed to compete with the record labels. If this alternative music delivery system survives the industry structure will become more competitive.

The promotional value to the music industry of radio broadcasting and concert promotion has been curtailed by deregulation in the telecommunications industry, which has transformed the telecommunications industry from monopolistically competitive to an oligopoly. The technological breakthrough of digital broadcasting may further entrench that oligopoly. Satellite radio has the potential to challenge the dominance of traditional land based radio broadcasting. However, the present duopolistic nature of that nascent industry is not conducive to expanding promotional opportunities for the music industry.

The following analysis elucidates the issues summarized above. It hinges on the creations of cable television, digital recording and peer to peer song swapping on the Internet, quantum leaps in technology that changed the industry structure and business model and generated changes in government policy.

AURAL/RECORDING/BROADCASTING DEVELOPMENT
1790s TO 1970s

The economic foundation of the music industry in the United States is copyright protection, created under Article 1, Section 8 of the Constitution of the United States. This law was enacted on May 31, 1790 and signed by President George Washington (Sanjek, 1988a). Under the law, songwriters became entitled to time delimited compensation when consumers used their product. The commercialization of music came about largely through the efforts of music publishers, who, due to the capital investment involved in engraving and printing, copyrighted and sold sheet music on a subscription basis. Their product was utilized in live performances of musical compositions by singers and/or musicians. Early in the nineteenth century, following the lead of the American book trade, music publishers abandoned the subscription business model in favor of an "off the shelf" model where consumers would purchase whatever music sheets they desired (Sanjek, 1988a). The nature of the industry was competitive. In Philadelphia, the early hub of music publishing, thirty-two music publishers entered and exited the music business by the turn of the nineteenth century (Sanjek, 1988a).

The patented inventions of the tin foil cylinder based "talking machine" by Thomas Edison in 1877, commercially manufactured by Columbia Phonograph Company, and the master recording based celluloid discs of Emile Berliner in 1888, marketed by the Victor Talking Machine Company, sparked commercial interest and competition in the mechanical reproductions of musical works (Sanjek, 1988a). The early recorded music business was both a duopoly and a joint venture based monopoly. A pact between Victor and Columbia pooled their disk patents and allowed the companies to operate under a cross-license agreement (Sanjek, 1988a). Their patents and high startup costs barred the entry of competitors.

The original basis for copyright protection in the music industry, Article 1, Section 8 of the United States Constitution, was written before the invention of recording technology, which became a significant factor in the commercialization of the industry. Disputes between music publishers and the recording companies over the copyright status of songs as well as legal conflicts over public performances of copyrighted music without permission led to the enactment of the Copyright Act of 1909. The Act granted songwriters and music publishers that registered and notified others of the copyright status of their works the exclusive right to perform the copyrighted work for profit. The Act also extended copyright protection to and established a royalty rate for mechanically recorded music (Moser, 2002).

With the inventions of talking motion pictures, radio and television (TV) broadcasting, three new outlets were created that allowed the music industry to capitalize on royalties granted it by the Act. By 1928, talking pictures had become the principal means for promoting popular music (Sanjek, 1988b). Although live and recorded performances were part of Amplitude Modulation (AM) radio broadcasting since its popularization in the 1920s, the development in the late 1930s of Frequency Modulation (FM) radio signal transmission reduced the static of radio broadcasts and greatly enhanced radios commercial viability (Sanjek, 1988b). Although record labels were not directly compensated for airplay of their recordings, a concession granted by the Act, the promotional value of frequent airplay of a "hit" song increased record labels' sales. Invented in the late 1930s, television became commercially viable after World War II and gave song writers and music publishers another royalty revenue source. Live and recorded musical compositions became part of the television content of the three major networks: National Broadcasting Company (NBC), Columbia Broadcasting System (CBS) and American Broadcasting Company (ABC) (Sanjek, 1988b). The "free" nature of radio and television broadcasts enhanced their popularity with listeners, which in turn promoted the songs that were heard in these broadcasts.

By the 1950s the major commercial outlets of musical compositions were in place: live performances, sheet music, recorded music, motion pictures, television and radio broadcasting. The major contribution that technology played in the commercialization of music for the next twenty years was incremental increases in recording technology. The introduction of Long Playing (LP) albums, transition from monaural to stereophonic sound, introduction of reel to reel, 8 track and cassette recordings all contributed to the growth of recording industry revenues as consumers upgraded their record collections to higher quality or more convenient to use sound recordings (Sanjek, 1988b).

Due to the substantive investment necessary to identify promising songwriters and recording artists as well as to produce and distribute commercially successful recordings, the industry began evolving toward an oligopoly. By the 1970s, five major record labels, Warner/Elektra/Atlantic, CBS, A&M, Capitol and RCA, controlled 52.6 percent of the *Billboard* top 25 album charts (Hull, 1998). However, the effect of industry convergence toward oligopoly on consumer musical choices and album prices as well as industry profits was more akin to monopolistic competition than monopoly, primarily because of competition between the major labels as well as the existence of smaller, less well known labels that had a substantial presence in the industry as evidenced by their 47.4 percent market share of the *Billboard* top 25 album charts.

The business model of the recording industry was founded on the sale of albums to consumers. The industry worked closely with the radio broadcast industry and concert operators to promote songs and artists, as well as with retailers to sell albums.

The introduction of low cost reproducible tape recordings allowed individuals to easily copy and distribute song recordings and led to piracy, a significant economic problem for the recording industry. By 1971 the volume of unauthorized tape sales was estimated at one-third the volume of legitimate recordings. The most egregious form of piracy was domestic and offshore bootleg recordings that were sold as originals. The recording industry responded by taking the perpetrators to court. A more insidious but equally formidable problem was individuals swapping recorded songs. Although a potential copyright violation, legal remedies for home recording were not particularly effective since the violators weren't easily identifiable, recording quality decreased with each subsequent copy and the distribution system was limited (Hull, 1998).

To clarify the status of technological developments since the passage of copyright law in 1909, Congress enacted the Copyright Law of 1976. This law dispensed with formal copyright registration (but not notification), extended copyright terms to the author's lifetime plus 50 years (the previous standard was a 28-year term starting with the creation of the work and renewable for an additional 28-year period) and formalized the fair use doctrine (Moser, 2002). In general, fair use entitles consumers of intellectual property to use and copy works in a reasonable manner for their own convenience. The fair use provision gives consumers music listening portability. It allows individuals to listen to legally obtained music when (time portability), where (space portability) and in whatever media format (format portability) they choose (Electronic Frontier Foundation, 2002).

The Copyright Law of 1976 did not contain a satisfactory remedy for piracy for two reasons. First, about the only effective way to curb piracy is for Congress to ban the possession by individuals of recording equipment in much the same way that it forbids the possession of an automatic weapon by unauthorized individuals. Congress was loath to take this or any other action that was perceived as violating individuals' right of fair use. Second, the broadcast of a song over the airwaves is technically a sound recording reproduction in the same manner that a tape copy is and would require radio broadcasters to make royalty payments to record labels. Not surprisingly, the radio broadcasting industry objected to this possible requirement (Hull, 1998).

Through the inclusion of copyright protection in the Constitution, the United States government policy fostered the commercial development of music since the nation's inception. Technology introduced new commercial outlets for music: recorded music, motion pictures, and radio and television broadcasting. Government policy advanced the commercialization of music by revising copyright law to address issues arising out of new uses for music created by technological development. The business model of the music industry was based on the sale of record albums and the "free" accessibility of music to consumers via radio and television broadcasts. Although industry structure was oligopolistic in terms of firms dominating the industry, due to competition among the major record labels as well as the major presence of independent labels, from the perspective of the consumer it was monopolistically competitive. The major economic issue facing the music industry was not industry structure but piracy. One of technology's offsprings, cheap reproduction of song recordings, enabled piracy and hurt industry revenue. Congress could not resolve this problem since there is no easily enforceable method of curbing piracy that does not encroach on individuals' right to fair use.

CABLE TV/DIGITAL RECORDING
AND TRANSMISSION/INTERNET–
1980s TO PRESENT

Three major technological developments emerged that drastically affected the commercialization of music: cable TV, digital recording and transmission, and the Internet. The first cable TV system was developed in the late 1940s as a method to increase television set sales in rural areas. The market for television sets could be expanded by installing an antenna on a mountaintop and retransmitting the signal via wire to households in the valley below. Recognizing the threat to their revenues, established television broadcasting companies persuaded the Federal Communications Commission (FCC) to limit the number of distant signals imported to a large market and prohibit cable systems from showing movies less than ten years old or televised sporting events less than five years old. These restrictions, coupled with costly barriers to entry, precluded cable TV industry development. The competitive environment changed in the mid-1970s with the easing of FCC restrictions and the introduction of magazine publisher Time Inc.'s Home Box Office (HBO), a pay TV-movie distribution organization. By the late 1970s competitors to HBO arose: Showtime, launched by the communications giant Viacom, and the Warner-Amex sponsored The Movie Channel. By the mid-1980s subscription based cable TV became the dominant program delivery system in the United States (Vogel, 1998). Although they could get air wave broadcast network programming for free, consumers valued the choices offered by cable TV enough to pay a monthly subscription fee. The business model of the television broadcasting industry had changed. A precedent was set for consumers' to pay a monthly subscription fee for entertainment.

The music industry benefited from the technological development of cable TV and corresponding expansion of television viewing choices for consumers. Revenues increased due to the use of music in the programs offered by cable TV. The promotional benefits to the recording industry associated with the introduction in 1981 of Music Television (MTV) reversed a downturn in demand for recorded music (Vogel, 1998).

Digital recording was the next major technological development to significantly impact the commercialization of the music industry. The introduction of Compact Discs (CDs) in the mid-1980s gave listeners a near studio grade sound quality. CDs originally had an added advantage for record labels in that easily affordable recording equipment wasn't available. Record sales increased and piracy was staved since consumers couldn't copy one CD to another CD. This windfall was short lived. Affordable recording equipment was introduced that allowed CDs to be burned (copied). Furthermore, unlike analog based tape recording technology, sound degradation under digital copying was nonexistent. In spite of legislation stiffening the penalties for piracy, and vigorous pursuit of violators by major record labels, piracy continued to plague the recording industry (Hull, 1998).

The convergence of digital recording (CDs) and the technological development of peer-to-peer Internet networking in the mid-1990s added another dimension, electronic delivery, to music distribution. Internet portals (web sites) evolved that allowed individuals to swap songs freely over the Internet (Sharman Networks, 2002). Software became available to download these songs to CDs for the marginal cost of a blank CD. Customers that formerly purchased CDs at retail outlets converted to freeloading peer-to-peer users, dramatically dropping music industry revenues (Goodlatte, 2002). Piracy was no longer constrained by physical distribution. Revenue losses through individual peer-to-peer song sharing became a larger problem for the music industry than either tape or CD based piracy (Lam and Tan, 2001).

GOVERNMENT POLICY RESPONSE TO PEER-TO-PEER PIRACY

The introduction of digital recording and peer-to-peer song swapping brought the same legal problem to the music industry in the 1990s as did the introductions of recording technology at the turn of the century and recordable tapes in the 1960s. Technology had developed that wasn't envisioned when the governing copyright law was enacted. Governmental response was similar. Congress passed the Digital Performance Right in Sound Recordings Act of 1995 (DPRSRA) and the Digital Millennium Copyright Act of 1998 (DMCA). DPRSRA entitled songwriters and recording artists to compensation for subscription and on-demand transmissions of digital music. Section 1201 of DMCA outlawed attempts to circumvent copyright protection technology for any purpose, including individual use. Congress took a much stronger stand towards piracy than they did with the passage of the 1976 law. Copyright protection became copy protection.

Although section 1201 of DMCA gave the recording industry definitive legal standing in pursuing piracy, it discouraged the purchase of digitally delivered music since music legally no longer had media format portability (i.e., individuals could not legally copy a digitally delivered song from their personal computer to a CD), which in turn precluded space and time portability (e.g., could not play their newly purchased music in their cars or at the beaches). The only legal way purchasers of digitally transmitted music could listen to music was on their personal computers.

The recording industry turned to technology to collect the revenues due them under the Acts by developing the Secure Digital Music Initiative (SDMI), a fee based encrypted electronic delivery system. SDMI was abandoned in its infancy because it wasn't secure (Taylor, 2002). Outside of the entertainment industry, technology developed that enabled users to swap movies as well as songs over the Internet, threatening the revenues of the movie industry (Taylor, 2002). The movie industry joined the recording industry in seeking relief in the courts by suing the major peer-to-peer Internet portals (*Metro-Goldwyn-Mayer Studios et al. vs. Grokster et al.*). The success of the entertainment industry in shutting down peer-to-peer Internet networks resembles the struggle between Hercules and Hydra in Greek mythology. Each time the industry won a court case shutting down a peer-to-peer network, two others sprang up.

In addition to prosecuting the Internet portals that allow song swapping, the recording industry intends to prosecute individual users of these portals. The Recording Industry Association of America is suing the Internet Service Provider Verizon for a list of users who have illegally posted songs on the Internet (Havighurst, 2002). One Arizona company has settled out of court with the

RIAA for $1 million after it was accused of allowing its employees to trade songs on an internal network. The RIAA is aware of other companies with unauthorized music files on their systems and several investigations are ongoing (Sherman and Smith, 2002).

Although the music industry received Congressional protection enforced by the courts, digital song trading continued unabated. "The vast majority of students would never shoplift a CD at a record store, but think nothing of accessing the same CD for free online. These computer users treat the Internet as their own digital jukebox" (Sherman and Smith, 2002, p. 2). Last year, piracy cost the United States 118,000 jobs, $5.7 billion in wages and $1 billion in lost tax revenue. Globally, one in three recordings is a pirated copy (Sherman and Smith, 2002).

In addition to enlisting Congress and the courts in its fight against copyright infringement, record labels have revised their approach to electronic song delivery. They have partnered with software developers and Internet Service Providers to offer subscription fee based streaming and downloading of "digitally wrapped" music. This fee based delivery system is equivalent to cable and satellite television broadcast delivery systems which have received widespread consumer acceptance. It is much cheaper than the distribution of physical media (tapes, CDs), lowering record label costs. Costs are also lowered because of its "on demand" delivery nature. Record labels do not have to deal with forecasting demand for an album and having physical CDs in the appropriate retail outlets in a timely fashion. An electronic copy of a song or album can be delivered to consumers when and where they want it.

This electronic delivery system has changed the structure of this sector of the industry from an oligopoly to a duopoly. Two major alliances have been formed: MusicNet, a joint venture of record companies BMG, EMI and Warner with America Online as the Internet portal, and Pressplay, an alliance of record companies Sony and Universal with Microsoft and Yahoo providing the user interface (Harmon, 2002). The record labels are reluctant to allow subscribers to burn CDs for fear of copying and the resulting free distribution on the Internet. This is at odds with the desires of consumers, who want portability. MusicNet doesn't offer CD burning while Pressplay allows limited burning of copy protected CDs that can be played on compatible portable devices. The effect of the change in industry structure from oligopoly to duopoly on consumer choices was negligible. The competition for consumers' dollars by five major labels was replaced by competition between two alliances of the labels.

Technology change has facilitated the development of peer-to-peer networking (P2P) which allows artists to offer their songs directly to consumers via the Internet (MusicCity, 2002). If consumers are willing to search multiple web sites for music that suits their tastes, this would supplant the song/artist screening function of record labels and mitigate if not obviate the need for the distribution network of record labels and retail outlets. Consumer acceptance of peer to peer networking would drive industry structure toward monopolistic competition from the perspective of product providers (firms). Consumers should benefit from more choices and lower prices since the alliances formed by the major labels, MusicNet and Pressplay, would have competition from peer to peer networks.

Technology change and legal issues, more so than industry structure, will have the major impact on the price of music and industry profits. Low cost electronic delivery promises more and cheaper music choices for consumers. However, technology based efforts by the recording industry to distribute songs digitally and collect revenues have met with limited consumer acceptance due to lack of portability. Government policy has supported advances in the commercialization of music. Through DPRSRA and DMCA Congress has affirmed both copyright and copy protection for digital song delivery. The legitimate commercialization of music has been harmed by digital/Internet technology primarily in the area of revenue losses via individual piracy. The recording industry has responded by using the courts to stop Internet peer-to-peer software providers and suing an Internet Service Provider for a list of individual violators.

Balancing copyright protection and individual fair use rights is the unresolved issue that has impeded full commercial realization of the benefits of digital/Internet technology regardless of potential industry structure changes. The business model could change from revolving around album sales to a subscription service if consumers had inexpensive portability. The music industry might provide portable digital songs if individual piracy were not so pervasive.

RADIO BROADCAST INDUSTRY

One of the major benefits that the radio broadcasting industry provides to the music industry is the ability to increase album sales through the promotion of new and existing artists via radio airplay. Those musical acts that the public identifies as popular via high chart ranking normally go on concert tours, which further promotes album sales. This benefit has been curtailed by governmental policy and technological innovation in the radio broadcasting industry. The Telecommunications Act (TCA) of 1996 removed the restraint that a company could own no more than 2 stations in any market and no more than 28 overall. One company in particular, Clear Channel Communications, has profited from the passage of the law and now owns over 1,200 stations and syndicates over 100 programs for 7,800 radio stations (ClearChannel, 2002). Other companies have followed suit, changing industry structure from monopolistically competitive to an oligopoly.

Consolidation in the radio broadcasting industry since the Act's passage has resulted in programming syndication and homogenization of play lists, making competition for air play very intense and decreasing the probability of airplay for a fledgling artist (Mathews and Ordonez, 2002). Unlike the oligopoly and duopoly structure of the music industry, which remained monopolistically competitive from the perspective of consumer choices, the oligopoly in the radio broadcasting industry resembles a monopoly in that consumer choices have been reduced. Although broadcast music is a "free good" for consumers, they pay the higher "price" of a monopoly produced good by having fewer songs and artists to listen to. This in turn can reduce record label revenues since the introduction of new artists and songs is a major source of growth.

The technology advance of digital radio is the biggest thing to happen to radio in a century. FM broadcasts will equal CD-quality sharpness and depth, while AM formats will sound like current FM analog stations. Stations converting to digital radio can expect to pay from $30,000 to $200,000 to upgrade, depending on the condition of their equipment (Vrana, 2002). Listeners won't have to subscribe to digital radio, but they will have to buy a receiver that will cost about $100 (Vrana, 2002). The primary beneficiaries of digital radio will be AM radio station owners, since their signals will be vastly improved (Lawson, 2002). Due to industry consolidation brought about by TCA, as well as barriers to entry due to the complications of getting a broadcast license from the Federal Communications Commission and the costs of establishing a broadcasting station, the technological advance of digital radio will primarily benefit existing radio station owners, further entrenching oligopoly in the industry.

Live concerts are another of recording artist and record label (through increased album sales) revenue sources that have been indirectly impacted by TCA. The concert industry has become a near monopoly, reducing consumer choices and increasing prices. The major radio broadcaster, Clear Channel, has a subsidiary that books approximately 70 percent of concert ticket revenue in the United States (Clear Channel, 2002). Through co-ordination of its radio and concert divisions, Clear Channel has the potential to exert control over an artist's career. "Premiere also broadcasts Clear Channel Entertainment concerts and new CD debuts, enhancing the synergies between divisions" (ClearChannel, 2002). Other major concert promoters include Concerts West and House of Blues (Waddell, 2002). Concert promoter consolidation further diminishes the ability of record labels to promote their promising artists, resulting in fewer musical choices for consumers and diminishing revenue growth opportunities for the industry.

Although telecommunications deregulation threatens to create an oligopoly in the radio broadcasting segment of the industry, technological innovation has spawned potential competition via webcasting and satellite radio. Webcasting allows listeners to hear recordings digitally broadcast over the Internet. Congress provided royalty protection for webcasting to the music industry through the Digital Millennium Copyright Act of 1998 (DMCA). Even though webcasting quality is poor due to the current technology, it may become a threat to airwave based radio broadcasting since there is no need for expensive transmission equipment. Technology innovation could spawn a monopolistically competitive webcasting industry structure. However, government policy may contribute to the creation of an oligopoly. The current government set royalty rate of 0.07 cents per listener per song is financially unfeasible for many small webcasters, but affordable for large ones (Briggs, 2002). Because consumers can currently only listen to webcasts on personal computers its lack of portability may inhibit its commercial development.

In addition to webcasting, another technological development, satellite radio, may eventually pose a threat to traditional radio. Unlike webcasting, satellite radio has substantial barriers to entry which has led to the current duopoly industry status. XM and Sirius satellite radio allows anyone who pays the monthly user fee and has the appropriate equipment to receive via satellite transmission any of its radio programs anywhere in the continental United States (XM Satellite Radio, 2002). This business model is the same as cable and satellite subscription fees services in the television broadcast industry. Satellite radio has deep pockets; the major automobile manufacturers have invested in it and are offering satellite radios as options on their products (Schaeffler, 2002). The main competitive advantage that satellite ratio has over webcasting is portability for the user. Even if satellite radio is commercially established, the benefits to the music industry may be minimal. Because of its duopolistic nature, play lists will be circumscribed, limiting their promotional value to the recording industry.

The effect of government policy via TCA on the commercialization of the radio broadcast industry has been to foster the creation of an oligopoly, reducing opportunities for record labels to introduce new talent that in turn would generate industry revenue growth. The technological advances of digital broadcasting and webcasting (through its rate structure) favor the strengthening of that oligopoly. Technology has created competition in the form of satellite radio, however, its duopolistic nature is not conducive to creating new growth opportunities for the recording industry. Regardless of the evolution of industry structure in the music broadcast industry (analog, digital and satellite radio, webcasting), the net effect on the music industry is fewer growth opportunities due to fewer consumer choices of artists and songs to listen to.

CONCLUSION

Government policy in the form of copyright protection allowed the music industry to develop commercially in the United States. Technological innovation spurred commercial growth in the music industry through the introduction of the recording, motion picture, radio and television broadcasting industries. Record labels were willing to risk the capital necessary to create industry growth through the discovery and development of new musical talent. Due to the costs associated with this endeavor, the music industry structure became an oligopoly. Competition among the major record labels, as well as the substantive presence of independent labels, rendered the industry monopolistically competitive from the perspective of consumer choices, recorded music prices and industry profits.

The availability of affordable recording devices reduced the revenues of the music industry. The marriage of digital recording and Internet song swapping particularly exacerbated the industry's piracy problems. A duopoly was created when members of the recording industry partnered with Internet portals to create an "on demand" subscription based music delivery model, a move which has the potential to impact the commercialization of music through increased demand and lower costs. However, consumers have not embraced this model due to its lack of portability as well as readily available "free" song swapping. Competing artist-to-consumer delivery systems have developed and if they survive may revert the industry structure to a monopolistically competitive one. Government policy has not been able to resolve the piracy issue because of the inherent conflict between copyright protection for the artisans and their business agents and individual's fair use privileges. The major impact on issues normally associated with industry structure, consumer choices, product prices and industry profits, have been driven primarily by technology changes and the legal issues surrounding "fair use" versus piracy more so than by the economic structure of the music industry.

Deregulation in the telecommunications industry transformed its structure from monopolistically competitive to an oligopoly and in the process diminished the effectiveness of two major promotional sources for the music industry: radio broadcasting and concert promotion. Technological changes in the industry brought about by digital radio tend to benefit existing radio station owners, further entrenching oligopoly in the industry. The royalty rates mandated by Congress for webcasters will tend to drive that sector of the industry toward oligopoly. Although satellite radio may challenge traditional radio broadcasting, its duopolistic nature is not conducive to the exposure and development of new artisans. Government policy and technological innovation in the music broadcast industry have been conducive to the formation of an oligopoly/duopoly, which in turn has reduced consumer listening choices, diminishing growth opportunities for the music industry.

REFERENCES

Baye, Michael R. and Richard O. Beil (1994), Managerial Economics and Business Strategy, Burr Ridge, IL: Irwin.

Briggs, John (2002), "Internet: Music Stops in Protest," *New York Times*, May 1, 2002, Section C, page 4, column 4.

Clear Channel Communications (2002), *Entertainment, Radio*, San Antonio, TX: Author, Retrieved October 21, 2002 from the World Wide Web: http://www. ClearChannel.com.

Electronic Frontier Foundation (2002), *Copyright and Fair Use FAQ*, San Francisco, CA: Author. Retrieved October 21, 2002 from the World Wide Web: http://www. eff.org/cafe/drmgame/copyright-faq.html.

Goodlatte, Bob (2002, May 27), "Stealing Entertainment," *The Washington Times*, page 1, Retrieved November 22, 2002 from the World Wide Web: http://asp. washtimes. com/.

Harmon, Amy (2002), "Technology Briefing: Internet: 2 Music Services Win Catalog Rights," *The New York Times*, November 14, 2002, page C7, New York.

Havighurst, Craig (2002), "Fight Continues Between File Traders, Music Industry," *The Tennessean*, October 8, 2002, page E1, Nashville.

Hull, Geoffrey (1998), *The Recording Industry*. Boston: Allyn and Bacon.

Lam, Calvin K. M. and Bernard C. Y. Tan (2001), "The Internet is Changing the Music Industry, " *Communications of the ACM*, 44(8), August, 62-66.

Lawson, Richard (2002), "Digital Radio on FCC's Agenda," *The Tennessean*, October 10, 2002, page E1, Nashville.

Mathews, Anna Wilde and Jennifer Ordonez (2002), "Music Labels Say It Costs Too Much to Get Songs on the Radio," *The Wall Street Journal*, June 10, 2002, page B1, New York.

Moser, David (2002), *Music Copyright for the New Millennium*, Vallejo, CA: ProMusic Press.

MusicCity.com (2002), *FAQ*, Franklin, TN: Author, Retrieved November 25, 2002 from the World Wide Web: http://www.musiccity.com/faq/html.

Sanjek, Russell (1988a), *American Popular Music and Its Business, Volume II From 1790 to 1860*. New York: Oxford University Press.

Sanjek, Russell (1988b), *American Popular Music and Its Business, Volume III From 1900 to 1984*. New York: Oxford University Press.

Schaeffler, Jimmy (2002), "GM's XM Support Could Boost Entire Industry," *Satellite News*, 36 (September), 1-3.

Sharman Networks (2002), *Sharing and the P2P Philosophy*, Los Angeles, CA: Author. Retrieved October 14, 2002 from the World Wide Web: http://www.kazaa.com/us/help/glossary/p2p_philosophy.html.
Sherman, Cary and Lamar Smith (2002), "Music Industry Says Internet Downloads Have Caused Losses, to Tune of Millions," *Knight Ridder Tribune Business News*, Oct. 12, 2002, pages 1-2, Washington.
Taylor, Chris (2002), "Burn, Baby, Burn," *Time*, May 20, 2002, 159(20).
Vogel, Harold L. (1999), *Entertainment Industry Economics*. Cambridge, UK: Cambridge University Press.
Vrana, Greg (2002), "Digital Radio Come Down to Earth," *Design News*; Nov. 4, 2002, pages 48-52, Boston.
Waddell, Ray (2002), "HOB Entertainment off the Market," *Billboard*, 45 November 9, 2002, page 20, New York.
XM Satellite Radio (2002), *How it Works*, Washington, DC: Author. Retrieved October 14, 2002 from the World Wide Web: http://www.xmradio.com/how_it_works/introduction.html.

LEGAL REFERENCES

Copyright Act of 1909, Title 17 U S Code S. 1.
Copyright Act of 1976, Pub. L. 94-553, Title I, Sec.101, Oct. 19, 1976, 90 Stat. 2541.
Digital Millennium Copyright Act of 1998, Pub. L. 105-304, Oct. 28, 1998, 112 Stat. 2860, United States House of Representatives, Washington DC.
Digital Performance Right in Sound Recordings Act of 1995, Pub. L. 104-39, Nov. 1, 1995, 109 Stat. 336, United States House of Representatives, Washington DC.
Metro-Goldyn-Mayer Studios et al. vs. Grokster et al., United States District Court for the Central District of California, Western Division, October, 2001.
Telecommunications Act of 1996, Pub. L. 104-104, Feb. 8, 1996, 110 Stat. 56, United States House of Representatives, Washington DC.
United States Constitution, Article I, Section 8 United States House of Representatives, Washington DC.

Government Programs
and Diffusion of Innovations in Taiwan:
An Empirical Study
of Household Technology Adoption Rates

Maxwell K. Hsu
Hani I. Mesak

SUMMARY. This research mirrors Olshavsky's (1980) groundbreaking study by using penetration data for household technologies in Taiwan. Results support conventional wisdom that adoption rates are increasing over time. Moreover, rates were negatively associated with price. This research goes beyond Olshavsky by examining whether adoption rates differ across geographical regions types (i.e., rural, townships, and metropolitan). Adoption rates for nine household technologies at the three geographical types are not significantly different. Support is lent to the proposition that government rural development policies promoting edu-

---

Maxwell K. Hsu, PhD, is Assistant Professor of Marketing, College of Business and Economics, University of Wisconsin-Whitewater, Whitewater, WI 53190 (E-mail: hsum@uww.edu).

Hani I. Mesak, PhD, is State Farm Endowed Professor and Professor of Quantitative Analysis, College of Administration and Business, Louisiana Tech University, Ruston, LA 71272 (E-mail: mesak@cab.latech.edu).

[Haworth co-indexing entry note]: "Government Programs and Diffusion of Innovations in Taiwan: An Empirical Study of Household Technology Adoption Rates." Hsu, Maxwell K., and Hani I. Mesak. Co-published simultaneously in *Journal of Nonprofit & Public Sector Marketing* (Best Business Books, an imprint of The Haworth Press, Inc.) Vol. 13, No. 1/2, 2005, pp. 91-110; and: *Government Policy and Program Impacts on Technology Development, Transfer and Commercialization: International Perspectives* (ed: Kimball P. Marshall, William S. Piper, and Walter W. Wymer, Jr.) Best Business Books, an imprint of The Haworth Press, Inc., 2005, pp. 91-110. Single or multiple copies of this article are available for a fee from The Haworth Document Delivery Service [1-800-HAWORTH, 9:00 a.m. - 5:00 p.m. (EST). E-mail address: getinfo@haworthpressinc.com].

Digital Object Identifier: 10.1300/J054v13n01_06   *91*

cation and physical infrastructures have fostered relatively equal diffusion patterns of household technologies in Taiwan. *[Article copies available for a fee from The Haworth Document Delivery Service: 1-800-HAWORTH. E-mail address: <docdelivery@haworthpress.com> Website: <http://www.HaworthPress.com> © 2005 by The Haworth Press, Inc. All rights reserved.]*

KEYWORDS. Diffusion of innovations, technology adoption, household technologies, technology transfer, rural development

INTRODUCTION

Technology transfer can be broadly defined as the set of activities and processes whereby technology is passed from a supplier to a receiver (Bessant and Rush 1993). Our current understanding of technology transfer suggests that it involves a communication process, with learning and change taking place on both sides. Rogers and Shoemaker (1971), for example, saw the process as one of communication of an innovation from a source to a receiver–an interpretation that highlights the role of the characteristics of the source and receiver, the message itself, and the environment in which it takes place in influencing the adoption of an innovation. This theory goes some way towards explaining why users adopt at different rates, even when they are located in the same population. According to Rogers (1983), the diffusion process of an innovation is like the spread of an epidemic and the rate of adoption is usually estimated by the length of time required for a certain percentage of the members of a system to adopt an innovation. One of the most important aspects of the diffusion of consumer technological innovations is how demand is stimulated for new products in different parts of the world or in different regions within a country.

Studies that explicitly examine the diffusion of innovations in international markets are scarce. Takada and Jain (1991) studied the diffusion processes of durable goods in four Pacific Rim countries (i.e., U.S., Japan, Taiwan, and South Korea). Other notable studies dealing with cross-national diffusion of innovations include those of Gatignon, Eliashberg, and Robertson (1989); Helsen, Jedidi, and DeSarbo (1993); Ganesh and Kumar (1996); and Kumar, Ganesh, and Echambadi (1998). It is noted that these studies generally focus their attention on the temporal (i.e., time-related) dimension as well as the spatial (i.e., geographical-related) dimension by analyzing the differences in diffusion processes *among* a number of countries. The national systems of innovation (NSI) approach typically focuses on a number of representative macro variables and aims to interpret national differences in technology development, such as regulatory legislation, natural geography, and regional and national historical experiences (Weber and Hoogma 1998). Little research, however, has been done to examine whether adoption rates vary *within* a country. The purposes of this study are to examine whether evidence that supports the acclaimed accelerating adoption rates can be found outside the United States and to test whether the mean adoption rates across three levels of government-defined jurisdictions (i.e., countryside, cities and townships, and metropolitan areas) in Taiwan are significantly different from each other. Results from this study could shed light on a compelling public policy question: do the efforts of the institutions of government, relative to technology transfer and rural development, play a role in the diffusion of household technologies?

GOVERNMENT TECHNOLOGY TRANSFER POLICY

Technology diffusion and technology commercialization are two sides of the same coin, and government policy on technology transfer could differ. At one end, government may intentionally adopt a policy to enhance technology imports and to develop the capacity for indigenous learning. At other end, government may adopt the no-intervention policy by giving no preferential treatment to the technology inflows or the technology commercialization. It is almost axiomatic that there are direct, powerful, symbiotic and causal relations between national policies (especially developmental) and the mechanics of communicating, transporting, moving and receiving people, goods and ideas.

Focusing on the regions once ruled by the Chinese in the Pacific Rim countries, China did not have a coordinated information technology (IT) policy until the mid-1980s. Recognizing its lag in the use of technology, China's central government in Beijing initiated a series of "Golden" projects in the 1990s which aimed at leaping forward in the development of information technology infrastructure and application (Burn and Martinsons 1999). Taiwan, also known as the "silicon island," initiated its government industrial strategy to complement the efforts of the IT industry by supporting research and development on the public sector side and transferring technology to the private sector. A number of such policies with the intention to speed up the development of the higher technological segment in Taiwan, such as the establishment of the Hsin-chu Science-based Industrial Park to provide a conducive environment for the high-tech industry and the expansion of the government-sponsored research institutes to serve as a technology transfer channel for the private sector, have been widely documented (Mathews 1997). Similarly, with the intention to be a regional center for scientific and technical knowledge, the Singapore government has implemented policies to enhance the diffusion of IT applications at all possible levels (Corbett and Wong 1999). In contrast, though Hong Kong has one of the most advanced telecommunications networks in the world and offers protection for intellectual property rights, it has no state-directed IT plan. Colin Greenfield, a former head of Hong Kong's Information Technology Services, stated "the use of information technology in Hong Kong is requirement-driven rather than coordinated and promoted by the Government, apart from promotion through its own consumption" (Burn and Martinsons 1999). It would be immensely interesting to study the spatial diffusion of technological innovations in China. Due to the lack of reliable sub-national data in China, however, this study attains the best possible by focusing its attention on the case of Taiwan.

Taiwan's economy "took off" in the 1970s. As its economy became increasingly stronger in the past three decades, various home appliances have made their way into Taiwanese homes. Economists and marketers have long been concerned with diffusion of household inventions among populations as indicators of modernization. Usually these inventions are long lasting and have been referred to as household durables. They represent modern inventions with commercial potential as they become distributed through populations via commercial market diffusion processes. In fact, household technological innovations such as air conditioners, clothes washers, color TVs, and personal computers have undoubtedly changed the way people allocate their time at home. In examining the possible linkage between the government technology transfer policy in Taiwan and the spatial diffusion of household technologies at the sub-national level, the present study attempts to fit ownership data into a logistic diffusion model.

DIFFUSION MODELS

In an attempt to evaluate the diffusion of innovations theory, researchers use diffusion models to "depict the successive increases in the number of adopters of a product and predict the continued development of a diffusion process already in progress" (Mahajan, Muller, and Bass 1990). Among the various aggregate diffusion models, the Bass (1969) diffusion model is a unique one in the sense that it incorporates both the innovative and imitative effects into its structure. The innovative effect explains the demand stimulated by external factors such as advertising while the imitative effect explains the demand stimulated mainly by internal factors such as word-of-mouth communication.

Though the Bass model has its advantages, both the pure innovative and the pure imitative diffusion models have their own merits. Indeed, one superior merit of the imitative model is that it "provides a means of projections for substitution in the absence of an adequate historical data base" (Lilien, Kotler, and Moorthy 1992). When reliable sales data are unavailable in a certain market, marketers can still take advantage of the pure imitative type of a diffusion model for the purposes of market projections, explanation of diffusion patterns, and policy making. With information about the historical market penetration and the anticipated ultimate level of penetration, for example, the adoption rate of an innovation can be estimated from the Mansfield-Blackman model (Blackman 1974).

With reasonable assumptions, Mansfield (1961) illustrated that the growth (over time) of the number of firms adopting an innovation should conform to an S-shaped logistic function. Similar to Mansfield's approach, Olshavsky (1980) applied a logistic innovation diffusion model to estimate the rate of adoption for 25 products sold in the United States. In Olshavsky's study, where time was hypothesized to be the driving force of the rate of adoption of consumer products, he found accelerating adoption rates for the studied 25 products. That is, a new product introduced in the market at a later time would have a larger adoption rate than another that has been introduced at an earlier date in the same market.

HYPOTHESES

Young (1964) analyzed sales data of 30 electrical home appliances, and his findings suggest that product life cycles are shortening due to the rapidly accelerating technological development. This intriguing claim later motivated Olshavsky's (1980) study to investigate whether the rate of adoption is accelerating over time using penetration data, which do not inherit the typical problem of sales data (i.e., repeat and multiunit purchases). Notably, many marketing researchers have cited Olshavsky's groundbreaking paper as a supporting evidence for the increasing adoption rates in the modern marketplace. This finding has been further examined by a number of studies. For example, Qualls, Olshavsky, and Michaels (1981) found that the length of the introduction and growth stages of home appliances are negatively correlated with the year of product introduction and thus confirming the earlier findings of Olshavsky (1980). Using first-time buyer sales data related to a number of home appliance products and consumer electronics sold in the United States, Bayus (1992) applied diffusion models to compute three measures of diffusion rates including the time to peak sales, the total contagion level, and the growth parameter in a logistic function. Afterwards, these multiple measures of the diffusion rate were compared across products by year of introduction. Since no statistically significant relationship was found between the variable "year of introduction" and the three measures of diffusion rate, it was concluded that diffusion rates have not been accelerating over time.

Mahajan et al. (1990) summarized the established diffusion of innovations literature and provided direction for future research. Though generally marketing scholars advocate that diffusion rates have been accelerating, the inconsistent empirical findings in the literature, implying that such rates are either accelerating or remain stationary over time, are notable and appear to suggest that this topic deserves to be revisited again using some new data. Because Taiwan's market system is similar to the one in the United States in that the central government of Taiwan generally encourages fair competition among rival firms and does not obtrude the marketplace unless it is extremely necessary, it is anticipated that the U.S. experience would apply to Taiwan's marketplace in this regard. We turn next to examining the relevant literature related to the variation in the adoption rate across geographical areas.

Differences in adoption rates at the cross-national level are well documented in the literature. For example, Swan (1973) suggested that the patterns of diffusion of synthetic rubber differ in eleven studied countries. Specifically, some countries had low long-run equilibrium shares of rubber markets while others had high long-run equilibrium shares. Stated another way, some countries are characterized by fast diffusion rates while others are characterized by slow rates. Likewise, Lindberg (1982) observed that in the case of television receivers, it took only 5 years to reach 25% of households in Sweden, while it took 12 years to reach that penetration level in France. However, only two studies have been found at the sub-national level to investigate the geographical variation of an innovation's adoption rate within a country.

Ormrod's study (1990) implicitly supports the notion that the adoption patterns of an innovation vary at the U.S. state level. Ormrod (1990) explicitly stated "the spatial distribution that emerges when an innovation diffuses typically reflects two primary influences, those of information flow and local acceptability." Information flow refers to the movement of information from one place to another, and it has been suggested that knowledge about an innovation is not likely to be evenly distributed–some places gain information about an innovation rapidly and others more slowly. Local acceptability refers to the level of local receptiveness toward an innovation, and it is an influential factor on the diffusion of innovations because the operating characteristics of an innovation could enhance or prohibit the diffusion process. For example, it appears reasonable to expect a faster adoption of snowmobiles in the state of Alaska than in the state of Texas. Similarly, Redmond (1994), investigating the diffusion rates of color TVs and VCRs at the U.S. regional level, found a substantial degree of variation across regions.

In the case of Taiwan, however, its unique socio-economic factors lead us to expect rather uniform adoption rates in different geographical areas. The size of Taiwan is slightly smaller than Maryland and Delaware combined, but it is characterized by its high level of population density,[1] high level of road density,[2] and nearly saturated media coverage (in 1997 in Taiwan, there were 190 news agencies, 3,491 publishing corp., and 33 broadcasting corp.[3]). According to Redmond (1994), the presence of a highly developed and coordinated marketing system, including national distribution chains and national media promotion, may prevent differential spatial diffusion from occurring in a developed economy. Moreover, since interpersonal connections and networking are of foremost importance to reduce uncertainty in the diffusion process that is specific to consumer technological innovations, high exposure to advertising messages everywhere in Taiwan and close interaction between people, allowed by geographical proximity of various governmental jurisdictions, may jointly foster a rather uniform diffusion pattern in Taiwan.

In addition, we argue that the rate of adoptions should be accounted by two diffusion-related dynamic externalities: demand-side linkages and human capital development. Demand-side linkages are influenced by a country's pattern of rural development, and it has been documented that "rapid rural income growth–farm and nonfarm–with equitable income distribution and the dominance of SMEs [*small- and medium-scale enterprises*] in industrial development" being two notable Taiwan's development experience (Park and Johnston 1995). Bourguignon, Fournier, and Gurgand (2001) pinpointed that one of the most striking features of Taiwan's rapid economic development is its very limited changes in the distribution of income. Active government policies (e.g., land reform) promoting equality result in relatively narrow discrepancies in income levels between people living in the urban areas and those who live in the rural areas, and, accordingly, enhance the diffusion pattern across geographical space. With respect to human capital development, Lucas (1988) developed a theory in which human capital plays a pivotal role in the economic growth process and Darrat, Hsu, and Zhong (2002) showed that both economic openness and human capital accumulation contribute significantly to Taiwan's economic growth in the post-World War II period. Moreover, in Taiwan the quality of rural labor was considered to be a critical factor in its success. The Taiwanese government's emphasis on rural education not only improves farmers' managerial capability but also their receptivity to technological innovations. It is worth mentioning that high and increasing wages reduce inequality and further improve workers' purchasing power. Together with government investments in transportation, electrification, and communications, these stable and encouraging rural economic development policies enroot a geographically balanced economic system in Taiwan and thus "narrowing" the gap in terms of living standards between metropolitan and rural areas.

Given that the physical infrastructures provide people with ample access to electricity, water, sewage treatment, roads, etc., and that the government rural development policy is in place, these factors jointly create a common cultural orientation toward consumer technologies. Thus, it seems reasonable to claim that informational effects on the diffusion of household technologies in Taiwan is likely to result in a faster, but geographically similar, adoption rates of household technologies in different government jurisdictions. Thus, the following hypotheses are formulated:

H1:   The rate of adoption is increasing over time in Taiwan.

H2:   The adoption rate of a household technological innovation does not differ by government-defined administrative jurisdictions in Taiwan.

METHOD

Given that no consensus has been reached in terms of which measure of adoption rate (e.g., the time to peak sales, the growth parameter in a logistic function) is superior to the others, we used a method similar to the one employed in Olshavsky's (1980) study. Specifically, we fitted the penetration data to the Blackman (1974) logistic diffusion function. The model is represented mathematically by the following form:

$$F(t) = \frac{\overline{F}}{1 + e^{-(a + bt)}} \tag{1}$$

where:

- $F(t)$ is the proportion of households adopting the innovation by time t,
- $\overline{F}$ is the ultimate penetration that the innovation can capture in the long run,
- a is a constant that positions the curve on the time scale,
- b is a growth rate coefficient, and
- t is measured in years from 1900.

For the Blackman model (1), parameter b is taken as a measure for the adoption rate. It can be shown, upon minor mathematical manipulations, that expression (1) takes on the following form related to the diffusion of a certain innovation j:

$$\ln\left(\frac{F_j(t)}{\overline{F}_j - F_j(t)}\right) = a_j + b_j t \tag{2}$$

where:

- $F_j(t)$ is the proportion of households adopting innovation j by time t,
- $\overline{F}_j$ is the ultimate penetration that innovation j can capture in the long run,
- a_j is a constant that positions the curve of innovation j on the time scale, and
- b_j is the growth rate coefficient of innovation j.

In the Olshavsky's study, the ceiling or the ultimate penetration value was estimated as the maximum penetration achieved. In the present study, $\overline{F}_j$ is chosen to be given by $Max\{F_j(t)\}$·. In other words, the maximum penetration, $\overline{F}_j$, is assumed to occur in^t 1999, the latest data point available to the authors. Regression model (2), after adding an error term, can be estimated using the method of ordinary least squares (OLS). The cross-innovation data can be pooled using dummy variables. After adding an error term, the model takes the following form:

$$Y_j(t) = \ln\left(\frac{F_j(t)}{\overline{F}_j - F_j(t)}\right) = \alpha_1 + b_1 t + \sum_{k=2}^{J} a_k D_k + \sum_{k=2}^{J} \beta_k D_k t + \varepsilon_j(t) \tag{3}$$

where:

- D_k is a dummy variable that takes on the value 1 if $k = j$ and 0 if otherwise,
- J is the number of studied household technological innovations,
- a_1, b_1, α_2 through α_J, and β_2 through β_J, are parameters to be estimated, and
- $\varepsilon_j(t)$ are independent random error terms, assumed to be identically normally distributed with zero mean and constant variance.

Regression model (3), designated as the *Full Model*, can be estimated using the method of ordinary least squares (OLS). It is worthy to note that the pooling procedure produces estimates for the parameters a_j and b_j, related to a particular innovation j, that are consistent with those that would have been obtained had the penetration data of that innovation been analyzed separately using expression (2).

To examine whether adoption rates across J products are equal, the *Reduced Model* (4), is derived from the full model (3) by setting $\beta_2 = \beta_3 = \cdots = \beta_J = 0$ to obtain

$$Y_j(t) = \ln\left(\frac{F_j(t)}{F_j - F_j(t)}\right) = a_1 + b_1{}' + \sum_{k=2}^{J} \alpha_k D_k + \varepsilon_j(t), \qquad (4)$$

The typical test statistic F^* could be calculated by the following formula:

$$F^* = [(SSE(R) - SSE(F))/(df_R - df_F)] \div [SSE(F)/df_F] \qquad (5)$$

where:

- SSE(F) is the sum of squares of errors for the full model,
- SSE(R) is the sum of squares of errors for the reduced model,
- df_F is the error degrees of freedom associated with the full model, and
- df_R is the error degrees of freedom associated with the reduced model.

The F test statistic is used to compare the *Full Model* with the *Reduced Model*. A significant F-statistic means that the adoption rates are significantly different across the examined products. Moreover, a significant F-statistic could shed light on the reasonableness of using parameter b to measure the diffusion process as this measure contains considerable information about the diffusion process of the studied products.

Hypothesis H1 can be assessed by regressing the estimated adoption rate b_j on the variable of time (i.e., the year of introduction). Specifically, the following regression model is used:

$$\hat{b}_j = \beta_{0j} + \beta_1 X_{1j} + v_j \qquad (6)$$

where:

- $\hat{b}_j$ is the adoption rate of innovation j, estimated using equation (2),
- X_{1j} is the year of introduction of innovation j,
- v_j represent independent error terms of innovation j, assumed to be normally distributed with zero mean and constant variance.

When examining the overall significance of the above model, a statistically greater than zero estimator for parameter β_1 will lead to the confirmation of an increasing rate of adoption over time. Because of a relatively small number of observations (i.e., 13 innovations), differences in the dependent and independent variables for all 78 pairs of the related innovations $(78 = (13 * 12)/2)$ will be used to increase the error degrees of freedom. It is noteworthy that Helsen et al. (1993) used a similar procedure in estimating their regression models.

With regard to Hypothesis H2, a one-way ANOVA is employed. A statistically insignificant related F-statistic would give support to our hypothesis that the adoption rate of a technological innovation does not differ by government-defined administrative jurisdictions in Taiwan. These levels are the government-defined administrative jurisdictions representing (1) countryside, (2) cities and townships, and (3) metropolitan areas.

For the first part of the analysis associated with H1, penetration data related to 13 household technologies ranging from 1975 through 1993 were collected from various issues of the *Research Report of Home Appliance Penetration in Taiwan*, published by the government-owned Taiwan Power Company. Namely, annual penetration data were secured for the following 13 products: air conditioners, clothes dryers, clothes washers, coffee makers, color TVs, electronic-magnet stoves, electronic stoves, juice makers (blenders), microwave ovens, personal computers, refrigerators, vacuum cleaners, and VCRs. For the second part of the analysis associated with H2, penetration data over a relatively smaller number of products were collected due to the data availability issue (i.e., penetration data must be available at the three studied jurisdictions over time). Specifically, penetration data across nine household technologies, run by electricity, are available for the following four years: 1991, 1993, 1997, and 1999, in various issues of the *Research Report of Home Appliance Penetration in Taiwan*. The examined nine household technologies are: air conditioners, clothes dryers, clothes washers, dehumidifiers, electric oven toasters, electric water heaters, microwave ovens, personal computers, and vacuum cleaners.

RESULTS

Using formula (5), the calculated F test statistic ($F(12,63) = 8.05$) was found to be statistically significant at the .001 level. This strongly indicates that adoption rates for the studied household technologies are different. As for each of the individual technologies, except for the case of coffee makers, significant t-statistics were observed for the adoption rate coefficients for the remaining 12 products. Using expression (2), the estimated parameters along with the P-values and the adjusted-R^2 values for each innovation are reported in Table 1. Our results suggest that the Mansfield-Blackman logistic growth diffusion model generally provides a good fit to the penetration data as was the case in Olshavsky's (1980) study.

When the estimated adoption rate variable is regressed against the year of introduction over the 13 innovations using their pair differences, the coefficient of the time variable is positive and significantly different from zero (see Table 2), which provides evidence to support the hypothesized accelerating rate of adoption over time. We further note that Olshavsky (1980) tested an ad hoc hypothesis that some of the variability might have been attributed to purchase price differences across these technologies, and he included a dichotomous price variable in equation (6) to capture the price variation. Similar to Olshavsky's approach, the present study employed a price dummy to classify the studied products into two categories (i.e., high and low) using an arbitrary cutoff point of $100 dollars. After the price variable was added to equation (6), the time variable remained significant and its directional impact did not change, while the coefficient of the price variable was found to be negative and significantly different from zero at the .05 level (see Table 2).

TABLE 1. Parameter Estimates and Goodness-of-Fit Measures of the Mansfield-Blackman Model for Thirteen Household Technologies

Household Technology	a	b	Adjusted-R^2
Air Conditioners	−26.57 (.00)	.32 (.00)	.78
Clothes Dryers	−50.82 (.00)	.58 (.00)	.97
Clothes Washers	−19.48 (.00)	.25 (.00)	.80
Coffee Makers	−66.85 (.25)	.75 (.25)	.71
Color TVs	−28.35 (.00)	.36 (.00)	.94
Electronic-Magnet Stoves	−110.1 (.01)	1.23 (.01)	.96
Electronic Stoves	−57.82 (.01)	.66 (.01)	.90
Juice Makers (Blenders)	−15.88 (.00)	.21 (.00)	.75
Microwave Ovens	−77.65 (.01)	.87 (.01)	.89
Personal Computers	−45.02 (.02)	.51 (.02)	.82
Refrigerators	−11.56 (.01)	.17 (.00)	.62
Vacuum Cleaners	−55.16 (.02)	.63 (.02)	.86
VCRs	−49.42 (.00)	.57 (.00)	.81

Note: P-values are reported within parentheses.

TABLE 2. Significance Levels of Time- and Price-Related Null Hypotheses

Time-Related Hypothesis (H1): H_0: Coefficient of the time variable is zero.			Time- and Price-Related Hypotheses (H1): H_{01}: Coefficient of the time variable is zero. H_{02}: Coefficient of the price variable is zero.		
Null Hypothesis	Estimated Coefficient	P-value	Null Hypotheses	Estimated Coefficient	P-value
H_0: $\beta_1 = 0$	.029	.000	H_{01}: $\beta_1 = 0$	.027	.000
			H_{02}: $\beta_2 = 0$	−.099	.049

Using a slightly different set of technologies, the adoption rates estimated using regression equation (2) for the studied nine home technological products relative to the considered three geographical levels in Taiwan are reported in Table 3. Specifically, Table 3 shows the adoption rate parameter estimates, a 95% confidence interval for every adoption rate parameter, and the adjusted R^2 values for each of the studied innovations. It is noteworthy that, for each of the studied household technologies, the computed 95% adoption rate confidence intervals were found to be overlapping across the three government jurisdictions. Table 4 reports the ANOVA F test statistic ($F(2, 24) = 2.56$, P-value = .10) and the nonparametric Kruskal-Wallis Rank Test statistic (Chi-square value = 4.19, P-value = .12). Taken together, all statistical measures fail to reject the assertion that the adoption rate of a household technological innovation does not differ by government-defined administrative jurisdictions in Taiwan.

DISCUSSIONS AND CONCLUDING REMARKS

In this article, we focus attention on penetration (ownership level) data relative to some representative, infrequently purchased household technologies, which means that the influence of replacement and repeat purchases upon results is negligible. The impact of time found in this research is in accordance with the notion of Olshavsky's (1980) paper, which provides support to the assertion of accelerating adoption rates over time in Taiwan. In a recent study, Hsu et al. (1999) found that the average adoption rate for lag countries (countries where the innovations are later introduced) is higher than the lead country's adoption rate in 13 European countries. It appears that a rapidly shortening product life cycle has become a trend in the global marketplace. The rapid technology adoption rates may reflect the importance of household technologies to the modern lifestyle in an established economic system where they are no longer

TABLE 3. Parameter Estimates and Goodness-of-Fit Measures of the Mansfield-Blackman Model for Nine Household Technologies Across Three Government-Defined Jurisdictions in Taiwan

Household Technology	Countryside Adoption Rate (95% C.I.) Adjusted R^2	Cities and Townships Adoption Rate (95% C.I.) Adjusted R^2	Metropolitan Areas Adoption Rate (95% C.I.) Adjusted R^2
Air Conditioners	.295 (.027 − .563) Adjusted R^2 = .99	.403 (−.688 − 1.494) Adjusted R^2 = .91	.238 (−.357 − .833) Adjusted R^2 = .93
Clothes Dryers	.482 (.205 − .759) Adjusted R^2 = .99	.458 (−.843 − 1.759) Adjusted R^2 = .90	.118 (.031 − .267) Adjusted R^2 = .98
Clothes Washers	.085 (−.128 − .298) Adjusted R^2 = .93	.043 (.016 − .070) Adjusted R^2 = .99	.038 (−.099 − .175) Adjusted R^2 = .85
Dehumidifiers	.214 (−.927 − 1.355) Adjusted R^2 = .70	.229 (−.673 − 1.131) Adjusted R^2 = .82	.143 (−.140 − .426) Adjusted R^2 = .95
Electric Oven Toasters	.229 (−.242 − .700) Adjusted R^2 = .95	.112 (.037 −.187) Adjusted R^2 = .99	.153 (−.575 − .881) Adjusted R^2 = .75
Electric Water Heaters	.209 (.147 − .271) Adjusted R^2 = .99	.409 (−.397 − 1.215) Adjusted R^2 = .95	.164 (−.169 − .497) Adjusted R^2 = .95
Microwave Ovens	.329 (.240 − .418) Adjusted R^2 = .99	.361 (−.137 − .859) Adjusted R^2 = .98	.239 (.080 − .398) Adjusted R^2 = .99
Personal Computers	.374 (−.134 − .882) Adjusted R^2 = .98	.396 (−.204 − .996) Adjusted R^2 = .97	.299 (−.291 − .889) Adjusted R^2 = .95
Vacuum Cleaners	.280 (−.160 − .720) Adjusted R^2 = .97	.332 (−.691 − 1.355) Adjusted R^2 = .89	.275 (−.670 − 1.220) Adjusted R^2 = .86

Note: A 95% confidence interval related to the associated adoption rate is reported within parentheses.

TABLE 4. *F*-Statistic and Chi-Square Statistic of Geographical Setting-Related Null Hypothesis

Testing Method: ANOVA		
Null Hypothesis (H2)	Test Statistic	P-value
H_0: $u_{countryside} = u_{cities\ and\ townships} = u_{metropolitan\ areas}$	$F(2, 24) = 2.57$	.10
Testing Method: Kruskal-Wallis Rank Test		
Null Hypothesis (H2)	Test Statistic	P-value
H_0: $u_{countryside} = u_{cities\ and\ townships} = u_{metropolitan\ areas}$	$\chi^2(2) = 4.19$	.12

considered as luxury items for an ordinary household in Taiwan. Perhaps the rapid speed of adoption is also partially attributed to the fact that all studied technologies have successfully passed the *innovators* stage (i.e., most of them are with a penetration rate between 16% and 50%) and thus exhibit significant learning and epidemic effects. As Olshavsky (1980) noted, such a trend "threatens the usefulness and meaningfulness of the distinctions drawn among 'innovators,' 'early adaptors,' 'early majority,' 'late majority,' and 'laggards.'" A rapid adoption rate may also challenge the new product entry strategy in terms of the decision of a waterfall (i.e., a phased launch) strategy.

While the price coefficient was positive but not significantly related to the adoption rate in Olshavsky's (1980) study, the price coefficient is found to be negative and significantly associated with the rate of adoption in the present study. Our results seem to reflect the conventional wisdom that a relatively more expensive innovation tends to be adopted at a relatively lower speed. However, we note that our research is subject to certain limitations. For example, given the limited set of technologies (both Olshavsky's (1980) study and ours concentrate exclusively on household appliances), it is hard to draw conclusions that are generally applicable to other types of products. As a result, further research is needed to validate this finding using a wider range of innovations in other countries.

Through the examination of the household technology diffusion in Taiwan, it seems clear that important conclusions can also be drawn in terms of government rural development and technology transfer policies from the presented findings. The rapid adoption rates found in Taiwan may be accounted for by a wide range of socio-economic forces, such as increased speed of communication and the effectiveness of personal network and advertising media in a populated economic system. In conjunction with these socio-economic phenomena, the Taiwanese government's rural economic development policy is thought to pave the road for a faster but geographically balanced diffusion of innovations.

Prior research suggests two contradictory viewpoints in terms of governmental regulations on the diffusion of technologies: one view suggests that the main driving force is regulations and standards enforced by government, while the other view criticizes regulations on the grounds that they are ineffective or distortionary in the diffusion of technological innovations (Cetindamar 2001). It is noteworthy that opponents of government intervention do not necessarily go against all of the services provided by the government. Rather, they hold the viewpoint that the society does not need more than minimal public-goods infrastructures that are sufficient enough to build a base for the private sector to act. Summarizing the arguments against and for government intervention, Kraemer, Gurbaxani, and King (1992) indicated that the key discriminator between the proponents and opponents is "how does one define the critical infrastructure requiring government involvement?"

In the case of Taiwan, the government did not enforce any specific policy to increase or impede the diffusion of household technologies prior to or during the sample period but it did formulate a technology transfer plan (i.e., the 10-year *Sectoral Development Plan for the Information Industry* in the 1980s). Taiwan's computing plan is aimed not only at the production of personal computers, monitors and PC motherboards, but also at the application of computing technology (Cherng and Lin 1990). Coordination and direction of computing technology development are provided by the *Institute for Information Industry* which promotes "the development of the local computer industry by training the high-level manpower needed by the industry, conducting research and development, introducing advanced techniques, providing capital for the industry, and giving preference to locally made computers in government procurements" (Kraemer et al. 1992).

A comparison of the personal computer's adoption rates to other studied household technologies' adoption rates in the three government-jurisdictions shows no clear evidence that the existence of a government technology transfer policy leads to a higher adoption rate of personal computers. personal computers at the three jurisdiction levels had the following adoption rates: Countryside: .374; cities and townships: .396; metropolitan areas: .299 (see Table 3). However, we found higher or equally comparable adoption rates related to other household technologies (e.g., countryside: .482 for clothes dryers; cities and townships: .403 for electric water heaters; metropolitan areas: .275 for vacuum cleaners) with no state-directed facilitating programs. Basically, the finding of "steady growth" across personal computers and other household technologies suggests that government technology transfer policy might have not made a difference in the adoption of personal computers. If government policy has made a difference, we would have expected to see an overwhelming larger adoption rate of personal computers than its counterparts. In summary, government involvement in rural development may be effective in lifting the overall population's standard of living and in creating a more economically balanced society. On the other hand, the effects of government computing promotion activities seem to be marginal and subtle in the case of Taiwan. Thus, the findings of this study favor the viewpoint that an economically invisible hand could have taken care of the household technologies' market demand, given the necessary infrastructures (e.g., reliable power provision, sound education system) have been made available by the government. This is not to say that policy does not matter. What it might indicate is that the government computing policy is more effective in promoting production (and hence exports) of computers on the supply side than the individual purchase of personal computers on the demand side.

Though no significant geographical difference in the category of household technologies was found, interestingly, the cities and townships appear to take the lead on six products (i.e., air conditioners, dehumidifiers, electric water heaters, microwave ovens, personal computers, and vacuum cleaners) in term of the adoption rate/speed. Additionally, except for the electric oven toasters, none of the studied products experiences the highest adoption rate/speed in the metropolitan areas. That is, the findings of the current study implicitly suggest that large metropolitan areas tend to experience a slightly slower adoption rate (though not statistically noticeable) than other less populated jurisdictions in Taiwan. This seems to echo the viewpoint that underlies the diffusion of technological process in the inter-firm empirical studies. It was suggested that "generally, technology diffuses faster in less concentrated markets (both from the supply and adoption sides), and that large firms tend to adopt innovations earlier than smaller ones" (Baptista 1999).

As a final remark, this study could also shed light on new product entry strategy in international markets. A conventional "waterfall" marketing strategy for international marketers is to enter a set of wealthy countries with GDP per capita above a predetermined level, and then to gradually enter other countries. When entering a specific country, international marketers often start from the metropolitan areas, which are characterized with a relatively high income and established distribution channel systems, and then penetrate to the suburban areas and the more distant countryside. Nonetheless, the findings in the present study suggest that, it may be wise for international marketers of household technologies to consider a "sprinkler strategy" by simultaneously targeting not only the metropolitan areas but also the less populated cities or university towns when they attempt to enter developed and populated countries like the Asian tigers. Further empirical research is needed to investigate diffusion patterns across a wider range of innovations, different geographical regions, and across different country contexts (e.g., diffusion patterns in the PRC and South Korea versus the European Union countries).

AUTHOR NOTE

Maxwell K. Hsu's publications have appeared in *Information & Management, Applied Economics Letters, International Journal of Advertising, International Journal of Business and Economics*, and *Service Marketing Quarterly*. Dr. Hsu's current work is in international diffusion of innovations, information technology, advertising, and service quality.

Hani I. Mesak's publications have appeared in *Management Science, Decision Sciences, Marketing Science,* and the *Journal of Service Research.* Dr. Mesak's current work is in advertising pulsation, and consumer adoption of technology innovations.

NOTES

1. Population density is 573 persons/Km^2 in Taiwan and 27.3 persons/Km^2 in the U.S. in 1992 (Table I in *Social Indicators in Taiwan Area of the Republic of China,* 1993 by Executive Yuan, Republic of China).
2. Paved road density is 91.1 Km/Km^2 in Taiwan and 68.5 Km/Km^2 in the United States in 1995 (*World Road Statistics,* 2000).
3. These statistics are available from Table 3-6 in *1999 Indicators of Taiwan Market,* 1999, edited by DENTSU Communication Inc. (the original data source is "Government Information Office, Executive Yuan, Republic of China").

REFERENCES

Bass, Frank M. (1969), "A New Product Growth Model for Consumer Durables," *Management Science,* 15(January): 215-227.

Baptista, Rui (1999), "The Diffusion of Process Innovations: A Selective Review," *International Journal of the Economics of Business,* 6(1): 107-129.

Bayus, Barry L. (1992), "Have Diffusion Rates Been Accelerating Over Time?" *Marketing Letters,* 3(3): 215-226.

Bessant, John and Howard Rush (1993), "Government Support of Manufacturing Innovations: Two Country-Level Case Studies," *IEEE Transactions on Engineering Management,* 40(1): 79-90.

Blackman, A. Wade, Jr. (1974), "The Market Dynamics of Technological Substitutions," *Technological Forecasting and Social Change,* 6: 41-63.

Bourguignon, F., M. Fournier, and M. Gurgand (2001), "Fast Development With a Stable Income Distribution: Taiwan, 1979-94," *Review of Income & Wealth,* 47(2): 139-163.

Burn, Janice M. and Maris G. Martinsons (1999), "Information Technology Production and Application in Hong Kong and China: Progress, Policies and Prospects," in *Information Technology Diffusion in the Asia Pacific: Perspectives on Policy, Electronic Commerce and Education,* Eds. Felix B. Tan, P. Scott Corbett, and Yuk Yong Wong, Hershey, PA: IDEA Group Publication: 7-35.

Cetindamar, D. (2001), "The Role of Regulations in the Diffusion of Environment Technologies: Micro and Macro Issues," *European Journal of Innovation Management,* 4(4): 186-193.

Cherng, C.J. and C.C. Lin (1990), "The Computerization Policy of the R.O.C. Government," paper presented at the First Annual Meeting of Country Experts on Government Policy and Information Technology, November 12-17, Singapore.

Corbett, P. Scott and Yuk-Yong Wong (1999), "Seeding the Clouds of Change: The Planned Evolution of Singapore into an Intelligent Island," in *Information Technology Diffusion in the Asia Pacific: Perspectives on Policy, Electronic Commerce and Education*, Eds. Felix B. Tan, P. Scott Corbett, and Yuk Yong Wong, Hershey, PA: IDEA Group Publication: 68-78.

Darrat, Ali F., Maxwell K. Hsu, and Maosen Zhong (2002), "Foreign Trade, Human Capital and Economic Growth in Taiwan: A Re-Examination," *Studies in Economics and Finance*, 20(1): 85-94.

Ganesh, Jaishankar and V. Kumar (1996), "Capturing the Cross-National Learning Effect: An Analysis of an Industrial Technology Diffusion," *Journal of the Academy of Marketing Science*, 24(Fall): 328-337.

Gatignon, Hubert, Jehoshua Eliashberg, and Thomas S. Robertson (1989), "Modeling Multinational Diffusion Patterns: An Efficient Methodology," *Marketing Science*, 8(Summer): 231-247.

Helsen, Kristiaan, Kamel Jedidi, and Wayne S. DeSarbo (1993), "A New Approach to Country Segmentation Utilizing Multinational Diffusion Patterns," *Journal of Marketing*, 57(October): 60-71.

Hsu, Maxwell K., Hani I. Mesak, Otis W. Gilley, Sean T. Dwyer, and Thomas L. Means (1999), "Spatial Variation in Diffusion of Technological Innovations at the Cross-National Level," in *Advances in Marketing*, eds. R. Keith Tudor, Sheb L. True, and Lou E. Pelton, Southwestern Marketing Association: 101-109.

Kraemer, Kenneth L., Vijay Gurbaxani, and John L. King (1992), "Economic Development, Government Policy, and the Diffusion of Computing in Asia-Pacific Countries," *Public Administration Review*, 52(2): 146-156.

Kumar, V., Jaishankar Ganesh, and Rai Echambadi (1998), "Cross-National Diffusion Research: What Do We Know and How Certain Are We?" *Journal of Product Innovation Management*, 15: 255-268.

Lilien, Gary L., Philip Kotler, and K. Sridhar Moorthy (1992), *Marketing Models*, Englewood Cliffs, NJ: Prentice Hall.

Lindberg, Bertil C. (1982), "International Comparison of Growth in Demand for a New Durable Consumer Product," *Journal of Marketing Research*, 19(August): 364-371.

Lucas, R.E., Jr. (1988), "On the Mechanics of Economic Development," *Journal of Monetary Economics*, 22: 3-42.

Mahajan, Vijay, Eitan Muller, and Frank M. Bass (1990), "New Product Diffusion Models in Marketing: A Review and Direction for Research," *Journal of Marketing*, 54(1): 1-26.

Mansfield, Edwin (1961), "Technical Change and the Rate of Imitation," *Econometrica*, 29: 741-765.

Mathews, John A. (1997), "A Silicon Valley of the East: Creating Taiwan's Semiconductor Industry," *California Management Review*, 39(4): 26-53.

Olshavsky, Richard W. (1980), "Time and the Rate of Adoption of Innovations," *Journal of Consumer Research*, 6: 425-428.

Ormrod, Richard K. (1990), "Local Context and Innovation Diffusion in A Well-Connected World," *Economic Geography*, 66(April): 109-122.

Park, Albert and Bruce Johnston (1995), "Rural Development and Dynamic Externalities in Taiwan's Structural Transformation," *Economic Development & Cultural Change*, 44(1): 181-208.

Qualls, W., R. Olshavsky, and R. Michaels (1981), "Shortening of the PLC–An Empirical Test," *Journal of Marketing*, 45(Fall): 76-80.

Redmond, William H. (1994), "Diffusion at Sub-National Levels: A Regional Analysis of New Product Growth," *Journal of Product Innovation Management*, 11: 201-212.

Rogers, Everett M. (1983), *Diffusion of Innovations*. 3rd Ed. New York: The Free Press.

Rogers, Everett M. and F. Shoemaker (1971), *Communication of Innovations*, New York: The Free Press.

Swan, Philip L. (1973), "The International Diffusion of an Innovation," *Journal of Industrial Economics*, 22(September): 61-69.

Taiwan Power Company, *Research Report of Home Appliance Penetration in Taiwan*, Republic of China (printed in Chinese).

Takada, Hirokazu and Dipak Jain (1991), "Cross-National Analysis of Diffusion of Consumer Durable Goods in Pacific Rim Countries," *Journal of Marketing*, 55(April): 48-54.

Weber, Matthias and Remco Hoogma (1998), "Beyond National and Technological Styles of Innovation Diffusion: A Dynamic Perspective on Cases from the Energy and Transport Sectors," *Technology Analysis and Strategic Management*, 10(4): 545-566.

Young, Robert B. (1964), *Product Growth Cycles–A Key to Growth Planning*, unpublished results, Stanford Research Institute, Menlo Park, California.

Telemedicine
from a Macromarketing Viewpoint:
A Critical Evaluation
with Proposed Licensing Strategies

Ashish Chandra

Charles E. Pettry, Jr.

David P. Paul, III

SUMMARY. Telemedicine is the practice of medicine from a distance, in which interventions, diagnostic and treatment decisions and recommendations are based on information transmitted through telecommunication systems. While telemedicine may be the wave of the future, many

Ashish Chandra, PhD, is Associate Professor of Health Care Administration, Graduate School of Management, Marshall University, South Charleston, WV 25303-1600 (E-mail: chandra2@marshall.edu). His research areas are in health care marketing.

Charles E. Pettry, Jr., JD, is a lawyer at Allen Law Offices, Charleston, WV. His research interests are in the field of legal affairs in medical technology. He may be contacted at P.O. Box 5308, Charleston, WV 25361 (E-mail: chuckpettry@ hotmail.com).

David P. Paul, III, DDS, PhD, is Assistant Professor of Marketing and Health Care Management, School of Business Administration, Monmouth University, 400 Cedar Avenue, West Long Branch, NJ 07764 (E-mail: dpaul@monmouth.edu). His research interests include health care marketing and service quality issues.

[Haworth co-indexing entry note]: "Telemedicine from a Macromarketing Viewpoint: A Critical Evaluation with Proposed Licensing Strategies." Chandra, Ashish, Charles E. Pettry, Jr., and David P. Paul, III. Co-published simultaneously in *Journal of Nonprofit & Public Sector Marketing* (Best Business Books, an imprint of The Haworth Press, Inc.) Vol. 13, No. 1/2, 2005, pp. 111-135; and: *Government Policy and Program Impacts on Technology Development, Transfer and Commercialization: International Perspectives* (ed: Kimball P. Marshall, William S. Piper, and Walter W. Wymer, Jr.) Best Business Books, an imprint of The Haworth Press, Inc., 2005, pp. 111-135. Single or multiple copies of this article are available for a fee from The Haworth Document Delivery Service [1-800-HAWORTH, 9:00 a.m. - 5:00 p.m. (EST). E-mail address: getinfo@haworthpressinc.com].

Digital Object Identifier: 10.1300/J054v13n01_07

hurdles must be removed to allow it to prosper. A major challenge is licensure. This paper evaluates the current medical licensure requirements in the United States, with an eye toward how government programs can impede or encourage the diffusion of new technologies into commerce, based upon the specific domain of telemedicine. There is a great degree of variation in telemedicine regulations from state to state. Mutual licensing strategies for telemedicine and ensuing challenges are proposed and discussed. The authors also identify various areas of potential applications of telemedicine and explore the implications for managers of health care organizations who would be interested in incorporating telemedicine into their organization. *[Article copies available for a fee from The Haworth Document Delivery Service: 1-800-HAWORTH. E-mail address: <docdelivery@haworthpress.com> Website: <http://www.HaworthPress. com> © 2005 by The Haworth Press, Inc. All rights reserved.]*

KEYWORDS. Telemedicine, medical practice, telecommunications, licensure, government regulation, health care

INTRODUCTION

Macromarketing has been defined as "the effect of marketing on society and vice versa" (Sirgy 2001, 134). Fisk (1971) suggested that marketing's social contribution could be examined through the major "publics" (biological survival/environmental habitability, business interest, government, and consumer sovereignty) whose goals typically are served by the marketing discipline. Thus, one of the domains of macromarketing is to evaluate the societal effects of the market system as it is regulated by governmental and other control systems (Nason 1994). Laws and regulations govern and control transactions in society (Panda and Dholakia 1992), impeding or assisting the interstate transfer of goods and/or services (Hollander and Popper 1994). Governmental regulations of the market do, however, produce significant rigidities and inefficiencies in the system, and it is incumbent upon marketing scholars to "catalog such frictions, explain their origins (so as to help predict their reoccurrence) and analyze their effect" (Hollander and Popper 1994, 69). Because these governmentally-induced frictions and inefficiencies significantly affect what is permissible practice in not only what are considered to be commercial enterprises, but also in the co-called "learned professions," including healthcare (Jost 1997), this paper examines how governmental regulations can impede the diffusion of a promising innovation in the healthcare industry: telemedicine.

The proper practice of medicine is critical to the well being of the populace, and conversely, "bad medicine" may involve substantial risk of injury or even death to individuals. Thus, regulation of the practice of medicine has long been seen as an appropriate function of governments. Telemedicine will be used as a particular case to illustrate how governmental policies, which are designed to regulate public welfare for the public good, can impact the diffusion of a particular technological innovation in a regulated industry.

The paper is divided into four major sections: an Overview of Telemedicine, Governmental Regulation, Potential Applications, and Conclusions. The Overview section includes a discussion of the background of telemedicine, medical regulation/licensure issues, reimbursements issues, important terms used in the field, and some historical developments. The Governmental Regulation section examines in depth the state laws regulating telemedicine, the problems resulting from these varying regulatory schemes, and the position of the major national physicians' medical body, the American Medical Association, on the situation. In the Potential Telemedicine Applications section, existing and potential applications of telemedicine in terms of the U.S. domestic market (use in correctional facilities, housecalls/home health, rural health clinics, regional telemedicine centers, and managed care applications of telemedicine), and also some non-domestic applications of telemedicine, through a discussion of both personnel (e.g., pharmacists) and general international telemedicine usage are reviewed. Finally, some future challenges and recommendations for the future, and implications for managers are presented in the Conclusions.

OVERVIEW OF TELEMEDICINE

Definitions

For a better understanding of the following discussion, it may be helpful to define certain terms as they are commonly used.

"Telemedicine"–Magenau (1997) defined telemedicine as "a wide range of medical services delivered from a remote site via electronic networks. It uses telecommunications networks to transmit medical data (i.e., x-rays, high-resolution images, patient records, and videoconference consultations) from on location to another. Such transmission occurs on the Internet, on corporate Intranets, using videoconferencing equipment, and on ordinary telephone lines."

The definition provided by the World Medical Association (Anonymous 1999) is similar: *"Telemedicine is the practice of medicine, from a distance, in which interventions, diagnostic and treatment decisions and recommendations are based on clinical data, documents and other information transmitted through telecommunication systems."*

A more recent definition explicitly takes into account the two-way interaction between physician(s) and patient: *"Telemedicine is . . . the use of advanced telecommunication technologies to exchange health information and provide health care services across geographic, time, social, and cultural barriers"* (Yallapragada and Paruchuri 2002).

These definitions have several things in common: a patient (perhaps accompanied by a local physician, perhaps not), for whom a medical opinion concerning diagnosis or treatment is rendered by a physician in another state or country, as result of the transmission of individual patient data by electronic or other means to the distant physician.

Although the use of telemedicine has expanded greatly in recent years, the issue of licensing has not been fully addressed as it relates to telemedicine in the United States. While most state statutes require a telemedicine practitioner to be licensed in the state in which the patient resides, we will propose model legislation that does not contain such a requirement. A number of states have included language in their statutory schemes that requires the patient to be attended by a local physician (e.g., Alaska Code 08.64.370). However, we would argue that this requirement essentially negates the concept of modern telemedicine, by requiring an attending physician to be physically present at the remote site to receive data from the telemedicine practitioner.

The authors believe that the significance of state and national boundaries should be de-emphasized so that the practice of telemedicine can develop free of restrictive licensing requirements. It is suggested that the current licensing requirements for telemedicine practitioners will inhibit the development of the practice of telemedicine throughout the country and may serve to restrict the purpose for which it was intended.

"Consultation"–This term is discussed at this juncture because most state licensing statutes exempt "consultations" from licensing requirements, and some would argue that such an exemption serves to provide a de facto exemption for telemedicine. However, it should be noted that the definitions often clearly require two physicians–one who attends the patient and is present at the remote site and another who receives the electronically transmitted patient data. It is submitted that such arrangement is not telemedicine in the modern sense, but merely a consultation where electronic patient data transmission has replaced a face-to-face conversation between peers.

Steadman's Medical Dictionary defines "consultation" as a meeting of ". . . two or more health professionals to evaluate the nature and progress of disease in a particular patient and to establish a diagnosis, prognosis and therapy." Again, it is submitted that the term "consultation" does not apply to telemedicine regardless of which of the above definitions is used.

Background

Telemedicine relies on the electronic transfer of patient information from a healthcare provider to and from the location of the patient, and may involve medical diagnosis, medical treatment, and/or medical education. Such electronic transmissions may include x-rays, high-resolution images, patient records, and/or videoconference consultations. The range of information highways available to transmit healthcare information includes the Internet, corporate Intranets, video conferencing equipment, radio waves to cellular dishes or satellites, and ordinary telephone lines (Magenau 1997). Telemedicine allows physicians throughout the United States and even the world to expand their expertise by sending and receiving electronic images and information. While telemedicine may often be delivered via two-way, full motion video, healthcare providers can also use it to transmit and store static images or videos for later review and consultation.

Thus, via telemedicine, healthcare information and services can be provided in previously underserved areas, particularly in geographic areas where there is an inadequate supply of specialists (Bisby 1998). Technological innovations bring patients, technology resources, and local and remote health care providers together in a virtual medical location. These innovations in technology allow healthcare providers to meet the needs of patients effectively and efficiently (Magel 1999). Telemedicine is being applied in cardiology, dermatology, dentistry, gynecology, internal medicine, neurology, pediatrics, trauma, radiology, surgery, and home health care (Reich-Hale 1999), and has been shown to improve continuity of care by improving both access and coordination of clinicians' activities (Balas et al. 1997).

In many areas, such as recruitment for clinical trials, purchasing prescription drugs, applications for health insurance, and even consultations with medical care providers, traditional health care activities are migrating to the Internet (Goldman and Hudson 2000). It is this area that drives the field of telemedicine, but the anticipated migration of telemedicine to the Internet will not take place unless the regulatory scheme now in place is changed significantly. The "patchwork" of state licensing laws presents an obstacle to the eventual practice of telemedicine across state lines (Broenden and Perry 2000), an issue which is addressed more fully later.

Medical Regulation and Licensure in the United States

The practice of medicine in the United States is highly regulated, and the complex licensing requirements vary in all 50 states. There are, however, common themes in the various licensing frameworks. Most licensing statutes were enacted years ago, and reflect a regulatory climate more suited to the practice of medicine in the days of the house call. As the practice of medicine changes, due at least in part to the increasing use of the technology, it is suggested that its licensing aspects must change as well.

Medical licensing is initially implemented by the definition of the "practice of medicine" contained in the various state codes. By way of example, the code of West Virginia defines the practice of medicine as the "diagnosis or treatment of, or operation or prescription for, any human disease, pain, injury, deformity or other physical or mental condition" (West Virginia Code, 30-3-4). If one performs a service encompassed by such definition, one is practicing medicine, and must be licensed to do so by the state in which the medical service is delivered. There are, however, certain "loopholes." Some states' codes exclude certain practices from the licensing provisions of the medical licensing statutes. Other states' codes exclude certain practices not from the licensing requirements but from the definition of the "unauthorized practice of medicine." For example, in West Virginia the exemptions from the "practice of medicine" are outlined in the latter manner, and as they relate to telemedicine, the pertinent statute (West Virginia Code, 30-3-13) provides, among other things, that certain listed activities do not constitute the unauthorized practice of medicine (i.e., do not require a medical license in the state). Among these listed activities are those of physicians licensed in other states who ". . . are acting in a consulting capacity with physicians . . . duly licensed in this state, for a period of not more than three months" (West Virginia Code, 30-3-13(b)(2)). This statue appears to exclude telemedicine from medical licensing requirements, but as discussed herein, the proposed definition of modern telemedicine would not include the "consult" exemption, as we refer to it herein, and hence would not be excluded.

The regulation of healthcare certainly falls within the purview of each state to protect the welfare of its citizens. However, the inherent distrust/mistrust of the competence of other states to perform this same task makes multi-state licensing of healthcare professionals quite difficult. Until this problem is resolved in some manner, the interstate diffusion of telemedicine and its benefits will continue to be impaired.

Reimbursement Issues

An additional noteworthy issue in telemedicine is the reimbursement issue (Coile 2000a; Edlin 1999a; Huston and Huston 2000; Krizner 2002). The Balanced Budget Amendment of 1997 requires Medicare to reimburse for telehealth consultations under certain conditions. Beginning in 1999, Medicare paid for telehealth services in rural counties designated as health professional shortage areas (Morris, Nickelson and Mahr 1998), including counties in California, Arkansas, Georgia, Iowa, Montana, Oklahoma, North Dakota, South Dakota, Virginia, West Virginia, Illinois and Kansas. Some argue that Medicare (and perhaps Medicaid) reimbursement may be the fuel that accelerates the implementation of telemedicine (Herrick 1998). The program was previously restricted to certain designated rural areas with a shortage of healthcare providers, but a provision in the Medicare-Medicaid funding bill recently signed expands the Medicare telehealth programs to all counties outside metropolitan statistical areas and, for the first time, allows reimbursement for telemedicine at urban Medicare demonstration sites (Versel 2001). These new rules took effect October 1, 2001, the beginning of the government's 2002 fiscal year. The changes broaden the Medicare telemedicine program beyond mere consultation with remote specialists to include coverage for office visits, drug maintenance, psychotherapy, and other services defined by dozens of specific CPT codes. Also, under the new regulations Medicare will not require a "telepresenter"–a physician or other qualified healthcare practitioner–to be present at the rural site with each patient during a consultation with the remote specialist. More recently, the Medicare Remote Monitoring Services Coverage Act of 2001, which required that Medicare reimburse physicians the same fee for remote monitoring of patients that they are paid for a face-to-face encounter, was passed. This law became effective on January 1, 2003 (Landa 2002).

Perhaps reimbursement, being largely driven at the Federal government level via Medicare, will result in a push toward greater flexibility of "movement" (via telemedicine) of physicians.

Telemedicine Development

Telemedicine is currently in its infancy and the cost for its implementation may be expensive. However, the required technology is both improving rapidly and decreasing in cost, and primary care physicians in the near future can be expected to provide medical consultations from affordable and convenient multimedia desktop systems, at least intrastate (Reich-Hale 1999). For example, Sprint Healthcare Systems, Sprint's healthcare-dedicated business unit, recently provided all of Florida Hospital facilities with videoconferencing/ telemedicine services. This kind of service allows a cost-effective approach for physicians to perform remote diagnoses and consultations. Ralph Randall, regional client manager for Sprint, says "We will implement technology that changes the way healthcare is delivered and improves the lives of both patients and healthy individuals" (Mycek 1997, 18). This service allows radiologists to read X-rays and includes electronic storage and transmission of ultrasound, electrocardiograms, and other scanned images. It is interactive and allows patients and physicians to work simultaneously on their records to provide necessary treatment (Mycek 1997).

In the international economy, telemedicine represents one of the primary applications of electronic commerce. Through its use, health care can be provided and patient information can be transmitted via computer networks. It can decrease or even eliminate the need for patients and health care providers (doctors, nurses, therapists, etc.) to be physically present in the same geographic location. Plock (1998, 42) postulates that a key benefit is that "by linking up with medical specialists in a developing country, a U.S. telemedicine provider can avoid many of the regulatory barriers it would encounter with an 'on the ground' operation." After many decades, telemedicine is finally claiming its place in health care, in countries as far ranging as the United States, Saudi Arabia, Ireland, Jordan, and Malaysia. But while technology has been developing rapidly, the response from the medical profession has been rather slow. Now in its early years, telemedicine will have to overcome the same legal, financial, personal and cultural obstacles that have confronted every other technological change in the history of medicine (Jarudi 1999).

Telemedicine allows the centralization of physicians and medical records for patients. This centralized data can then be made available to health care providers at off-site or other geographic locations to address patient needs. In 1993, the Telemedicine Center at Yale University School of Medicine linked over 400 U.S. and Russian health care providers to allow them to participate in clinical conferences (Fishman 1997). However, full implementation remains restricted by the current licensure requirements.

Continuing developments in electronic commerce, especially the continuing development of the Internet and other telecommunications approaches, will certainly make telemedicine a more cost effective option than physical travel of doctor(s) and/or patient(s) when long-distance consultations and/or discussions are required.

GOVERNMENTAL REGULATION

Regulatory Scheme

Telemedicine represents a new source of anxiety for state-based regulators who are concerned about healthcare professionals practicing medicine across state boundaries without appropriate state license(s) (Coile 2000). Licensure is the legal process whereby permission is granted to individuals to engage in a particular occupation or use a particular title. Thus, individuals not licensed in a particular field are legally forbidden from performing any of those tasks or duties which are permitted to members of the specified profession. Licensure to practice medicine is granted at the level of the individual state, not at a regional or national level. Achievement of a state medical license is thus is an absolute prerequisite for the practice of medicine in any given locality.

As noted above, the statutes of the various states address the licensure of telemedicine in different manners. Eight states currently address the issue directly (West Virginia Code, 30-3-13; Mississippi Code, 73-25-34; Hawaii Code, 453-2(b)(3); Missouri Code, 334.010; Montana Code, 37-3-34, 342, 343; North Carolina Code, Chapter 90, Section 90-18(c)(11); Oklahoma Code, 36-6801 et seq.; Vernon's Ann. Civ. St. 151.056 [Texas Code]), although the statutory provisions and practical effects differ. Forty-two states have yet to address the issue. Of the eight states that have addressed the issue, the statutes vary widely.

Hawaii's statute (Hawaii Code, 453-2(b)(3)) recognizes telemedicine, and exempts the practice of telemedicine from state licensure, but only in circumstances of consultations with physicians licensed in Hawaii.

The Mississippi statute (Mississippi Code, 73-25-34) recognizes telemedicine, but specifically requires licensure for its practitioners, unless in the area of defined consultations. Missouri recognizes telemedicine in its statute (Missouri Code, 334.010), but requires licensure, except for consultations. Montana recognizes telemedicine, and requires a "telemedicine certificate" for all practitioners (Montana Code, 37-3-34, 342, 343).

North Carolina exempts telemedicine from licensing requirements, but only when consultations are conducted on an irregular basis with a licensed physician or medical school personnel (North Carolina Code, Chapter 90, Section 90-18(c)(11)). Oklahoma has enacted the Oklahoma Telemedicine Act (Oklahoma Code, 36-6801 et seq.) which defines telemedicine, provides for payment therefore by state Medicare managed care programs and private insurers, and requires specific informed consent by the patients of telemedicine practitioners.

In 1999, Texas enacted legislation (Vernon's Ann. Civ. St. 151.056 (Texas Code)) to provide that telemedicine practitioners are subject to licensing requirements, except those who provide "only episodic consultation services on request to a physician licensed in this state who practices the same medical specialty." It is noteworthy that the Texas statute addresses an issue previously not addressed–a requirement that the telemedicine practitioner and the local physician practice the same specialty to remain exempt from the licensing requirements. Such language appears to emphasize consultation between equals to remain a "consultation" and hence exempt from licensure. A consultation between physicians not practicing the same specialty would not be exempt from licensure. It may be the intent of the statute to discourage such arrangements.

The West Virginia statute defines telemedicine and specifically requires a telemedicine practitioner to be licensed in that state (West Virginia Code, 30-3-13) (see Table 1).

Although it is unlikely that state-by-state licensure will be replaced by national licensure, some additional coordination of state licensure agencies is becoming increasingly important (Jost 1997). Telemedicine is likely to force increased cooperation among the states with respect to at least some aspects of licensure of healthcare professionals, and should result in the decreased ability of states to restrict the services of a healthcare professional licensed in another state (Grandade 1996; Prager 1995).

The regulatory scheme for the practice of medicine in general, and telemedicine in particular, is at best a "patchwork quilt" of conflicting and often confusing state regulations. These state-by-state restrictions, varying in what is allowed and what is forbidden, and under what conditions, present a significant barrier to the efficient diffusion of telemedicine and its benefits to the general populace of the United States.

Licensure as a Major Problem

A number of problems hinder the growth of telemedicine, including patient privacy, medical malpractice, quality of care, informed consent, inadequate reimbursement, coverage by Medicare and private health plans, inadequate infrastructure in rural areas, costs of the required technology, informed consent, intellectual property, corporate practice of medicine, and governmental regulation in the form of licensure (Braender and Perry 2000; Conhaim and Page 2003; Edelstein 1999; Heich-Hale 1999; Krizner 2002). While each of these problems represents a significant hurdle, this paper will concentrate on perhaps the most contentious of the public sector difficulties, licensure. Much of the difficulty between telemedicine and licensure of healthcare providers

TABLE 1. Licensing Aspects of Telemedicine–Survey of State Codes–2002

STATE	CODE SECTION	DESCRIPTION
Alabama	34-24-51,50	no exemptions for telemedicine
Alaska	8.64.370	exempts consults w/physician
Arizona	32-1421	exempts "single/infrequent consult"
Arkansas	17-95-203	exempts "occasional" consult
California	Bus & Prof 2060	exempts "consultations"
Colorado	12-36-106(3)(b)	no mention of "telemedicine," exempts consults
Connecticut	Title 20,Sec 9	describes telemedicine, requires license
Delaware	24-1703	no exemptions; license required
Florida	58.303(b)	exempts consults w/physician when "meeting" physician
Georgia	43-34-20	no exemptions
Hawaii	453-2(b) (93)	exempts telemedicine for actual consults with licensed Hawaii physician
Idaho	54-1804	no exemptions for telemedicine
Illinois	225 ILS 60/49.5	no exemptions
Indiana	25-22.5-1-2	exempts consults w/Illinois licensed physician
Iowa	148.2 Io. Code Ann.	no exemptions
Kansas	65-2867,2869	exempts consultations
Kentucky	311.560	exempts consultations
Louisiana	37-1291 LRS	no exemptions
Maine	32-3270 MRSA	no exemptions
Maryland	14-302	exempts consultations
Massachusetts	112-2,6	no exemptions
Michigan	333.17001	no exemptions
Minnesota	147.09	exempts consultations
Mississippi	73-25-34	Recognizes telemedicine, requires license, unless it is a consult
Missouri	334.010	includes telemedicine in definition of "practice" requires license, but excepts consults;
Montana	37-3-34,342,343	requires "telemedicine certificate"
Nebraska	71-1,102	no exemptions
Nevada	NRS 630.020	no exemptions
New Hampshire	329-21	no exemptions
New Jersey	45:9-5.	no exemptions
New Mexico	61-6-17	provides for/requires "telemedicine license" (Note: statute repealed effective 7/1/2004)

TABLE 1 (continued)

STATE	CODE SECTION	DESCRIPTION
New York	16-6526	no exemptions
North Carolina	Ch 90, Sec90-18(c)(11)	exempts telemedicine on irregular basis in consult w/phys or w/medical school personnel
North Dakota	43-17-01,02	no exemptions
Ohio	4731.36	exempts consults
Oklahoma	36-6801	Oklahoma Telemedicine Act–recognizes and defines telemedicine; license required; exempts consults
Oregon	677.060	exempts consults when "meeting" w/Oregon licensed physician
Pennsylvania	63PS422.22	no exemptions
Rhode Island	5-37-12	no exemptions
South Carolina	40-47-140	no exemptions
South Dakota	36-4-10	no exemptions
Tennessee	63-6-201	no exemptions
Texas	151.056	includes telemedicine in "practice" and license requirements, except for "episodic" consults w/phys.
Vermont	26-1313	exempts consults w/Vermont licensed physician
Utah	58-6-7	no exemptions
Virginia	54.1-2901(7), (15)	exempts consults by Virginia physician with emergency medical technicians
Washington	18.71.030(6)	exempts practice by person licensed by State, provided such person does not open an office or appointment place to meet patients
West Virginia	30-3-13	defines telemedicine; requires license
Wisconsin	448.03	no exemptions
Wyoming	33-26-103	exempts consults by "physicians called into this state."

stems from the fact that healthcare licensure resides at the state level (Jost 1995; Shryock 1967).

A major problem hindering the expansion of telemedicine is the question of whether it is seen as bringing the clinician to the patient, or the patient to the clinician. Most state medical licensing boards seem to take the approach that it is the latter that is occurring in telemedicine, and require a practitioner to have a medical license in the state in which the patient resides. Alternatively, one could argue that it is the patient whose image/information is being transmitted, which may alleviate the need for the consultant to obtain another medical license. A third viewpoint holds that national licensure for telemedicine practitioners is necessary (Strode, Gustke, and Allen 1999).

A survey of the pertinent state statutes in the United States reveals that a clear majority of the states have not specifically addressed this issue. Forty-two states have not yet addressed the specific licensing issue of telemedicine, although the issue may be indirectly addressed in the area of "consultations." While not specifically excluding the practice of telemedicine from licensing requirements, the licensing statutes of a number of states provide an exemption/exclusion for consultations with locally licensed physicians. Physician-to-physician consultations have generally been allowed under "consultation exceptions" of state licensing laws, but now, with the advent of telecommunications between physicians and patients, a physician in one state could easily treat a patient–and thereby practice medicine–in another state. And that, almost without exception, requires a license (Hoffman 1998).

It is important to note that the concept of "consultation" necessarily includes a local physician who has already established a physician/patient relationship, and accordingly would not include a circumstance in which a patient or a health care worker was in contact with a physician in another state or country. If one were to adhere to the modern concept of telemedicine, where a health care worker–not a physician–electronically transmitted data to a physician in another state, territory or country, then the "consultation exemption" described herein would not apply. Under this more restrictive approach, it is fair to say that the practice of telemedicine is prohibited in all states of this country unless the telemedicine practitioner is licensed in the state in which the patient resides. Notably, while Mississippi recognizes telemedicine, it requires licensure of telemedicine practitioners except in the "consultation" circumstances.

In addition, most states not only prohibit the unlicensed practice of telemedicine but additionally make such practice a criminal offense. For example, the West Virginia statute provides than one convicted of engaging in the practice of medicine shall be guilty of a misdemeanor and fined not more than $10,000.00 and sentenced to not more than 12 months incarceration (West Virginia Code, 30-3-13(b)(2)). While not suggesting that medical licensing boards are interested in mindlessly prosecuting telemedicine practitioners, it nonetheless is an issue that needs addressed, and may in some measure be dependent upon the impetus from local medical societies and regulatory bodies. The Medical Board of California recently fined six physicians whose practices are located in Rhode Island, Tennessee, Arizona, and Florida a total of $48 million for prescribing drugs to California citizens over the Internet (Adams 2003).

The Federation of States Medical Boards of the United States ("FSMB"), whose membership comprises the medical boards of the United States, the District of Columbia, Puerto Rico, Guam, the Virgin Islands and 13 state boards of osteopathic medicine, was founded in 1912. In 1996 the FSMB recognized that despite its advantages, telemedicine also presented a licensing problem to practitioners. It seemed obvious that states would have to agree on some sort of modification(s) to their medical practice acts which would allow healthcare practitioners to diagnose across state lines, or else the practice of telemedicine (and the benefits resulting thereof) would be seriously restricted (Relman 1997). In April of 1996, the FSMB published a proposed model act that provided that physicians who practiced telemedicine would be required to obtain a special license from the state licensing board, limited to the practice of medicine across state lines. It would not allow the physician to enter the state physically for the purpose of engaging in the practice of medicine. The proposed act further provides an expedited licensing procedure for the telemedicine practitioner. Unfortunately, the proposed FSMB telemedicine act does not appear to have been adopted by any state(s) to date in the U.S.

Licensure remains a major hurdle for the efficient diffusion of telemedicine and its benefits, at least domestically, as most states determine the circumstances under which an out-of-state practitioner can utilize telemedicine to confer with an in-state colleague. Under current state regulations, a medical practitioner in one state could legally examine a patient in another state via telemedicine, but might face considerable legal difficulties if the physical locations of the patient and physician were reversed.

The Position of the American Medical Association

The American Medical Association (AMA) has monitored the growth of telemedicine for a number of years, and has developed a well-defined policy. The AMA current policy (www.ama-assn.org/apps/pf_online/pf_online?f_n=browse&doc=policyfiles/HOD/H-480.974.htm) calls on the AMA to:

- Evaluate relevant federal legislation.
- Encourage broad federal support for telemedicine that improves access to care for the underserved.
- Urge funding of demonstrations to evaluate impact on costs, quality, and physician-patient relationship.
- Work with payers to develop test reimbursement procedures.

- Encourage the Current Procedural Terminology editorial board to develop codes or modifiers for telemedicine.
- Develop a means of providing appropriate continuing medical education credit for educational consultations.
- Urge specialty societies to develop practice parameters for telemedicine and guide quality assessment and liability issues.
- Work with Federation of State Medical Boards and state and territorial licensing boards to develop licensure guidelines for telemedicine across state boundaries.

AMA position papers have noted that a physician seeking multiple state licenses may find the current system burdensome, time consuming and expensive. The licensing fees range from $1,108 in California to $20 in Pennsylvania, with a national average of $339. In addition, most states require a physical appearance at some point in the licensure application process, adding significant time and expense (Robertson 2002).

Recently, due largely to the growth of telemedicine, medical residents (physicians who have graduated from medical school but are still in post-graduate training) approved a resolution requesting that the American Medical Association abandon its historic opposition to national licensure and adopt a national standard of care (Greene 2000).

The AMA's position with respect to this thorny question remains mired in its position that licensure is a state issue. Thus, the AMA continues to "evaluate, encourage, urge, work with, and develop." In the end, the AMA's position appears to largely be one of "wait and see," rather than a leadership one.

EXISTING AND POTENTIAL TELEMEDICINE APPLICATIONS

Telemedicine can have widespread application in the coming years, and serve to bring quality medical care to historically underserved areas of this country and other countries. Several potential applications are as follows:

Correctional Centers

Correctional facilities have begun to avail themselves of telemedicine practice, and have experienced significant cost reductions and reduced security exposure from transporting prisoners to remote health care facilities (Chin 2000; Huston and Huston 2000). Previous research has indicated that, "Prison administrators welcome the use of telemedicine in order to cut down on the time, expense and security headaches of transporting patients, since few prisons maintain a full time medical staff" (Rogak 1999, 42). In 1998, telemedicine accounted for 30% of the medical consultations in prisons (Chin 2000).

House Calls/Home Health

The median age is the U.S. is expected to increase for the next 30 years, as the "Baby Boomers" continue to mature. Associated with this aging process will be an increase in chronic diseases, such as arthritis, high blood pressure, coronary heart disease, etc. (Knickman 2002). These types of health problems and patients are difficult to monitor in a cost-effective, traditional manner, but telemedicine presents an alternative to the typical "office visit" (Edlin 1999b; Jerant et al. 1998). Telemedicine allows physicians to make house calls in remote locations, which will not only be helpful in saving lives but also can help in restoring economic viability of remotely located health centers (Isenberg 1993). After patients and their homes are outfitted with appropriate technological monitoring equipment, they can be monitored on a regular schedule by nurses, or (if required) physicians. Telemedicine can thus be used to augment traditional home health care, not necessarily replace it (Josey and Gustke 1999). The up-front costs of the technology should be more than offset by the savings achieved by preventing costly hospitalizations which result from medical problems becoming acute (Holewa 1999).

Rural Health Clinics

Rural health care clinics could be established in areas that historically have not been served by a physician. Such clinics could be staffed by health care workers trained to interview patients and to perform clinical diagnostic testing, the results of which testing can be electronically transmitted to physicians located in other states. The physicians can then transmit diagnoses, prescribe medication, or make referrals. This practice could bring 21st century health care to rural areas of this country. The need here is great. There are no primary care physicians in more than 166 U.S. counties with 10,000 or fewer residents, and almost half of existing rural physicians are 55 or older. While physician assistants and advanced nurse practitioners can help "take up the slack" for some rural areas lacking physicians, many of these alternate providers choose to work in urban areas of the U.S. (Coile 2000b).

Regional Telemedicine Centers

Regional telemedicine centers could be established by existing regional health care centers, such as The Cleveland Clinic, The Mayo Clinic, M.D. Anderson Cancer Institute, etc. Under such an arrangement, health care workers in rural areas of the country could electronically transmit patient data to such regional centers, where physicians could evaluate the data, make diagnoses and make referrals if necessary. Under the present regulatory scheme, this practice would be proper only if the physician were licensed in the state where the patient was located or if a locally licensed physician made the data transmission.

Managed Care Applications

In the interest of reducing cost and providing specialty care, managed care companies could look to groups of specialists or large hospitals for telemedicine consultations–or even treatment–rather than using less technologically equipped local facilities. If telemedicine costs less, a for-profit company may "send" their patients–virtually, that is–to less expensive specialists in other localities (www.mmhc.com/hcbd/articles/HCBD9711/telemedicinewhatsbeyondth.html).

Remote Health Professionals

Many rural communities do not have resident health care professionals (such as doctors and nurses); however, they may still have a small pharmacy. Pharmacies located in these communities could serve as the rural site for telemedicine practitioners to render diagnosis and treatment to patients. Again, health care workers/paramedics could transmit electronic patient data to remote physicians, who then could transmit their diagnosis/prescription to the pharmacist in the rural location. This form of telemedicine can play a major role in the enhancement of the image of the pharmacy, and may also be an alternate revenue-generating arrangement for the pharmacist. The revenue generated via this form of health care delivery might even encourage more pharmacies in rural setting from shutting down. Healthcare providers should note, however, that aiding a non-licensed provider in the practice of medicine could result in the practitioner's facing civil fines, suspension of license to practice, and/or revocation of his/her medical license. Appropriate changes in the state laws covering telemedicine are mandatory if pharmacists are to become more fully involved in telemedicine without dangerous possible consequences to both healthcare practitioner and pharmacist.

In addition to remote rural areas in the U.S. where there is no primary health care provider, there are also many countries where there is a severe lack of well-trained specialists in certain fields of health care. In the absence of these specialists, the remotely located pharmacy or health care institution in the foreign country can be used as viable settings for a telemedicine center where most of the equipment needed for telemedical care can be installed. The patient could come to these centers and the pharmacist or the local general practitioner can help in attaching the appropriate equipment to the patient for diagnostic purposes. These health professionals can also assist the physician at the other end in making decisions regarding appropriate drug therapy based on the drug availability in that location. Pharmaceutical and other health care products and services marketers should not ignore this sort of utilization of telemedicine. These marketers can sponsor some of the equipment and the airtime needed to provide appropriate care. This sort of activity on the part of the marketers has the potential of gaining valuable publicity for them. This value-added service on the part of pharmaceutical companies can also be used by them in their promotional materials where they can show to the consumer and foreign governments that they are giving something back to the community. In an era when pharmaceutical and health care companies are considered by many as only money making enterprises, this type of activity will give a positive boost to their image.

International Applications

Telemedicine can be used to link U.S.-based physicians and hospitals to practitioners in developing countries (Coddington et al. 2000; Oyewole 2001). In a recent article in the *Lancet*, a telemedicine project in remote areas of southern India was described, in which a private-sector hospital chain established a 50-bed telemedicine center equipped with an operating theatre, a computed axial tomographic (CAT) scanner, x-ray facilities, and an integrated laboratory. Using special hardware and software, doctors at the telemedicine center can scan, convert, and send data images via satellite to teleconsultant stations several hundred kilometers away at specialty hospitals in Chennai and Hyderabad. These villagers previously had to go to a teaching hospital more than 100 kilometers away for these services (Sharma 2000).

Telemedicine has expanded its realm into even countries such as Jordan, Malaysia, and Panama that are considered by many as underdeveloped nations (Adams 2002; Jarudi 1999). The doctors in these countries have effectively utilized the concept and technology of telemedicine as a valuable channel of obtaining professional advice from their colleagues who are practicing medicine in countries such as the United States. In future, it will be interesting to see the role and level of participation of non-physician healthcare professionals in underdeveloped and developing nations in the dissemination of telemedicine.

According to Frasher and McGrath (2000), the Internet has an increasing presence in sub-Saharan Africa, evidenced by the availability of Internet services in 53 of 54 African countries in 2000, compared to 12 countries in 1997 (Fraser, Harnish, and McGarth 2000). For example, SatelLife, a Boston-based charitable organization, utilizes a low-earth satellite and phone lines to provide email services to some 10,000 health care workers in 140 countries. This allows electronic transmission of digital photograph, electro-cardiograms, and x-ray films, allowing rural health care workers and their patients to avail themselves of physicians/specialists who would not otherwise be accessible to persons living in rural, isolated areas.

Recently, British Midland has become the first airline in the world to install telemedicine technology in their aircraft making the long journey across the Atlantic. The equipment is "the first medical monitoring device designed for non-expert use on board an aircraft," and is capable of monitoring blood pressure, pulse rate, temperature, heart rate/rhythm (EKG), and blood oxygen and carbon dioxide levels (Anonymous 2002, 47). This data can be sent to a physician who can then advise the crew of what action(s) they should take to best assist the ill passenger.

Telemedicine applications range from the currently successful (e.g., use in correctional facilities and across national boundaries) to others which show future potential (e.g., house calls/home health, rural health clinics, etc.). Financial and legal considerations, perhaps brought to the forefront by managed care, will potentially be a major factor in the continuing development of telemedicine and its possible applications.

CONCLUSIONS

Challenges

Standards for privacy of medical information have been established by the Federal government (e.g., the Health Insurance Portability and Privacy Act), which also bars fraud, deceptive advertising practices, anticompetitive behavior, refusal to render emergency treatment, and (in some cases) remuneration for referrals. Historically, it has been state governments which have licensed and overseen medical practice, creating a myriad of legal regulations tied to state boundaries, creating the requirement that health care professionals must be licensed in each and every state in which they practice (Miller and Derse 2002).

Clearly, under the present statutory scheme, any telemedicine practitioner in this country must be fully licensed to practice both in the state in which he/she resides, as well as the state(s) in which the patient(s) reside, before engaging in the practice of telemedicine, even if the practitioner is merely serving in a consultation capacity. If providing patient care via a non-physician health care worker, the practitioner must definitely be licensed in the state where the patient resides.

In the eight states that have addressed the issue, it still remains necessary for the telemedicine practitioner to be licensed in the state where the patient resides. Although most states recognize telemedicine, they all require either specific telemedicine licensing or a consultation status with a locally licensed physician.

As long as states impose licensing requirements upon telemedicine practitioners, such requirement will serve as an impediment to the use of modern telemedicine to deliver medical care to underserved areas in the United States. Organizations offering healthcare services across country borders may encounter geographic and legal barriers to the implementation of their services. For example, in the U.S. there are concerns about physicians who regularly consult in states where they do not have a license to practice. Currently, a number of states require out-of-state physicians consulting across state lines to have a license in the state where the patient is located. Similar medical licensing difficulties may be encountered when providing telemedicine healthcare services across country borders in the global economy.

However, the authors believe that none of these issues could be addressed and accomplished in a satisfying manner under the current regulatory scheme. Clearly, the current licensure situation in the United States for telemedicine may, and almost certainly does, impede the spread of good medical care. In light of a recently published RAND study, described as "one of the most comprehensive undertakings on the subject of health-care adequacy" (Pereira 2003, D3), which found that Americans receive the appropriate diagnosis and treatment for many diseases only about 55% of the time (McGlynn et al. 2003), these barriers to consultation and treatment via telemedicine should be of grave concern to health care policy makers. With evidence that nearly half of U.S. residents are nor currently receiving appropriate medical care, the advantages of decreasing barriers to diffusion of telemedicine should be quite apparent.

Recommendations

The previous discussion indicates little coordination among the states and their respective medical licensing boards. Rather than addressing the need for uniform licensing of telemedicine practitioners, states have taken a seemingly parochial approach, ranging from full licensure to telemedicine licensure to ignoring the issue entirely. It appears that the most progressive approach is one that defines "telemedicine" but does not require a telemedicine practitioner to be licensed in the state in which the patient is being treated, although a currently valid medical license at the point of transmission would still be required. There are currently no states that have such a statutory scheme.

In order to make use of telemedicine to deliver medical care to underserved areas, it will probably be necessary for licensing boards and/or state legislatures to permit telemedicine practitioners to deliver health care through the use of non-physician health care workers. By requiring the use of locally licensed physicians–the "consult" requirement–the use of telemedicine to deliver health care to medically underserved areas is effectively prohibited. If state licensing authorities and/or legislatures approve the use of local non-physician health care workers/technicians to assist telemedicine practitioners, effective health care can be delivered to underserved areas of this country. A related issue is the willingness, or lack thereof, of local physicians, medical societies and state licensing authorities to "loosen" licensing requirements to allow the use of modern telemedicine. It may be that the current regulatory scheme is not the result of benign neglect, but rather the product of a preference for the status quo.

We suggest an approach that does not require local licensing of the telemedicine practitioner, but rather requires only that the telemedicine practitioner to be licensed in the state where he or she normally practices, and recognizes that the licensing in the practitioner's home state is essentially no different than licensing in any other state. To hold otherwise, as all states now do, is to suggest that the practice of medicine differs from state to state and that one state's licensing requirements are better/worse than another's.

The suggested approach is best described as the "mutual" licensure approach in the literature. The AMA defines "mutual" as "a system in which the licensing authorities voluntarily enter into an agreement to legally accept the policies and licensure processes of a licensee's home state. This approach has been adopted by the European Community and Australia to enable the cross-border practice of medicine. It has been successfully utilized by the Veteran's Administration, U.S. military, the Indian Health Service and the Public Health Service. The licensure based on mutual recognition is comprised of three components: a home state, a host state, and harmonization of standards for licensure and professional conduct deemed essential to the health care system. The health professional secures a license in his/her home state and is not required to obtain additional licenses to practice telemedicine in other states (Robertson 2002).

Implications for Managers

Given the current regulatory and statutory framework, it is important for managers to keep in mind the restrictive nature of the licensure issue. Despite advances in technology and communications, the 19th century restrictions on medical practice across state lines still exist. The practice of medicine still operates in a medico-legal environment, and until fundamental changes are enacted in both the medical and legal fields, licensure issues will continue to restrict the domestic growth of telemedicine. Even with appropriate software, hardware, willing physicians and appropriate reimbursement strategies, telemedicine practitioners must still be licensed in each state where a remote site is operated. Given the time, expense and varying requirements of multi-state licensure, the licensing issue is, and will continue to be, a major impediment to the growth of telemedicine. Managers must be aware of this limitation and develop appropriate strategies to operate with in the parameters imposed by these licensure issues.

On another front, managers should stand ready to work with physicians, licensing boards, professional societies, and state legislatures to bring about the changes needed to eliminate the restrictions on the practice of telemedicine across state lines and national boundaries. Unlike the practice of law, which is based upon knowledge of statutes that are, at least in part, state specific, the practice of medicine knows (or should know) no borders. Other than the medico-legal environment, the practice of medicine is no different in North Carolina than in Oklahoma. Nevertheless, current licensure statutes suggest the opposite. Managers need to be aware of this element, and be prepared to work proactively for needed change.

It is clear that the current AMA position strongly opposes any change in state-based licensure. One must also presume that this position reflects the opinion of a majority of its members. Managers must understand that the status quo of licensure is strongly favored by a majority of physicians. As with many fundamental changes, the change must be incremental. Perhaps the example of dental licensure, where regional licensing agencies now exist (Paul 2000), will serve as a model for more widespread medical licensure in the future. Managers must be prepared to demonstrate to physicians that the removal of barriers to the practice of telemedicine across state lines will eventually improve the health status of the country and will enhance the practice of medicine by providing greater access to health care by persons from traditionally underserved areas. In any event, if telemedicine is to develop further, some changes in medical licensure requirements will certainly be necessary.

REFERENCES

Adams, Damon (2002), "Telemedicine: A Forum for Learning for All Involved," *www.ama-assn.org/sci-pubs/amnews/pivk_02/prsd0502.htm*.

Adams, Damon (2003), "California Fines Out-of-State Doctors for Prescribing," *www.ama-assn.org/sci-pubs/amnews/pick_03/prsb0303.htm*.

Anonymous (1999), "World Medical Association Statement on Accountability, Responsibilities and Ethical Guidelines in the Practice of Telemedicine," *www.wma.net/e/policy/17-36_e.htm*.

Anonymous (2002), "Telemedicine Flying High," *Professional Engineering*, (15), 9, 1.

Balas, E. Andrew, Farah Jaffrey, Gilad J. Kuperman, Suzanne Austin Boren, Gordon D. Brown, Francesco Pinciroli, and Joyce A. Mitchell (1997), "Electronic Communication with Patients: Evaluation of Distance Medicine Technology," *Journal of the American Medical Association*, 278 (2), 152-159.

Bisby, Adam (1998), "Health Care Market Goes the Distance," *Computer Dealer News*, *www.findarticles.com/cf_0/m3563/n29_v14/21076876/p1/article.jhtml*.

Braender, Lori J. and Kara McCarthy Perry (2000), "Making a Virtual House Call," *National Law Journal*, 22 (52), C1, C16.

Chin, Tyler (2000), "Telemedicine Use Growing, but Slowly," *www.ama-assn.org/sci-pubs/amnews/picl_00/tesb0731/htm*.

Coddington, Dean, Keith Moore, Elizabeth Fischer, and Richard L. Clarke (2000), *Beyond Managed Care*, Jossey-Bass Inc., Publishers: San Francisco, CA.

Coile, Russell C., Jr. (2000a), "E-Health: Reinventing Healthcare in the Information Age," *Journal of Healthcare Management*, 45 (3), 206-210.

Coile, Russell C., Jr. (2000b), "Rural Health: Growth, Development (and Survival) of Rural Healthcare," Chapter 6 in Russell C. Coile, Jr., *New Century Healthcare: Strategies for Providers, Purchasers, and Plans*, Health Administration Press: Chicago, IL.

Conhaim, Wallys W. and Loraine Page (2003), "Is the Doctor Online?" *Information Today*, 20 (2), 29-31.

Edelstein, Scott A. (1999),"Careful Telemedicine Planning Limits Costly Liability Exposure," *Healthcare Financial Management*, (December), 63-69.

Edlin, Mari (1999a), "Several Barriers Inhibit Telemedicine," *Managed Healthcare*, 9 (8), 36.

Edlin, Mari (1999b), "Get Ready for Fine-Tuning: Telemedicine Is on the Way," *Managed Healthcare*, 9 (8), 35-39.

Fishman, Dorothy J. (1997), "Telemedicine: Bringing the Specialist to the Patient," *Nursing Management*, 28 (7), 30-32.

Fisk, George (1971), "New Criteria for Evaluating the Social Performance of Marketing," in *New Essays in Marketing Theory*, George Fisk (Editor), Allyn and Bacon: Boston, MA.

Fraser, Harnish S. F. and St. John D. McGarth (2000), "Information Technology and Telemedicine in Sub-Saharan Africa," *British Medical Journal*, 321 (7259), 465-466.

Goldman, Janlori, and Zoe Hudson (2000), "Virtually Exposed: Privacy and E-Health." *Health Affairs*, 19 (6), 140-148.

Grandade, P. F. (1996), "Implementing Telemedicine on a National Basis: A Legal Analysis of the Licensure Issues," *Federation Bulletin*, 83 (1), 7-17.

Greene, Jay (2000), "Residents Press for National Doctor Licensure," *www.ama-assn. org/sci-pubs/amnet/pick_00/prse0103.htm*.

Herrick, Timothy (1998), "Rural Clinicians Get Wired to the 'Virtual Clinic,'" *Clinician News*, 2 (2), 1.

Hoffman, Allen (1998), "TELEMEDICINE: What's Beyond the Hype?" *Health Care Business Digest, www.mmhc.com/hcbd/articles/HCBD9711/telemedicinewhatsbeyondth. html*.

Holewa, Lisa (1999), "Virtual House Call," *AMNews, www.ama-assn.org/sci-pubs/ amnews/pick_99/biza1122.htm*.

Hollander, Stanley C. and Katheen M. LaFrancis Popper (1994), "Balkanization of America: Lessons from the Interstate Trade Barrier Experience," *Journal of Macromarketing*, 14 (1), 62-72.

Huston, Terry L. and Janis L. Huston (2000), "Is Telemedicine a Practical Reality?" *Communications of the ACM*, 43 (6), 91-95.

Isenberg, Doug (1993), "Digital Doctors," *Atlanta Business*, 22 (9), 58-61.

Jarudi, Larma (1999), "Doctors Without Borders," *Harvard International Review*, 22 (1), 36-40.

Jerant, Anthony F., Loretta Schlachta, Ted D. Epperly, and Jean Barnes-Camp (1998), "Family Practice Management," *www.aafp.org/fpm/980100fm/lead.htm*.

Josey, Paula and Susan Gustke (1999), "How to Merge Telemedicine with Traditional Clinical Practice," *Nursing Management*, 30 (4), 33-36.

Jost, Timothy S. (1995), "Oversight of the Quality of Medical Care: Regulation, Management, or the Market?" *Arizona Law Review*, 37, 825-868.

Jost, Timothy S. (1997), "Introduction: Regulation of the Healthcare Professions," in Timothy S. Jost (Editor), *Regulation of the Healthcare Professions*, Chicago, IL: Health Administration Press.

Knickman, James R. (2002) "Futures," in *Jonas and Kovner's Health Care Delivery in the United States*, Seventh Edition, Anthony R. Kovner and Steven Jonas (Editors), 453-475.

Krizner, Ken (2002), "Telemedicine Still Looks for Inroads to Total Acceptability," *Managed Healthcare Executive*, 12 (5), 44-45.

Landa, Amy Snow (2002), "Telemedicine Payment Expansion Sought," *www.ama-assn.org/sci-pubs/amnews/pick_02/gysc0708.htm*.

Magel, Judith S. (1999), "Consolidation in the Health Care Sector," *Journal of Health Care Finance*, 25 (3), 22-28.

Magenau, Jeff L. (1997), "Digital Diagnosis: Liability Concerns and State Licensing Issues Are Inhibiting the Progress of Telemedicine," *Communications & the Law*, 19 (4), 25-43.

McGlynn, Elizabeth A., Steven M. Arsch, John Adams, John Keesey, Jennifer Hicks, Alison DeCristofaro, and Eve A. Kerr (2003), "The Quality of Health Care Delivered to Adults in the United States," *New England Journal of Medicine*, 348 (26), 2635-2645.

Miller, Tracy E. and Arthur R. Derse (2002), "Between Strangers: The Practice of Online Medicine," *Health Affairs*, 21 (4), 168-170.

Morris, T., D. Nickelson, and T. Mahr (1998), "Incremental Steps Lead to Reimbursement Success," *Telemedicine and Telehealth Networks*, 4 (1), 26-9.

Mycek, Shari (1997), "Providing More than Just Phone Service," *Healthcare Forum Journal*, 40 (6), 18.

Nason, Robert W. (1994), "Globalization and Macromarketing," *Journal of Macromarketing*, 14 (2), 1-3.

Oyewole, Philemon (2001), "Prospects for Developing Country Exports of Services to the Year 2010: Projections and Public Policy Implications," *Journal of Macromarketing*, 21 (1), 32-46.

Pandya, Anil and Nikhilesh Dholakia (1992), "An Institutional Theory of Exchange in Marketing," *European Journal of Marketing*, 26 (12), 19-41.

Paul, David P., III (2000), "The Potential Impact of the North American Free Trade Agreement on American Dental Licensure: A European Community Model," *Health Marketing Quarterly*, 18 (1/2), 87-98.

Pereiera, Joseph (2003), "Study Sets Correct Patient Care at 55%," *Wall Street Journal*, June 26, D3.

Plock, Ernest D. (1998) "Telemedicine is Emerging as a Cost-Effective Healthcare Alternative," *Business America*, (January), 42.

Prager, Linda O. (1995), "Medical Board Plan Would Speed Telemedicine Licensing," *American Medical News*, 30 (40), 4.

Reich-Hale, David (1999), "Technology Gains Fuel Telemedicine Growth," *National Underwriter*, 103 (9), 35-42.

Relman, Arnold S. (1997), "Regulation of the Medical Profession: A Physician's Perspective," in Timothy S. Jost (Editor), *Regulation of the Healthcare Professions*, Chicago, IL: Health Administration Press.

Robertson, Janice (2002), "Physician Licensure: An Update of Trends." *www.ama-assn.org/ama/pub/category/2378.htm*.

Rogak, Lisa (1999), "Telemedicine Spreading as Benefits Become Apparent," *Physician Financial News*, 17 (4), 42.

Sharma, D.C. (2000), "Remote Indian Villages to Benefit from Telemedicine Project," *Lancet*, 355 (9214), 1529.

Sirgy, M. Joseph (2001), *Social Indicators Research Series: Handbook of Quality of Life Research*, Boston, MA: Kluwer Academic Publishers.

Shyrock, Richard (1967), *Medical Licensing in America, 1650-1965*, Baltimore, MD: Johns Hopkins University Press.

Strode, Steven W., Susan Gustke, and Ace Allen (1999), "Technical and Clinical Progress in Telemedicine," *Journal of the American Medical Association*, 281 (12), 1066-1068.

Versel, Neil (2001), "Rural Reward," *Modern Physician*, 5 (2), 7.

Yallapragada, RamMohan R. and Madhu R. Paruchuri (2002), "Telemedicine in the Rural Areas," *Proceedings of the 2002 Business & Health Association Annual Meeting*, 295-296.

Patents, Copyrights, and Trademarks in the Early Twenty-First Century: The New Relevance of Some Old Concepts of Intellectual Property Rights

Thomas S. O'Connor

SUMMARY. After laying an historical foundation of the concept of "promonopoly" legislation in the face of a U. S. public policy devoted, since 1890, to a generally antimonopoly position, this paper examines the relevance of recent changes in the three types of domestic promonopoly law: patent law, copyright law, and trademark law. During the last thirty years of the twentieth century, and especially during the last decade of that period, all three aspects of American promonopoly legislation were strengthened. The term of copyright was extended from a maximum of fifty-six years to the lifetime of the last-living author of a work plus seventy years and coverage was broadened to extend to recordings–by a number of means–of performances; patents were lengthened in duration from seventeen years to twenty; and the law of trademarks was modified to include protection for trademark holders against "dilution," or loss of

---

Thomas S. O'Connor, PhD, is Senior Professor of Marketing, Department of Marketing, College of Business Administration, University of New Orleans, New Orleans, LA 70148 (E-mail: toconnor@uno.edu).

[Haworth co-indexing entry note]: "Patents, Copyrights, and Trademarks in the Early Twenty-First Century: The New Relevance of Some Old Concepts of Intellectual Property Rights." O'Connor, Thomas S. Co-published simultaneously in *Journal of Nonprofit & Public Sector Marketing* (Best Business Books, an imprint of The Haworth Press, Inc.) Vol. 13, No. 1/2, 2005, pp. 137-150; and: *Government Policy and Program Impacts on Technology Development, Transfer and Commercialization: International Perspectives* (ed: Kimball P. Marshall, William S. Piper, and Walter W. Wymer, Jr.) Best Business Books, an imprint of The Haworth Press, Inc., 2005, pp. 137-150. Single or multiple copies of this article are available for a fee from The Haworth Document Delivery Service [1-800-HAWORTH, 9:00 a.m. - 5:00 p.m. (EST). E-mail address: getinfo@haworthpressinc.com].

value of trademarks due to the acts of noncompetitors. The implications and rationales of these changes are examined in the context of their effect on business life at the beginning of the third millennium. While it is too soon since the majority of these changes in law went into effect to draw definitive conclusions concerning their ultimate consequences, it does appear that, in general, their thrust is to increase the value of intellectual property to its proprietors. *[Article copies available for a fee from The Haworth Document Delivery Service: 1-800-HAWORTH. E-mail address: <docdelivery@haworthpress.com> Website: <http://www.HaworthPress.com> © 2005 by The Haworth Press, Inc. All rights reserved.]*

KEYWORDS. Intellectual property, property right, patents, copyrights, trademarks

INTRODUCTION: GRANTS OF MONOPOLY BY GOVERNMENT

The Genesis of Patents

Growing out of a history over five hundred years long, the concept of patents on scientific advances and unique applications of the mechanic arts is an established policy of most governments. The first known patent in western Europe was granted by King Henry VI in 1449 (Her Majesty's Patent Office, 2000). The concept was institutionalized some twenty-five years later by the Venetian Patent Ordinance of 1474 (Europäisches Patentamt, 1997), and was recognized in the U. S. Constitution (Art. I, Sec. 8, Cl. 8). Patents have always had as their essential purpose the grant of monopoly to the creator of an invention, usually defined as an advance in the mechanic arts (engineering) or a scientific discovery. The logic behind "grants of letters patent," as they were originally known, is that the individual or his or her assignee who has developed or discovered the innovation, quite likely at substantial cost, should be allowed to recover that cost and reap the benefit of the advancement it represents while protected from competition–but not forever.

Copyrights on Written Works

Reservation of the rights to his or her written work to an author, a similar but philosophically different concept to that of patents, harks back to the reign of British Queen Anne in 1710 (Swarbrick, 2003), and was recognized in the U. S. Constitution by the same article recognizing patents. Copyrights protect, not an advancement in the sciences or engineering, but bodies of creative work, historically those that could be reduced to writing. As will be discussed later, recent changes in U. S. law have extended the protection of the copyright to performances recorded on film, tape, or by other means.

The objective of the law and of the related concept remains the same as with the patent: to reserve the exclusive right to his or her work to the creator of that work for a period of time.

Trademarks Arise from Common Law

Protection of the rights to one's business "mark," on the other hand, is a common-law right long ensconced by custom though specifically recognized in this country by the Act of March 3, 1881, and by the U. S. becoming a signatory of the Paris Convention of 1883. Trademarks do not necessarily, in and of themselves, reflect creativity. What they do is to "identify and distinguish the source of goods and services of one party from those of another" (USPTO, 1996). Trademark rights may arise out of a person's use of the mark or they may arise out of filing a registry application with the U.S. Patent and Trademark Office. In either case, the general rule of law is that the first person to use a trademark in business and to continue its use retains it. There is no specific limitation on the length of time for which a trademark may be used.

As a tool of differentiation, the underlying philosophy of trademarks is that people identify their preferences in goods and services by the names and symbols that those goods and services bear, hence the exclusivity of proprietary ownership from first use and the unlimited duration of the life of the trademark. As a device capable of conferring competitive advantage on its owner, a trademark may have substantial value. As such, piracy of trademarks has never been an unusual phenomenon and protection against such piracy a primary thrust of trademark litigation.

MONOPOLY RIGHTS AND FEDERAL ANTITRUST/ANTIMONOPOLY POSTURE SINCE 1890

Beginning in 1890, the thrust of U. S. legislation concerning competition has been stridently pro-competition and anti-trust–except where patents, copyrights, and trademarks are concerned.

The Sherman Act of 1890 (15 USC Ch. 1, §§1-37), Clayton Act of 1914 (amending 15 USC Ch. 1, §§12-27; 29 USC Ch. 1, §§52-53), Federal Trade Commission Act of 1914 (15 USC 2, Subch. 1, §§41-58), and Robinson-Patman Act of 1936 (amending 15 USC 1, §13), often thought of as the foundation stones of the national law of business for the U. S., all take a staunch stand against any action that reduces competition, fosters monopoly, or in any way restrains trade. Each of them, taken chronologically, tightens the existing strictures against monopolistic behavior.

On the other hand, the Trademark Law (15 USC Ch. 22, §§1111-1129), Patent Act (35 USC Parts I-IV), and Copyright Act (17 USC Chs. 1-13) have, during the same period of time, become increasingly pro-monopoly. To illustrate, consider that the provisions of the trademark laws have, in recent years, been liberalized so that "dilution" of trademarks, which might be described as only the most indirect type of competition between one mark and another, has become federally illegal under the provisions of the Trademark Dilution Act of 1995 (replaces 15 USC Ch. 22, §§1125, 1127); under the patent law, the granting of monopoly for new inventions was extended from seventeen years to twenty by a 1995 amendment (replacing 35 USC Part II, Ch. 14, §154); and copyrights on text materials, once good for twenty-eight years renewable for another twenty-eight, are now valid for the life of the last-living author of the work plus seventy years or up to 120 years for an anonymous work or work-for-hire by the terms of the Copyright Term Extension Act of 1998 (replaces 17 USC Ch. 3, §§302-304).

CHANGING PATENT CONDITIONS

Implications of the Patent Law

The logic behind all the "pro-monopoly" laws is, on its face, clear. Their premise is that intellectual creators of whatever sort should have the fruits of their labors preserved to their use. The duration of that protection, though at first seemingly arbitrary, reflects the government's–and, one might infer, society's–perception of the value of that right to the creator balanced against the need of society as a whole. Thus patents, where engineering technique or scientific advance if held for a long period of time by an individual might inhibit economic progress, are of relatively short duration.

Recent Changes in Patent Conditions

The terms of patents were lengthened in 1995, but only by three years, hardly a substantial advance. This is only the fourth extension of the term of patent in U. S. history, the sum total amounting to a six-year increment (from 14 to 16 to 17 to the present 20 years). One might suspect that this relatively slight extension, in addition to expressing the perception that scientific and engineering advances have relatively high public value, is also related to the fact that there are alternatives to patents and positive and negative consequences to patent acquisition. In order to acquire a patent on an invention or scientific advance, everything about the invention or advance must be divulged in the application–in detail. The patent itself then becomes public record so that a copy may be purchased by anyone and the exact details of the patented item and how it is created is known. A very old principle of engineering, author unknown, quite likely equally applicable to science in general, says that "Whatever one engineer designs another engineer of equal talent can design around."

Many inventors and scientists, recognizing this, decide to keep their inventions and discoveries secret or "proprietary," so the risk of losing the benefit of their creativity is lessened. Thus, the formula for Coca-Cola concentrate and the process by which Cracker Jacks are made remain trade secrets today, after 100 years or more. Even the most successful patents, such as those granted to Edwin Herbert Land for polarization apparatus and the "Polaroid" system of in-camera film development, ultimately either expire or are designed around, in the latter case resulting in an equivalent product becoming available long before the extant patents run out. Defending one's patents can also be an expensive process, so the issue of obtaining them is sometimes best simply avoided. Since patents apply to scientific and engineering phenomena, they are relatively easily designed around–a process often called "reverse-engineering"–but, by the same token, the discovery of whether such activities constitute an infringement of a patent may be very difficult to undertake in a court of non-scientists and non-engineers.

COPYRIGHTS THEN AND NOW

Extension and Expansion of Copyright Coverage and Protection

Copyrights are of such duration that the creators of the copyrighted work will be dead seventy years before their protection runs out. What is the logic of this point of view by comparison with the logic of patent duration? What good does a copyright of this duration do for the original authors? Here, the creation is entirely on the table. Once a work is written, recorded, or performed, there is nothing about it that is not known to those who wish to know it. Thus, what is protected is literally the unique act of creation, not a scientific revelation or engineering breakthrough or the name by which a product is known. A lengthy duration of copyright creates a highly merchantable product on which substantial sums can be made–provided the work retains some popularity–for a very long period of time without the threat of public domain publication. In other words, a book that has become a classic can still draw royalties for many, many years after the death of its last surviving author. The owners of the copyright, whether corporate, family, or of some other nature, still have something of value.

Moreover, copyright coverage was extended by the Copyright Act of 1976 (new 17 USC Ch. 1, §§101-120) to not merely the written word, but mechanically recorded (which covers practically any type of recording medium) performances and it is in this area that the lengthening of the term of copyright may have its most significant effect. There exist many recorded performances–films, sound recordings, and combinations of the two that are well above sixty years of age and still going strong. *Birth of a Nation* was made in 1914 and is still shown today. It is easy to imagine films like *The Wizard of Oz* or television series like *The Honeymooners* being shown fifty years from now–a hundred years from their making. While none of the mentioned performances are covered by the new law because they were made and copyrighted well before its effective date, their persistence is indicative of what may happen to future "classic" performances. Here, too, it is obvious who the beneficiaries of the extension will most likely be–corporations and estates owning the rights to these old performances–and they quite likely significantly more so than the performers who appeared in them.

Among the problems not yet totally resolved under the copyright laws are rights to computer software. In the early going of applications software–such as the word processor on which this document was composed–it was not at all unusual for competitors to access the source code of the products of others and re-verse-engineer a clone of the copyrighted program. Thus for *Lotus 1-2-3* there were the identical-appearing *As Easy As, Quattro*, and several other clones. The underlying problem with these "carbon-copies" was the issue of what was copyrighted, the appearance of the program when run or the underlying computer code that allowed it to run, or both. Obviously, Ashton-Tate, the proprietors of *Lotus*, would have preferred that it be both. The developers of the clones would have vastly preferred that it be the source code, the computer language that they had modified with the intention of avoiding accusations of copyright violation, that was protected. The Copyright Law allows for protection of "original works of authorship that are fixed in a tangible form of expression" (U.S. Copyright Office, 2000). Eight categories of copyrightable work are allowed, with computer programs being considered "literary works." As such, the appearance of the program on a computer monitor appears to be devoid of protection! Further, one cannot protect ideas, procedures, methods, systems, processes, concepts, or principles by copyright. Thus, the technology of the computer seems to have created a type of form of expression that is essentially "virtual" rather than tangible and, for the present, must remain in some of its aspects intellectual property that cannot be protected by traditional means.

TRADEMARKS–DURATION IS NOT THE ISSUE

Trademarks as Differentiators and Protectors of Image

Trademarks do not wear out, nor do they become "dated." A well-conceived trademark can conceivably last a millennium, and some are getting very close. One of Germany's oldest brands of beer, Weihenstephan, from the Bavarian State Brewery at Freising, claims to have been in existence since the year 1040 (Weihenstephan: älteste Brauerei der Welt, 2003). The author remembers having visited farms in Normandy that bore signs that proclaimed them not only to have been around under the same name since the eleventh century, but to have been in the same family the whole time.

Thus, unlike the other government-sponsored monopolies (or quasi-monopolies), trademarks can persist forever if properly managed, and as a result, are prime targets for copycats who seek to trade on the preference the original mark has built for itself among a segment of the consuming public. One might even make the case that in this instance the grant of monopoly over the covered mark is pro-competitive, rather than anti-competitive. As such, the recent broadening of the base of protection for the trademark owner is of major significance as a competitive tool for both for-profit and not-for-profit organizations.

Since there is no limit on the duration of trademark protection, broadening the protection provided by trademarks cannot be accomplished, as was the case with patents and copyrights, by lengthening their lives. Instead, a different device has been chosen, that of the concept of defense of a trademark against "dilution."

History of the Issue of Trademark Dilution

Trademarks have long been protected from direct infringement, or "piracy," direct copying of the trademark owner's mark on products similar to those made by the mark's proprietor. Until recently, however, the area of trademark dilution was, though evident in common law litigation since at least the late nineteenth century, "fuzzy" in its interpretation and application. Basically, dilution of a trademark takes place when the action of someone NOT the trademark owner results in a loss of value of the trademark of that owner *without confusion* on the part of consumers (Tyvser, 2000; Hall et al., 2002; Kaufman and Hudis, 1997).

The earliest known case of trademark dilution to the introduction into the market of "Kodak" bicycles in the late nineteenth century. The Eastman firm, proprietors of the trademark "Kodak" for photographic goods, brought suit against the Kodak Cycle Company (15 Rep. Pat. Cas. 105, 1898), and prevailed, the court holding that, despite the utter lack of likelihood of confusion, the Kodak mark should be protected against the harm it might suffer–whatever harm that might be–by being used as a name for bicycles.

New Trademark Legislation and Its Effect

The federal legislature was subsequently content to leave issues of dilution up to the courts to litigate, but during the course of the twentieth century about half the states of the American Union passed an array of trademark dilution statutes. Needless to say, these statutes were not identical, nor were they applied or interpreted the same way, but they did exist. And there the matter stood until 1995, when the Federal Trademark Dilution Act was passed. An amendment to the Lanham Act (15 USC Ch. 22, §1125(c), 15 USC Ch. 22, §1127), the new law prohibits the use of a famous trademark by any third party that causes dilution of the distinctive quality of the mark, that is, some diminution of its value to its proprietor. The underlying logic seems to be that the public derives benefit from protection of the famous mark and that reduction of the value of such a mark by an interloper is somehow wrong, even if the public is free of confusion.

The 1995 law carefully defines dilution as "the lessening of the capacity of a famous mark to identify and distinguish goods or services, regardless of the presence or absence of competition between the owner of a famous mark and other parties or likelihood of confusion, mistake, or deception" (15 USC Ch. 22, §1127) arising out of their use thereof. This definition does not obstruct the function of existing state laws. Many of the state statutes go further than does the federal statute, waiving, for example, the requirement that a mark be famous to be protected against dilution.

The issue of the "fame" of a mark calls for some discussion. Having used this term in its definition of dilution, Congress suggested in the Act (15 USC Ch. 22, §1125(c)1) eight factors which might be considered as criteria by which fame might be measured. These are:

a. the degree to which the mark is "distinctive"–either because of the mark itself or through acquisition of the characteristic;
b. the duration (in time) and extent (in space) of the use of the mark in connection with the particular goods and services for which its use is claimed;
c. the duration and extent of advertising and publicity of the mark;
d. the size of the geographic area in which the mark has been used;
e. the usual "channels of trade" by which the products are distributed;

   f. the degree to which the mark is recognized in the markets of its proprietor and the alleged diluter;

   g. the nature and extent of the use of the mark by third parties; and

   h. whether the mark was federally registered and if so, under which law. (There were laws of 1881 and 1905 precedent to the current law.)

There is obviously some area for debate as to the meaning of these criteria. What, for example, constitutes sufficient duration of use or advertising to create "fame"? How large, geographically, must be the extent of distribution of a mark to establish its fame? And, perhaps most interesting of all, what constitutes "distinctiveness" and how much of that must be present for dilution to occur?

It is beyond the scope of this paper to explore what will obviously be a very long and litigious process of quantifying each of the above criteria, which is perhaps why the criteria were used. Nonetheless, it might logically be suggested that certain broad conceptual rules might apply. One such might be that "confected" marks, through their inherent uniqueness, would have an easier task of proving fame than marks composed of pre-existent words or syllables. Harking back to the first mention of specific dilution in this paper, the name "KODAK," a completely confected term, makes an excellent example of the case for a distinctive mark. Today, more recent confected marks such as VERIZON, CYBEX, and FUBU (which is actually an acronym) and foreign–hence unusual–marks such as KUBOTA, MAHINDRA, and DAEWOO fill that bill. On the other hand, how distinctive are marks such as Best Western, Best Buy, and Best Products? That remains for the courts to determine.

The Public Policy Component of the 1995 Law

Recalling that the original basis for the U. S. trademark law was to protect the trademark's owner from direct competition and the public from fraud on the part of an infringer, the rationale of the need for a dilution statute is significantly weaker. It grows out of the thesis that proprietors of registered trademarks should derive a unique benefit from their registration regardless of whether or not there is confusion by the public as to the identity of the mark's proprietor. Dilution does not require confusion or its likelihood. It is for this reason that there was a tendency, even before the passage of the new Act, to add a dilution claim to all trademark litigations in states that had dilution laws (Supnik, 1997). Thus, if the trademark claim failed, dilution still remained as a basis for injunctive relief (*McDonald's Corporation v. Arche Technologies*, 17 U.S.P.Q. 2d 1557; N. D. of Cal., 1990). This case, as one might suspect, involved the use of an arch symbol by a computer vending company that McDonald's thought compromised their "golden arches." The claim of dilution rested on the purported uniqueness and distinctiveness of McDonald's symbol.

The existence of a federal dilution law may ultimately have the effect of rendering existing trademark laws irrelevant, broadening the basis for litigation to such an extent that any perceived similarity between two marks provides grounds for suit by an aggressive litigator. Referring to a case involving, in part, provisions of the then-new California dilution statute (California Business and Professional Code Section 14330), a federal court, having taken a dilution case under review some 25 years ago, said, "Until this statute is interpreted more fully by a California court, we feel constrained not to give it an overly broad interpretation lest it swallow up all competition in the claim of protection of trade name infringement." (*Toho v. Sears, Roebuck and Company*, 654 F. 2d. 788, 210 U.S.P.Q. 547, 552; Ninth Circuit, 1981).

So to what extent does real public policy exist concerning competitors' use of others' trademarks where there is not likelihood of confusion on the part of the public? There has certainly been a tendency, for a number of years, to protect trademarks which have achieved some prominence against use by others who would trade on their differentiating capacity even though no harm was done to consumers by the usurping users. Most academics in the marketing arena take a pro-trademark stance, viewing firms' trademarks as assets, part of their intellectual property rights, evidence of an earned monopolistically competitive position in the marketplace not to be sullied by use by others. The common term applied to the value of a successful mark is "brand equity" (Lavin, 1994; Keller, 2000).

On the other hand, at one point in the late 1970s, the Federal Trade Commission asked the Trademark and Patent Office to cancel the FORMICA trademark on the grounds that it had become a generic mark and no longer benefited the public, restraining competition. The FTC's claim was that the public was likely to be misled as to the function of the word, purchasing the allegedly "generic" brand without giving consideration to similar products of competitors because the competitors' products might appear to be other than the real product they sought (Supnik, 1997). In other words, the success of the proprietors of the FORMICA mark in differentiating and creating strong brand equity for their mark was to be punished by depriving them of the rights to that mark! The requested action was not, however, taken, and FORMICA remains a trademark rather than a generic name today. It is not so much that a petition to cancel the mark was filed, it is that the petition was filed, not by a competitor but by the government agency charged with maintaining competitive conditions in the U.S. marketplace. Public policy, if indeed it does exist, thus seems somewhat confused as to which actions and what sorts of trademark protection are appropriate so as to best serve the interests of the public at large.

The eight criteria mentioned earlier effectively deny protection to a large number of trademarks. New marks–those only recently created or only recently registered–are probably not eligible for protection under this statute. Moreover, though injunctive relief is in theory available under the law, the existence of the eight criteria for establishing the fame and distinctiveness of a mark and the necessity to test the achievement of those criteria argue against the likelihood of many summary judgments–judgments without the test of trial–being handed down in actions involving this issue.

It appears that we may be watching a case of gestation of public policy, rather than its implementation. The phrasing of the criteria advanced by the statute is extraordinarily subject to interpretation, and that is one of the purposes of the federal court system. As litigation proceeds under the statute, the various circuit and appellate courts will produce their written interpretations and decisions concerning the concepts and criteria involved in the issue of dilution. New law will evolve as common law is invoked and previously unthought-of arguments brought to the courtroom.

Bases for Litigation Under Recent Law

Since the new statute took effect, several rather interesting cases have been filed that seemingly carry the dilution argument further than might have been imagined in 1995. As might have been expected, the first several cases under the dilution law had to do with fairly straightforward and historically developed issues which can be classed into two categories: "blurring," use of a trademark such that the connection in consumers' minds between the claimant's mark and the claimant's goods or services is weakened; and "tarnishment," the use of a trademark by someone not its owner in an unsavory or unwholesome fashion, or in connection with inferior products (Tysver, 2000).

The concepts of tarnishment and blurring have seen wide application in that newest area of marketing, the on-line venue. Domain names on the Internet, since they did not exist until fifteen years ago, have proven a fertile area of exploitation for trademark diluters. The use of the domain name candyland.com, for example, to promote pornographic materials on the Internet, brought legal action from Hasbro, Inc., proprietors of the children's game of the same name (*Hasbro, Inc. v. Internet Entertainment Group, Inc.*, 40 U.S.P.Q. 2d. 1479; W.D. Wash., 1996) as a case of tarnishment. Hasbro prevailed.

Blurring cases also devolved from the Internet, as in the case of *Intermatic, Inc., v. Toeppen* (947 F.Supp. 1227; N.D. Ill., 1996; Davis and Stockton,

1998). In this instance, Mr. Toeppen had registered the trademarked name of the plaintiff as a domain name on the Internet (Intermatic.com), with the apparent intent of attempting to sell it to its normal proprietor or to others. This action (and Toeppen's related action of similarly registering over 240 other famous marks as domain names!) was found to blur the distinctiveness of the Intermatic mark (Strote, 1997).

Exemptions to the Usual Rules

Some uses of trademarks by nonowners are, incidentally, specifically exempted from the provisions of the law as free speech rights. They include "fair use" of a mark in the context of comparative commercial advertising or promotion, non-commercial uses of the mark, such as parody, satire, and editorial commentary, and all forms of news reporting and news commentary (15 USC Ch. 22, §1125(c)(4).

Further Expansion of the Interpretation of the Law

Additional interesting new cases have to do with the apparent expansion of the coverage of the law by court interpretation. One aspect of dilution left unclear in the original statement of the law had to do with the issue of "exact copying." In other words, was it necessary that the alleged infringement be as the result of an exact rendering of the words or graphics of the original trademark to be dilution? Several cases have seemingly set this issue to rest. In *Anheuser-Busch, Inc. v. Andy's Sportswear* (1996 U. S. Dist. Lexis 11583; N. D. Cal. 1996), the court agreed with plaintiff that use of the term "Buttwiser" on T-Shirts parodying the "Budweiser" beer trademark was a diluting alteration of the original use. This, and the contemporaneous decision in a case in Pennsylvania involving contention by the proprietors of the "famous mark" WAWA for a chain of over 500 convenience stores against the proprietor of a single convenience store styled HAHA (*WAWA v. Haaf*, 40 U.S.P.Q. 2d. (BNA) 1629; E. D. Pa. 1996) in which the WAWA interests were upheld, apparently overturns the previous appellate decision in *Mead Data Central, Inc., v. Toyota Motor Sales*, Inc. (702 F. Supp. 1309; S. D. N. Y. 1988), reviewed 875 F. 2d. 1026 at 1028-29; 2d Circuit 1989), which found that the terms LEXIS and LEXUS were not sufficiently similar (not identically similar) and thus not subject to an interpretation of dilution under the then-existing trademark protection statute.

Finally, in the more recent case of *Icee Distributors, Inc., v. J & J Snack Foods Corp. and WalMart Stores, Inc.* (U.S. District Court, W. D. La., CV 99-0850S), (tried in late 2001, presently under appeal), a federal district court in Louisiana held that the owner of a trademark had diluted the rights of its licensee by pursuing a brand extension that applied the trademark it owned to a line of products designed for sale in a different market segment than that in which the licensee operated. This decision seemingly flies in the face of the provision of the law that specifically extends the protections of the statute to the owners or trademarks, not their licensees.

CONCLUDING COMMENTS

It has never been said that the legal arena is simple, nor that the written language of the law is necessarily consistent with the interpretation to which it will be put by the courts. The Patent and Copyright Law changes made during the last thirty years have effected substantial changes in the protection of those aspects of intellectual properties. The much newer Trademark Dilution Act seems to be having a rather significant effect on the quantum of protection provided to trademarks and related rights, particularly on the Internet and in other areas where the application of new technology has left gaps in the protection previously afforded by the law. Whether this protection is preemptory of the original letter and intent of protecting trademarks against unreasonable competition remains yet to be established. However, in this context perhaps it is well to notice that the thrust of protection for intellectual property rights as embodied in patent and copyright statutes has been toward the duration of protection. Lacking the capacity to extend the duration of protection of trademarks, the Trademark Dilution Act seems to have done the next best thing, to broaden the interpretation of competition in the trademark arena. If one adds to this recent legal interpretations of the concept of "trade dress"–the unique components of the packaging and "presentation" of products–and the apparent extension of much broader coverage to that aspect of product composition, it would appear that the effect of the legislative and judicial components of the protection of property rights has been to significantly broaden those rights, perhaps at the expense of a substantive muddying of the waters concerning the nature, extent, and implications of that broadening.

AUTHOR NOTE

Thomas S. O'Connor received the PhD in Business Administration from the University of Alabama. He has published a number of books, monographs, and articles in the fields of consumer behavior, research methods, diffusion of innovation and technology transfer, and protection of intellectual property.

REFERENCES

Note: Case and law citations were given in full in the text. Each is in the form conventional to its type. Thus, U. S. Federal law is cited as the section of the U. S. Code in which it appears, then the fact that it is part of the U. S. Code, then the Chapter of the Code, and finally, the relevant sections, parts, and subparts, thus: 15 USC Ch. 22, §25(a)(4). Cases are cited in terms of the names of the litigants, plaintiff first, then defendant, then the source, as: *Jones v. Smith*, (931 F.Supp. 1227; N.D. Ill., 1996) where the material in parentheses refers to volume 931 of the *Federal Supplement*, page 1227, northern district of Illinois, 1996.

The citations below refer to facts, articles and opinions appearing in various publications and other venues.

Johnson, Robert A. and Sean O'Donnell, "Trademark Dilution Proof in Flux," *New York Law Journal*, Law Journal Publishing Company, 2000.

Kaufman, Jeffrey H. and Jonathan Hudis, "The Federal Dilution Act Celebrates its One Year Anniversary," www.oblon.com/Pub/display,php?Kaufman-601.html. Oblon, Spivak, McClelland, Maier, and Neustadt, Attorneys, 13 pp.

Keller, Kevin Lane, "The Brand Report Card," *Harvard Business Review*, January-February 2000, Harvard Business Publishing Reprint R100104.

Lavin, Douglas, "A Cloudy Issue: Will ATM Users be Confused by a Car Called Cirrus?" *The Wall Street Journal*, 16 February 1994, p. B1.

Strote, John R., "The U. S. Federal Trademark Dilution Act of 1995 and the Internet," www.iael.org/newsletter/d6dilute.html

Supnik, Paul D., "Mark of Distinction–A New Federal Law Protects Distinctive Trademarks Against Dilution Through Unauthorized Use." *Los Angeles Lawyer*, May 1997, pp. 14-22.

Swarbrick, David, "The Statute of Anne," www.swarb.co.uk/acts/1710AnneStatute.html

Tysver, Daniel A., "Trademark Dilution," *Bitlaw: A Resource on Technology Law, 2000,* www.Bitlaw.com/trademark/dilution.html, 2 pp.

_____*Copyright Basics*, United States Copyright Office, 2000, 20 pp.

_____*Federal Trademark Dilution Law in the United States.* www.ladas.com/BULLETINS/1996/FederalDilution.html, Ladas and Parry, Attorneys, 1996.

_____"Historical Dates," *Facts and Figures 1997*, Europäisches Patentamt, www.european-patent-office.org/epo/facts_figures/hist_en.htm

_____*The Law of Trademarks in the U.S.A.*, United States Patent and Trademark Office, 1996, 6 pp.

_____"Victor/Victoria–Supreme Court Hears Trademark Dilution Case," *Adlaw by Request*, Dickler Kent Goldstein & Wood LLP, 2002) www.adlawbyrequest.com/inthecourts/Victoria112502.shtml.

_____"Weihenstephan: älteste Brauerei der Welt," www.weihenstephaner.de/index.php.

Government and Market Mechanisms
to Provide Alternatives
to Scrap Tire Disposal

Nancy J. Merritt

William H. Redmond

Michael M. Pearson

SUMMARY. The current study examines disposal and recycling of scrap tires with a discussion of the role of government policies and an empirical study of tire dealers' preferences for market and government solutions to disposal problems. Preference for government intervention

Nancy J. Merritt, PhD, is Associate Professor of Marketing, Department of Marketing, Bowling Green State University, Bowling Green, OH 43403 (E-mail: nmerrit@cba.bgsu.edu). Her current research interests are in the areas of channel management and business-to-business marketing.

William H. Redmond, PhD, is Professor of Marketing, Department of Marketing, Bowling Green State University, Bowling Green, OH 43403 (E-mail: wredmon@cba.bgsu.edu). His current research interests are in the areas of new product diffusion and market structure.

Michael M. Pearson, PhD, is Professor of Marketing, College of Business, Box 15, Loyola University, New Orleans, LA 70118 (E-mail: pearson@loyno.edu). His current research interests are in the areas of interaction among channel members and computer modeling for the classroom.

[Haworth co-indexing entry note]: "Government and Market Mechanisms to Provide Alternatives to Scrap Tire Disposal." Merritt, Nancy J., William H. Redmond, and Michael M. Pearson. Co-published simultaneously in *Journal of Nonprofit & Public Sector Marketing* (Best Business Books, an imprint of The Haworth Press, Inc.) Vol. 13, No. 1/2, 2005, pp. 151-178; and: *Government Policy and Program Impacts on Technology Development, Transfer and Commercialization: International Perspectives* (ed: Kimball P. Marshall, William S. Piper, and Walter W. Wymer, Jr.) Best Business Books, an imprint of The Haworth Press, Inc., 2005, pp. 151-178. Single or multiple copies of this article are available for a fee from The Haworth Document Delivery Service [1-800-HAWORTH, 9:00 a.m. - 5:00 p.m. (EST). E-mail address: getinfo@haworthpressinc.com].

Available online at http://www.haworthpress.com/web/JNPSM
Digital Object Identifier: 10.1300/J054v13n01_09

was related positively to dealer perceptions of environmental concerns, consumer concerns and the severity of the problem, but related negatively to dealers' assessments of manufacturer effectiveness in disposal solutions. Results show the need for more information from manufacturers and government agencies and show dealer support of specific remedies in developing new uses, technologies, and markets for scrap tires. Market development and commercialization of recycled products may be facilitated by cooperation of involved business and government agencies. Particular government actions to stimulate markets include enforcement of disposal restrictions; provision of technological and capital assistances; participation in setting industry standards for recycled rubber products; and state procurement policies. *[Article copies available for a fee from The Haworth Document Delivery Service: 1-800-HAWORTH. E-mail address: <docdelivery@haworthpress.com> Website: <http://www.HaworthPress.com> © 2005 by The Haworth Press, Inc. All rights reserved.]*

KEYWORDS. Scrap tire disposal, recycling, dealer preferences, government programs and policies, market development

INTRODUCTION

Awareness and concern about waste disposal and environmental issues have been at high levels among consumer, business, and government groups and show every sign of growing in the future (Marcus, Geffen, and Sexton 2002). A key issue is the assignment of responsibility for remediation and restructuring of disposal and recycling activities. Should these activities fall mainly to government or market mechanisms? What specific remedies will be supported? Are market and public policy responses providing adequate solutions to disposal problems and aiding development of technologies and markets for the commercialization of recycled products?

Activities characterized as government or market solutions to disposal problems may take numerous forms, such as technological and market development, commercialization of recycled products, provision of services or information, and application of taxes and controls (Harris and Carman 1984). Businesses, public interest groups, governments, and others are divided about the methods needed to resolve environmental and recycling problems, and a variety of responses from businesses and governments have been proposed and tested (Apaiwongse 1991; Farrell 1999; Marcus et al. 2002; Owen 1998; Scrap Tire Management Council 1991a).

The current research setting is the disposal and recycling of scrap tires. According to the Rubber Manufacturers Association (RMA), responsibility for scrap tires has been shared among manufacturers, dealers, processors, and disposal site operators, as well as government agencies (RMA 2003). Government activities are principally at the state level in the U.S., at the province level in Canada, and at the national level in Europe (Fedchenko 2001; Moore 1998). The U.S. has no federal policy for scrap tires, and states differ widely on tire regulations and their support of market development (Scrap Tire Management Council 1991b; RMA 2002b). In the period from the mid-1980s to the present, state government activities included taxes on new tires to fund disposal alternatives and clear tire stockpiles; regulation of tire haulers, disposal sites, and processing firms; and provision of incentives for the development and commercialization of recycled products (Farrell 1999; Fedchenko 2002; Moore 2001; RMA 2002a, 2002b).

Market-based solutions to scrap tire disposal have primarily taken the form of recycling through scrap tire conversion to tire-derived fuel, asphalt paving and other construction materials (Miller 2002; Owen 1998; RMA 2002a). Market solutions to scrap tire management are not without impediments, including technological problems, funding challenges, generation of other environmental concerns, and variability in government regulation and support (Diefendorf 2000; Farrell 1999; Noga 2001; RMA 2002a).

Tire manufacturers, dealers, and processors share concerns about disposal of scrap tires but have different priorities for framing scrap tire regulations and programs and for funding product and market development (Moore 1998, 2001; Rowbotham 1990; RMA 2003). Varying priorities may reflect, for example, that tire manufacturers produce and sell tires across states, while tire dealers operate within particular states and are more directly connected to and influenced by local customers and their state and local communities.

Tire dealers play an immediate and key role in the sale of new, used, and retreaded tires; collection of state-imposed fees; and collection, storage, and disposition of scrap tires. The dealers are the direct contact with consumers and also contract with manufacturers, retreaders, and scrap tire haulers and processors. Lack of support by tire dealers for a solution to scrap tire disposal or to particular solution mechanisms is likely to elicit minimal cooperation and effectiveness. However, little is known about the diversity of opinion among dealers regarding government or market-based solutions to scrap tire disposal or factors that affect their support for particular solutions.

The current study examines tire dealers' preferences and support for government and market solutions. Following a description of the research setting, hypotheses regarding tire dealer preferences for government and market solutions to scrap tire management are developed. A national survey of dealers is used to test the hypotheses. Based upon the findings, we propose the need for cooperative efforts among government, tire dealers, and other industry participants for the successful remediation of scrap tire disposal and the development of viable products from scrap tires.

SCRAP TIRE DISPOSAL AND MARKET DEVELOPMENT

Estimates from the U.S. Environmental Protection Agency (EPA), the RMA and other sources indicate that at least 300 million used tires are generated per year in the U.S., amounting to more than one tire per person each year, a figure that has remained relatively constant in the past decade (U.S. EPA 1991, 2002; RMA 2002a; Scrap Tire Management Council 1991a). Prior to 1985, the absence of laws restricting scrap tire disposal and the absence of programs for other uses of tires led to two dominant practices: landfill disposal and stockpiling of whole tires (RMA 2002a).

Disposal of Scrap Tires

Hundreds of millions of non-biodegradable tires were dumped each year in landfills, tire stockpiles, or illegal locations, resulting in space, aesthetic, health, and other environmental concerns (U.S. EPA 1991). For example, tires float to the surface of landfills and provide breeding grounds for mosquitoes and rodents; tire stockpiles are fire hazards, and the resulting fires produce toxic smoke and release oil that pollutes the soil, ground water, and wells (Owen 1998; U.S. EPA 2002).

By 1990, all but two states had regulations or at least minimal scrap tire management programs, but less than 20 percent of used tires were being reused, retreaded, or exported; only 15 percent were being put to alternative uses; and the vast majority were still being dumped in landfills and stockpiles (Scrap Tire Management Council 1991a). Over the course of the decade, however, substantial results were produced by state governments that restricted disposal, licensed haulers and processors, and supported stockpile abatement with fees imposed on tire sales (Noga 2001; RMA 2002a). While industry participants developed uses for scrap tires, the most successful state programs stimulated these alternative markets by providing grants, research, and state procurement for construction projects (Farrell 1999; RMA 2002b).

Tire disposal was reduced to approximately 10 percent of the annual generation of scrap tires nationally by 2001. Tire stockpiles were reduced from over 3 billion scrap tires to estimates ranging from 300 and 800 million tires still stockpiled, with the improvement attributed to state efforts, to growth in scrap tire markets, and, unfortunately, to large tire fires (Noga 2001; *Tire Business* 2001; U.S. EPA 2002). However, a limited federal regulatory stance on stockpiles and other tire disposal created a patchwork of state policies, programs, and results (Farrell 1999; Moore 1998; Noga 2001; RMA 2003). Abatement efforts are still needed, particularly in the nine states (Alabama, Colorado, Connecticut, Michigan, New York, Ohio, Pennsylvania, Texas, and West Virginia) holding approximately 85 percent of stockpiled tires (Moore 2002; RMA 2002a). Further, several state programs for stockpile abatement have changed, including expiration or diversion of tire fees that were or could be used to reduce stockpiles (Farrell 1999; Noga 2001; Truini 2002). As noted in *Tire Business* (2001), "anything less than strict vigilance by state and local officials, tire dealers, and tire makers could cause the tire piles to mount again."

Alternatives to Tire Disposal

With the goal of reducing waste disposal, the U.S. EPA proposed a hierarchy of solutions: source reduction, recycling, and incineration (U.S. EPA 1991). The first alternative to tire disposal, source reduction by extension of the useful life of tires and through sales of used and retreaded tires, has produced limited results. The useful life of tires was extended from approximately 20,000-60,000 miles over several decades principally through the use of steel wire embedded in the tire (Miller 2002; Northwest Product Stewardship Council 2002). While further extension of the useful life of tires is possible, little occurred during the 1990s, and the embedded steel wire in tires complicates processing for other uses, including retreading of tires and production of recycled products (Farrell 1999). The number of tires that were retreaded for reuse doubled in the 1990s, indicating some source reduction (Miller 2002; Tire Retread Information Bureau 2003). However, sales of used and retreaded tires still account for less than 20 percent of used tires and continue to be limited principally to truck tires and small export markets due to competitive new tire prices and consumer perceptions that used and retreaded tires are unsafe (RMA 2002a; U.S. EPA 2002; Wallace 1990).

The second and third alternatives to tire dumping are recycling and incineration. Tire recycling processes use grinding or shredding to reduce tires into smaller pieces for civil engineering applications using shredded, ground, or crumb rubber, for further processing to produce higher value materials (e.g., athletic surfaces, mats, and carpet padding), and for incineration and energy production with tire-derived fuel (Farrell 1999; Miller 2002; Owen 1998). Ground rubber and civil engineering applications increased to approximately 12 and 14 percent, respectively, and use in tire-derived fuel increased to about 41 percent of scrap tires generated annually (RMA 2002a).

New applications for ground and shredded rubber emerged during the 1990s, including mixing crumb rubber in asphalt, adding scrap rubber in the manufacture of new tires, and using shredded tires in landfill, septic system, road construction and other civil engineering projects (RMA 2002a). Several companies entered the scrap tire processing business, Canadian firms exported ground rubber to the U.S., and new processing technologies emerged (Farrell 1999; Fedchenko 2002; Moore 2001; Owen 1998). These alternatives to dumping scrap tires, however, met economic and non-economic barriers to their use, including variability in state regulations, high costs and inadequate funding, technological and environmental constraints, oversupply of scrap tires, and customer and business reluctance (Farrell 1999; Owen 1998; RMA 2002a; Truini 2002).

Demand for ground rubber was met in 1990 by scrap rubber from the part of tires removed in preparation for retreading, and equipment for processing whole tires into ground or crumb rubber was in developmental stages (U.S. EPA 1991; Scrap Tire Management Council 1991a). In 1991, the U.S. Congress mandated use of ground tire rubber in asphalt paving for federally-funded highway projects and commissioned further study of the costs, useful life, and possible environmental concerns associated with asphalt rubber (U.S. EPA 1991). As noted in a recent report of the RMA (2002a), the U.S. legislation produced optimism in scrap tire industries but opposition in the paving industry and in state departments of transportation. Several companies entered the business of processing scrap tires into ground rubber; testing programs were started in some states; but most state departments of transportation refused to comply with the mandate. The Federal Highway Administration indicated that it was unlikely to monitor or punish states that did not comply, so little rubber asphalt paving was used. Congress repealed the mandate by 1993, with devastating effects for ground rubber producers.

Oversupply of ground rubber caused price reductions, marginal producers were forced out of business by 1996, and established producers were weakened. Positive side effects of the 1991 legislation were that some states tested and continued to use asphalt rubber paving and that state and university experiments focused on possible uses of ground or shredded scrap tires in highway and other construction projects (Farrell 1999; Owen 1998; RMA 2002a).

Shredded tires were being used as alternatives to materials such as rock or sand in embankments and retaining walls in landfill, septic system, and road construction projects by the early 1990s (U.S. EPA 1991). A major setback in these civil engineering applications occurred in 1995, however, when two large road embankments built with shredded tires began to heat. Rumors, although untrue, circulated widely that the asphalt road itself ignited, and planned shredded-tire applications were halted. Research led to guidelines for the depth of tire shred usage in construction applications, but civil engineering applications were limited for several years to alternative daily cover in landfills (RMA 2002a).

The largest market for scrap tires is the controlled incineration of whole, shredded, or ground tires to produce energy and an alternative fuel (Moore 2002; Truini 2002). Tire-derived fuel expanded considerably in U.S. pulp and paper mills, cement kilns, and power industries from 1990-1995, although the second half of the decade revealed a volatile market and lower growth rates (Owen 1998; RMA 2002a). In the economic boom of the later 1990s, cement kilns reduced their use of tire-derived fuel, citing a reduction in cement-making capacity with its use. Some pulp and paper mills discontinued use, citing increased zinc emission levels. Deregulation and potential sale of power plants in the utility industry produced fear that use of alternative fuels would create disincentives to potential buyers of the plants; several companies stopped using tire-derived fuel. Between 1998 and 2001, while no recovery was shown in the paper and utility industries, a dramatic increase in tire-derived fuel consumption by cement kilns was implemented to reduce production costs and to reduce nitrogen oxide emissions from burning coal (RMA 2002a; Truini 2002).

Emerging Markets, Technologies and Concerns

Expansion of major markets for scrap tires occurred in the past few years, and new applications, technologies, and concerns are emerging. New markets for scrap tires include extruded rubber products, production of carbon black and pyrolysis oil, and other products from ground or crumb rubber (Jupiter Consultancy 2002; Miller 2002; Onorato-Jackson 2001; RMA 2002a). By the late 1990s, new technologies were emerging for tire recycling industries and for dealing with problems in producing ground rubber and other products recycled from tires (Owen 1998).

Grinding processes raise blistering problems due to oxidation and generate heat that can alter the properties of recycled rubber (Owen 1998). For example, ground recycled rubber can bond mechanically but not chemically with other materials, unlike natural rubber. To make recycled rubber perform more like natural rubber, processes such as devulcanization to delink chemicals in the material are needed. Without devulcanization, many commercial applications require much smaller crumb rubber pieces with increased processing costs. The U.S. Air Force has experimented with micro-organism devulcanization and commercial trials were conducted. Others are developing cryogenic processes (cooling with liquid nitrogen) to derive crumb rubber without the problems of surface oxidation or alteration of the materials properties (Farrell 1999; Owen 1998).

Some of the new applications and processes, however, are considered emerging markets that may take a long time to develop (Jupiter Consultancy 2002; RMA 2002a; Truini 2002). The largest scrap tire markets, tire-derived fuel, may be impacted by changes to emission standards. Recently, the U.S. EPA began discussing new emission standards, with the possibility of classifying tire-derived fuel as a solid waste, which would subject combustion facilities to more stringent requirements and severely curtail the use of tire-derived fuel (RMA 2002a).

Industry experts are increasingly concerned about scrap tire product and market development in light of EPA emission standards, the economic climate, and a lack of uniform quality standards and supply continuity for crumb rubber (Farrell 1999; Owen 1998; RMA 2003). Some observers note a lack of systematic state policies and programs to enforce regulations and support market development through capital investments and purchasing policies (e.g., Farrell 1999; Truini 2002); others note state diversion or elimination of tire fees that were or could be used to support scrap tire industries and abate stockpiles (e.g., Noga 2001; RMA 2002b).

DEVELOPMENT OF THEORETICAL PROPOSITIONS

While business interests are commonly stereotyped as holding free market positions, particular government regulations, legislation, and programs are supported by many business firms (Cunningham, Sethi, and Turicchi 1982; Marcus et al. 2002). An individual will typically hold different preferences for a variety of government and market mechanisms and also hold a more global disposition regarding government versus market approaches, generally (cf. Harris and Carman 1983, 1984). These global preferences are conceptually and statistically distinct from preference for specific mechanisms (Aaker and Bagozzi 1982; Mills 1984), and the set of preferences are subject to situational or contextual considerations. Global preferences for government solutions to disposal problems are examined in three hypotheses regarding: (1) legitimacy of the government role; (2) market system constraints; and (3) newness and severity of the problem. A fourth hypothesis proposes an ordering of preferences for specific market and government mechanisms to remedy the scrap tire disposal problem.

Legitimacy of the Government Role

Waste disposal, in general, may be classified as a marketing externality. That is, disposal of the used product or packaging is not considered a primary concern during the initial marketing exchange, but results in unforeseen effects on the transacting parties as well as third parties (Nason 1986). The process of consumption is in reality a process of transformation associated with a lowering of the residual economic value of the product or packaging (Reidenbach and Oliva 1983). The usual options are to tap some of the remaining economic value through recycling or to discard the item. However, most disposal programs do not fully pay for themselves, and societal costs are incurred on economic as well as aesthetic dimensions. This particular form of market externality has been classified as a market failure (Harris and Carman 1983). The terms of exchange do not fully account for the consequences of production and consumption, and the extra costs must be borne by third parties or broadly allocated across society as a whole.

In the case of environmental concerns, businesses may take a strong bystander position that the government is morally responsible to act on behalf of the environment (cf., Granzin and Olsen 1991). Individual judgments that tire disposal is a social, environmental concern are predicted to lead to support for government intervention:

H1$_a$: Dealers who view tire disposal as a social, environmental problem will have higher preferences for government intervention.

Support for the first hypothesis is also found by considering the nature of exchanges among government and scrap tire channel participants. Scrap tires are collected by dealers from consumers, and dealers then contract with haulers and/or processors for removal of tires. The channel continues with haulers and processors distributing (1) whole or shredded tires to disposal sites; (2) retreaded tires to tire dealers who sell to consumers; or (3) processed tires and recycled products to manufacturers for use in producing new tires or to markets such as construction industries. In some respects the channel for scrap tires reflects the complex exchanges of other distribution channels, but considering government participation, the channel is more likely to be characterized by restricted and generalized exchanges (Bagozzi 1975; cf. Marshall 1998). Government regulation and program activities are principally directed to haulers, processors, and disposal site operators, with most direct benefits accruing to processors in such forms as grants for technology to process tires and government purchases of processed scrap tires. Thus, government agencies and processors are more directly contracting for benefits, that is, in two-party restricted exchanges (Bagozzi 1975). On the other hand, exchanges among government agencies, tire dealers and customers may be characterized as generalized exchanges that involve three or more parties in indirect, univocal, reciprocal transfers. Such generalized exchanges are often characteristic of situations involving public policy and support of public programs to provide benefits to the community at large (Marshall, Piper, and Micich 2002).

Tire dealers and consumers are asked to support government programs as voters and influencers of public policy, but do not typically achieve direct utilitarian benefits from taxes or other government mechanisms for scrap tire solutions. Social and other indirect benefits accrue to dealers and customers in the form of quality of life enhancements by solving aesthetic and environmental problems of scrap tire disposal. Dealers may also indirectly benefit from supporting government programs by enhanced public relations and future patronage, and both dealers and consumers may benefit from the government's support of scrap tire industries (e.g., longer lasting asphalt paving used by governments in constructing highways). Dealer willingness to support taxes and other government programs may be based upon these indirect benefits.

Legitimacy of government intervention may also be attained when individuals believe governments perform specific tasks well (Mills 1984), particularly individuals who are direct beneficiaries of public programs, but also those in generalized exchanges with indirect social benefits (Marshall 1998). Further, the availability of both government and market solutions to social problems suggests the use of a political economy and institutional approach (Grewal and Dharwadkar 2002). A key issue involves the basis for selecting among various political and market institutions, and the choices depend, at least in part, on judgments of efficiency or effectiveness (Arndt 1981, 1983). We expect varying judgments of the effectiveness of federal and state government efforts to solve scrap tire issues, particularly given variability in state legislation and regulations (Farrell 1999; RMA 2002b). To the extent that government branches are believed to be effective in handling disposal and supporting markets for scrap tires, they may be seen as the legitimate agents of scrap tire management:

$H1_b$: Dealers who perceive governments to be effective regarding tire disposal issues will have higher preferences for government intervention.

Market System Constraints

The second hypothesis is based upon channel, competitive, and consumer constraints in the market system that lead to failure of markets to absorb external costs of waste disposal, and thus justify government intervention. First, since businesses are typically organized for specific production and exchange tasks, these same organizations may not be easily modified or economically suited to perform other tasks. The possibility of channel conflict or inadequate compensation for disposal alternatives may also present serious costs to intermediaries in vertical recycling situations (Barnes 1982). When solutions from channel partners in the affected industry are perceived as unsuitable or ineffective, government intervention is likely to be preferred (Arndt 1981). This position corresponds to the "reconstructivist" perspective in which individuals are both concerned about the prospects for institutional change and skeptical about market solutions to environmental problems (Crane 2000). We hypothesize:

H2$_a$: Dealers who perceive that tire manufacturers are ineffective in disposal solutions will have higher preference for government intervention.

Second, the internalization of societal costs is difficult to justify in cases where such actions leave the firm at a cost disadvantage relative to competitors. A serious problem exists for socially concerned businesses if their competitors are not similarly inclined toward market solutions (Nason 1986). One role of trade associations is to encourage joint action among competitors in such areas of mutual concern (Rowbatham 1990). Effectiveness of trade associations, however, requires that members embrace social obligations, norms, and behaviors. Regulatory interventions may be more effective in imposing or inducing competitors to accept equivalent sacrifices in the internalization of social costs (Grewal and Dharwadkar 2002). To the extent that trade associations are ineffective in convincing competitors to accept equivalent sacrifices, government interventions may be preferred:

H2$_b$: Dealers who perceive that trade associations are ineffective regarding disposal solutions will have higher preference for government intervention.

Third, increasing customer concerns regarding disposal might motivate businesses to take a proactive market approach, if it is believed that the approach leads to higher demand or enhanced customer loyalty (Mills 1984). However, customers may be reluctant to pay increased prices or otherwise support businesses working directly to solve disposal issues. In the case of scrap tires, consumer reluctance is cited as a constraint to market solutions, such as buying used or recycled tires or products derived from scrap tires

(Friedlander 1990; Wallace 1990). Consumer concerns have been cited in support of government intervention in scrap tire disposal (Noga 2001; RMA 2002a). Thus, we hypothesize:

$H2_c$: Dealers who perceive increasing consumer concerns regarding tire disposal will have higher preferences for government intervention.

Newness and Severity of the Problem

Many observers take a contingency approach to the role of governments in business matters; one such contingency is newness and a second is severity of the problem. The contingency of newness suggests that government intervention is appropriate in order to assist the start up of innovative activities, after which the intervention may no longer be necessary (Marcus et al. 2002). The reasoning is parallel to the infant industries' argument that desirable businesses may need protection or subsidies for a period of time (Grether 1966).

The problem of tire disposal itself is not new, but government, environmental, and industry groups have shown increasing concern and attention in the past few years. From 1999 to 2001, scrap tire legislation was the second most discussed disposal and recycling issue in state legislatures (RMA 2002b). The impact on tire dealers and processors increased as a result of changes to regulations, expanded disposal costs, and emergence of disposal alternatives (Fedchenko 2002; Friedlander 1990; Moore 2001). Alternative markets for scrap tires, particularly the creation and commercialization of recycled products, however, are still developing industries (Farrell 1999; Onorato-Jackson 2001; RMA 2002a). The absence of industry standards of practice and consistent outlets for scrap tires may produce dealer uncertainty and adaptive efforts to cope with a high rate of change. In light of such uncertainty and modifications to dealer operations, government intervention may be desired:

$H3_a$: Dealers who are modifying operations as a result of tire disposal issues will have higher preference for government intervention.

A second contingency is the severity of the problem. Individual judgments of the seriousness of environmental problems may lead to preference for government intervention by consumers (Aaker and Bagozzi 1982; Granzin and Olsen 1991). Similarly, if business people judge disposal activities as a serious problem compared to other business problems, government intervention may be preferred to market solutions. Uncertainty stemming from severity of direct impacts on tire dealers is predicted to be associated with higher preferences for government intervention:

H3$_b$: Dealers who regard disposal issues as serious business challenges will have higher preference for government intervention.

Preference for Specific Remedies

While the previous hypotheses suggest preferences for government intervention versus market solutions in general, specific mechanisms may be differentially preferred. Government responses may embrace legislative and regulatory activities including government controls, taxes or fees, service provision, and support of new technologies for using or disposing of the product. Market solutions may include business or channel absorption of disposal and associated costs, as well as development of new industries to use or recycle the product.

Differential preferences for particular government responses are predicted based upon Harris and Carman's (1984) ordering, which follows three principles: (1) most to least compatible with markets; (2) private to public control; and (3) least to most coercive. From high to low preference, government responses are information provision, standards of practice, taxes or subsidies, controls on collective action, direct controls to allocate resources or restrict production or exchange, and government provision of services.

We expect to find preferences of tire dealers to reflect the order given by Harris and Carman, but hypothesize differences based upon their general preferences for leadership from the government or from within the industry. Those preferring industry leadership in general would favor the first half of the ordering, since those mechanisms are more compatible with markets, allow more private control, and are less coercive. The government-preference group, on the other hand, is likely to favor the second half of the ordering, namely government controls on collective action, direct controls, and provision of services. Specifically, we hypothesize:

H4: Dealers who prefer industry leadership will more strongly favor market development, manufacturer services, and government mechanisms that involve information, standards, and tax or subsidy remedies; those who prefer government leadership will more strongly favor controls and government provision of services.

METHODOLOGY

A mail survey of U.S. tire dealers was used to assess preferences, concerns, and activities regarding scrap tire disposal. Academic and industry professionals developed and revised a questionnaire that was sent to dealership owners or managers. A sample of 750 independently-owned tire dealers was randomly selected from a list of U.S. tire dealers provided by a national tire manufacturer.

Description of the Sample

Responses were received from 222 of the 750 dealers for a response rate of 30 percent. Table 1 shows the profile of the location and tire business of responding dealers. Respondents represent 40 states and nearly equal proportions of urban, suburban, and rural markets. The 10 states not represented in the sample are in the lower quartile of state population size. While these state populations are small, we do not interpret this to mean that tire disposal is not an issue.

The sample closely matches a profile provided by the manufacturer in terms of dealership size, sales, and geographic location. Sales ranged from less than $500,000 to over $10 million, but 42 percent of the dealers reported annual sales volume below $1 million. An average 682 used tires per month were generated by the responding dealers, with over half of the used tires stored at the dealership. Overall, disposition of used tires was estimated to be 11 percent retreaded or sold for reuse, 19 percent sent to recycled product markets,

TABLE 1. Profile of Responding Tire Dealers

Trade Area		Percent	Tire Business		Percent
Location	Urban	32%	Primary product	Automobile tires	53%
	Suburban	34		Light truck tires	19
	Rural	34		Heavy truck tires	22
City Size	Less than 10,000	24%	Primary business	Retail	55%
	10,000-34,999	26		Wholesale	20
	35,000-99,999	21		Commercial	24
	100,000-999,999	21			
	1,000,000 or more	7	Annual sales	< $500,000	18%
				500,000-999,999	24
				1,000,000-4,999,999	43
				5,000,000 or more	15

and 42 percent taken by haulers to storage or disposal locations. Although dealers indicated they did not know the disposition of approximately one-quarter of the tires, the final disposition of tires taken by haulers to storage or disposal locations was not indicated, suggesting dealers may not know the fate of a majority of used tires from their dealerships.

Measures Used to Test Hypotheses

Measures are reported in the appendix, and, except as otherwise noted, used a 5-point Likert scale ranging from (1) strongly disagree to (5) strongly agree. First, dealer preferences for solutions to scrap tire disposal were assessed. Responses to two items regarding federal and state government leadership in solving scrap tire issues were averaged to form an index of government preference as the dependent variable in Hypotheses 1-3 (GPREF; alpha = .74). An additional item measured preference for manufacturer leadership (MPREF), reflecting a market-sector preference. Dealers with GPREF scores higher than, equal to, or lower than MPREF were categorized into three respective groups: government (n = 49), equal (n = 87), and manufacturer preference (n = 78). Preferences for 19 specific mechanisms included five market, five manufacturer, and nine government solutions.

Second, environmental and business concerns were measured with a 5-point Likert scale. To establish the relative salience of scrap tire disposal as an environmental issue, dealers indicated their agreement with each of five issues posed as a major environmental issue: scrap tire disposal (ENV1), wetlands, ozone layer, landfill capacity, and scrap tires in landfills. An additional item measured the belief that scrap tire disposal is more of an environmental issue than an economic issue (ENV2). Business concerns included scrap tire disposal as a serious problem (PROB), restrictive regulations (REGS), information needed (INFO), and consumer concerns (CCONC and CBUY). Dealers were asked whether they had changed disposal methods in the past three years (CHNG) and whether they were investigating new methods (INVST).

Third, dealer perceptions of the effectiveness of government and market-sector responses to scrap tire issues were assessed with items posing federal government, state government, manufacturers, and trade associations as doing a good job of working on solutions to scrap tire disposal. Federal and state government scores were averaged to form an index of effectiveness (GEFF; alpha = .78). Manufacturer and trade association items were used as separate measures of the effectiveness of two market organizations (MEFF and TEFF).

RESULTS

Dealer Concerns

Dealers indicated salience of scrap tire disposal as a major environmental issue. Table 2 shows average responses on a 1-5 scale to items of environmental and business concerns, as well as effectiveness of trade associations, manufacturers, and governments. First, of five issues posed as major environmental concerns, landfill capacity and scrap tire disposal (ENV1) ranked as the highest and were not significantly different in ratings (t = 1.35, p = .178). Scrap tire disposal as an environmental concern (average = 4.23) was rated significantly higher than environmental concerns of the ozone layer (p = .057), preservation of wetlands (p < .001), and tires in landfills (p <.001).

Second, dealers' strongest business concerns were the need for more information regarding tire disposal alternatives and seriousness of the problem (averages = 3.98 and 3.51, respectively). Responses suggested, however, that dealers did not perceive restrictive regulations, strong consumer concerns, or that consumers were likely to buy tires based upon whether the manufacturer

TABLE 2. Dealer Concerns Ranked by Averages

Type	Concern	Measure	Mean*	Standard Deviation
Environmental	Landfill capacity	ENV5	4.32	.84
	Scrap tire disposal	ENV1	4.23	.98
	Ozone layer	ENV4	4.06	.97
	Wetlands	ENV3	3.85	1.03
	Tire landfill	ENV6	3.77	1.21
Business	Need more information	INFO	3.98	.89
	Serious problem	PROB	3.51	1.07
	Consumer concern	CCONC	2.81	1.08
	Restrictive regulations	REGS	2.77	.89
	Consumer buying	CBUY	2.68	1.06
Effectiveness	Trade association	TEFF	3.29	1.13
	Manufacturers	MEFF	2.09	1.08
	State government	SGEFF	1.98	.94
	Federal government	FGEFF	1.84	.85

* Average using a 5-point scale of (1) strongly disagree to (5) strongly agree.

had a disposal program. Dealers indicated some effectiveness of trade associations in disposal solutions (average = 3.29), but did not view manufacturers, state government, or federal government as effective in general (averages = 2.09, 1.98, and 1.84, respectively).

Preference for Government Intervention

Hypotheses 1-3 were tested using correlation and regression analyses. Correlations among the dependent variable, preference for government intervention (GPREF), and all independent variables are shown in Table 3; results of regression analyses are presented in Table 4, with GPREF as the dependent variable. Collinearity indices are less than 30, suggesting multicollinearity is not a threat to stability and interpretability of the regression coefficients (Belsley, Kuh, and Welsh 1980).

Regarding the first hypothesis, significant, positive correlations between preference for government intervention (GPREF) and environmental concern measures (ENV1, $r = .22$, $p < .01$; ENV2, $r = .21$, $p < .01$) support $H1_a$. Regression analysis also supports government preference as a function of environ-

TABLE 3. Correlations Among Variables Used to Test the Hypotheses

ITEM	GPREF	ENV1	ENV2	REGS	GEFF	MEFF	TEFF	CBUY	CCONC	CHNG	INVST	CHNG
GPREF	1.00											
ENV1	.22[1]	1.00										
ENV2	.21[1]	.27[1]	1.00									
REGS	.07	−.01	.01	1.00								
GEFF	−.08	.02	.04	−.09	1.00							
MEFF	−.15[2]	−.22[1]	.02	.10	.29[1]	1.00						
TEFF	−.07	−.17[2]	.07	.07	.27[1]	.34[1]	1.00					
CBUY	.21[1]	.19[1]	.01	.17[2]	.02	−.05	.14[2]	1.00				
CCONC	.14[2]	.22[1]	.10	.24[1]	.03	.01	−.02	.26[1]	1.00			
CHNG	.04	.11[3]	.05	.06	−.03	−.13[3]	.14[2]	.06	.07	1.00		
INVST	−.01	−.03	−.12[3]	.10	−.20[1]	−.14[2]	−.13[3]	.05	.01	.14[2]	1.00	
INFO	.14[2]	.23[1]	.04	.14[2]	−.09	−.13[3]	−.17[2]	.22[1]	.12[3]	.08	.22[1]	1.00
PROB	.25[1]	.26[1]	.14[2]	.27[1]	−.07	−.07	−.04	.06	.07	.15[2]	.10	.25[1]

[1] Correlation significant at the $p < .01$ level.
[2] Correlation significant at the $p < .05$ level.
[3] Correlation significant at the $p < .10$ level.

TABLE 4. Regression Analyses with Government Preference (GPREF) as the Dependent Variable

Model/ Hypothesis	Independent Variables	Unstandardized Coefficient	Standard Error	Standard Coefficient	T Value	Prob.[1]
1	ENV1	.26	.08	.21	3.06	p < .01
	ENV2	.13	.07	.13	1.8	p = .07
	REGS	.07	.09	.05	.8	p = .42
	GEFF	−.10	.10	−.07	−1.04	p = .30
	Constant	1.92	.50		3.82	p < .01

[1] F = 4.53; $p < .01$; adjusted R^2 = .06; largest condition index = 15.87.

Model/ Hypothesis	Independent Variables	Unstandardized Coefficient	Standard Error	Standard Coefficient	T Value	Prob.[2]
2	MEFF	−.16	.08	−.15	−2.08	p = .04
	TEFF	.02	.08	.02	.24	p = .81
	CBUY	.21	.08	.19	2.66	p < .01
	CCONC	.10	.08	.09	1.27	p = .21
	Constant	2.88	.39		7.47	p < .01

[2] F = 4.09; $p < .01$; adjusted R^2 = .06; largest condition index = 11.88.

Model/ Hypothesis	Independent Variables	Unstandardized Coefficient	Standard Error	Standard Coefficient	T Value	Prob.[3]
3	CHNG	−.03	.16	−.01	−.16	p = .87
	INVST	−.15	.16	−.06	−.93	p = .35
	INFO	.14	.09	.10	1.47	p = .14
	PROB	.26	.08	.24	3.39	p < .01
	Constant	2.07	.41		5.07	p < .01

[3] F = 4.24; $p < .01$; adjusted R^2 = .06; largest condition index = 13.29.

mental concerns (shown by the first model in Table 4). Higher ratings of scrap tire disposal as a major environmental concern (ENV1) and recycling as more of an environmental than an economic issue (ENV2) are predictive of higher preference for government intervention ($p < .01$ and $p = .07$, respectively). However, neither correlation or regression analysis support $H1_b$ that higher preference for government intervention is associated with perceived government effectiveness. Measures of restrictive regulations (REGS) and government effectiveness (GEFF) are not significant predictors of government preference (GPREF) at the p < .10 level.

Hypothesis 2 states that preference for government intervention will be higher for dealers who assess manufacturers and trade associations as ineffective in scrap tire disposal issues (H2$_a$ and H2$_b$) and who perceive increasing consumer concerns (H2$_c$). Correlations shown in Table support the hypothesis regarding manufacturer effectiveness (MEFF, r = −.15, p < .05), consumer preference for manufacturers with disposal programs (CBUY, r = .21, p < .01), and increasing consumer concerns (CCONC, r = .14, p < .05). Trade association effectiveness (TEFF) is not significantly correlated with government preference at the p < .10 level, however. Regression analysis shown in the second model of Table 4 also supports that government preference (GPREF) is a negative function of manufacturer effectiveness (MEFF; p < .05) and a positive function of consumer preference for manufacturers with disposal programs (CBUY; p < .01). Variables of trade association effectiveness (TEFF) and consumer concerns (CCONC) are not significant predictors at the p < .10 level. Thus, H2$_a$ regarding manufacturer effectiveness is supported by the analysis, H2$_b$ regarding trade association effectiveness is not supported, and H2$_c$ is supported considering the consumer preference measure.

The third hypothesis predicts higher government preference for dealers who are modifying their operations (H3$_a$) and perceive disposal as a serious problem (H3$_b$). Neither of the two dichotomous measures of modifications (CHNG and INVST) are significantly related to government preference, as indicated by correlations or regression coefficients, failing to support H3$_a$. Perceived information need (INFO) and business problems (PROB) regarding tire disposal are correlated positively with government preference (GPREF) the p < .05 level (r = .14 and .25, respectively), supporting H3$_b$. The third regression model in Table 4 also supports government preference as a function of the seriousness of the problem (PROB, p < .01), although information need is not shown as a significant predictor of government preference.

Preferences for Specific Mechanisms

Preferences for 19 specific solutions to the problem of scrap tires are divided into: (1) market uses and development; (2) manufacturer service provision; and (3) government remedies. Table 5 shows average responses for each solution, contrasted for dealers with preference for government, equal, or manufacturer leadership. The most preferred solutions by all three groups are market uses and development, with the exception of retread and reuse of tires; all averages are above 4.0 on a 5-point scale. Information provision by manufacturers and government is also preferred above other manufacturer or gov-

TABLE 5. Average Specific Preferences by General Preference Groups

1. Market Uses and Development	Government (n = 49)	Equal (n = 87)	Manufacturer (n = 78)	F	Prob.[1]
Find new uses for tires	4.50	4.37	4.45	0.56	.57
Innovate with new technology	4.29	4.29	4.26	0.04	.96
Burn tires for energy	4.29	4.27	4.11	1.08	.34
Use for road construction	4.16	4.31	4.04	2.39	.09
Retread and reuse tires	3.06	3.11	2.97	0.21	.81

[1] Hotellings Multivariate F = 1.13; p = .34.

2. Manufacturer Services	Government	Equal	Manufacturer	F	Prob.[2]
Information provision	3.39	4.08^2	4.05^2	12.01	< .01
Rebates for disposal	2.04	2.80^2	2.82^2	7.61	< .01
Finance shipping of scrap tires	1.96	2.56^2	2.62^2	6.08	< .01
Buy back scrap tires	1.88	2.24	2.09t	2.01	.14
Clean landfills and stockpiles	1.53	1.94^2	2.05^2	4.54	.01

[2] Significantly higher than government group; Hotellings Multivariate F = 4.36; p < .01.

3. Government Remedies	Government	Equal	Manufacturer	F	Prob.[3]
Information provision	4.16	4.07	3.85	1.81	.17
Standards: license scrap haulers	3.88	3.82	3.91	0.12	.89
Controls: penalize illegal disposal	3.88	4.19	3.99	1.41	.25
Controls: restrict landfill disposal	3.53	3.52	3.60	0.09	.92
Controls: store only in stockpiles	3.56	3.47	3.35	0.50	.61
Service: clean landfills, stockpiles	3.14	3.12	2.99	0.32	.72
Excise tax to fund disposal alternatives	2.40	2.52	2.42	0.19	.82
Road tax to fund disposal alternatives	2.14	2.47	2.16	1.84	.16
Gasoline tax to fund disposal alternatives	1.94	2.31	2.00	2.51	.08

[3] Hotellings Multivariate F = .87; p = .61.

ernment remedies. The ranking of government remedies parallels Harris and Carman's (1984) ordering, with the exception of the category of taxes and subsidies. Tax mechanisms were the least preferred solutions by responding dealers, perhaps reflecting the measures specification of taxes to fund disposal alternatives without any specification of subsidies per se.

Hypothesis 4 predicted that dealer preference for manufacturer leadership, a market-sector solution, is associated with higher specific preferences for market uses, manufacturer service provision, and government remedies of information provision, standards, and taxes. For those dealers generally preferring government leadership, Hypothesis 4 predicted higher specific preferences for controls and service provision by the government. Averages for the specific solutions were compared with MANOVA, and results are reported in Table 5. Multivariate effects are found for mechanisms under the heading of manufacturer service provision (F = 4.36, p < .01), supporting Hypothesis 4 with respect to information provision, disposal rebates, shipping of scrap tires, and cleaning landfills and stockpiles. Dealers classified as preferring manufacturer leadership or having equal preference for manufacturer and government leadership have significantly higher preferences for manufacturer service provision than those with preference for government leadership. However, differences among the groups were not supported relative to the five market uses (F = 1.13, p = .34) or government remedies (F = .87, p = .61). Thus, Hypothesis 4 was supported only with respect to manufacturer service provision.

Summary of Survey Results

Tire dealers play a central role in the disposition of scrap tires through their direct consumer contacts, collection of used tires, and contracts with manufacturers, haulers and processors. Further, dealers operate within the states that regulate disposal and support alternative markets for scrap tires, a factor that is particularly important given the varying content and extent of state government interventions. The central role of dealers makes their participation vital to the success of programs to reduce scrap tire disposal through alternative market uses and commercialization of recycled products.

Dealer preferences include both government and manufacturer-sponsored programs. Regardless of the dealer's overall preference for government or manufacturer leadership, information provision and development of new uses, technologies, and markets for scrap tires were the most preferred solutions. Dealers indicated that they do not know the final disposition of many tires collected at their dealerships, and that consumers do not appear to be very concerned about scrap tires. Both manufacturer and government sources of information were seen as preferred solution mechanisms, and the role of industry and government sources in education and, perhaps, promotion of scrap tire issues and solutions is clearly suggested.

Results of the current study also indicate dealers have strong environmental and business concerns regarding scrap tire disposal, and their perceptions of the severity of these problems is associated with preference for government leadership in disposal solutions. However, dealers did not judge federal or state governments or manufacturers to be effective in dealing with scrap tire issues. Industry trade associations were the only group perceived as even moderately effective. The effectiveness of these sources may be strongly tied to their provision of information, enforcement of regulations for tire haulers and disposal sites, and support in developing technologies and uses for scrap tires.

DISCUSSION

Several aspects of the scrap tire disposal problem appear to be amenable to a wider role for government-sponsored programs and the cooperation of dealers, manufacturers, processors, trade associations and government agencies. Improved dialog and coordination among manufacturers, dealers, processors, trade associations and both federal and state government agencies is suggested by the study of tire dealer preferences, as well as industry and government sources regarding scrap tire issues and environmental industries in general. Cooperative programs including manufacturers, government, and other related industries contribute to Japan's success in using or exporting nearly 90 percent of its scrap tires (RMA 2003).

Systematic Solutions Involving Industry and Government

In environmental industries, European and Asian models emphasize business-government cooperation (Marcus et al. 2002). For example, the Dutch emphasize abatement of pollution sources along the entire production-consumption-disposal chain. The government provides technological assistance and tax mechanisms to support business profitability, while polluters pay for abatement. Profitability may be enhanced by cost reductions from use of new technologies and generation of revenue from new products. However, cost reductions, technological success, and market development are long-term strategies, and U.S. industry frequently demands more immediate results. Additional constraints include bureaucracy in government programs and in businesses, uneven enforcement of regulations, and a lack of coordination with suppliers and customers (Diefendorf 2000; Farrell 1999; Marcus et al. 2002).

Greater success in developing and commercializing market uses for scrap tires in the U.S. may result from cooperative programs that involve tire dealers, scrap tire processors, and manufacturers of tires and other products, as well as federal and state policies and programs. Mechanisms such as the Small Business Innovation Research Program and the Federal Technology Transfer

Program can prove instrumental in improving and sharing solutions to scrap tire disposal (U.S. EPA 2003). The purpose of the Federal Technology Transfer Program and state programs are to make scientific and technological developments accessible to private industry and other government agencies. Users are supposed to develop new products or processes to enhance competitiveness of the U.S. industries and/or to improve quality of life. Examples in the scrap tire industry include U.S. Air Force research and commercial trials for devulcanizing processes (Owen 1998).

Premus (2002) cites case studies of entrepreneurs developing environmentally friendly technologies to highlight three governmental policies that would improve market development and commercialization: (1) start up and early expansion capital; (2) environmental standards for and taxes on pollution; and (3) procurement policies including products derived from new technologies. Lack of equity capital is a major barrier to growth in environmental industries, which have struggled to attract venture capital and to deal with other barriers to market entry and development, such as market fragmentation and varied regulation, permitting processes, and enforcement across states (Diefendorf 2000). In addition, a track record of market volatility caused by federal and state governments withdrawing subsidies and incentives makes these markets even less attractive (RMA 2002a).

Specific to scrap tires, Farrell (1999) and the RMA (2002a, 2002b) argue for capital in the form of start up grants and time-limited fees to support development of recycled products, rather than subsidies or reimbursements for collection, processing, or use of scrap tires. Grants for research and processing technologies and equipment, for example, have jump started alternative markets for scrap tires that can be self sustaining, as compared to the false economies created when short-term subsidies are given to industry participants simply for using scrap tires (Farrell 1999). As states remove fees, subsidized markets fail.

More success has been shown in states that use a longer-term strategy of stimulating technology and market development in scrap tire and other environmental industries (Marcus et al. 2002; RMA 2002b; U.S. EPA 2003). The State of North Carolina funded research by Continental Tire to boost use of recycled rubber from 3 to 25 percent of the content of new tires (RMA 2002a). Grants from the State of Illinois to purchase processing equipment aided the use of tire-derived fuel (Farrell 1999). Success could also be facilitated by development of technologies to produce ground or crumb rubber, oil, and carbon black (Jupiter 2002; Onorato-Jackson 2001; Owen 1998).

The second suggestion by Premus (2002) to support market development and commercialization of recycled products is government policy in setting environmental standards and taxes. Federal regulation and standards from the

EPA, in cooperation with industry, have been suggested for environmental testing and permitting in general (Diefendorf 2000). In the case of scrap tires, tire-derived fuel markets are very dependent upon EPA emission standards, as well as standards for the tire-derived materials used for the fuel (e.g., whole or ground tires). Standards from the American Society for Testing and Materials are a step toward making tire chips a commodity in tire-derived fuel (RMA 2002a). Owen (1998) suggests large-scale market development for crumb rubber products is unlikely without uniform standards or at least differential grades of crumb rubber. Processors and equipment manufacturers in the crumb rubber market guard cost data and operate without uniform quality standards. Customers appear to be dependent upon single suppliers with proprietary products. Owen (1998) notes the greatest opportunities for high value crumb rubber markets are in cooperative ventures between processors and potential large users, such as tire manufacturers and plastic compounders. However, government and industry cooperation is needed to set crumb rubber standards that would facilitate such market uses.

Finally, Premus (2002) suggests procurement policies including products derived from new technologies. Clearly, scrap tire use in civil engineering and construction projects has been and could be further facilitated by federal and state procurement policies. The federal program for rubber asphalt use in road construction in the early 1990s was not only unsuccessful but devastating to industry participants when states refused to comply (RMA 2002a). The federal program was premature, but industry and state-supported research has since that time shown the value and produced standards for use of rubber asphalt and tire shreds in construction and civil engineering applications. However, Florida is one of only two states to systematically address all four component markets of asphalt, civil engineering, tire-derived fuel, and crumb rubber applications. A significant part of this effort has been accomplished through purchasing policies (Farrell 1999).

A comprehensive solution to scrap tire disposal will require continued development of recycled product markets and the cooperation of a variety of business and government parties. The current study reveals the need for federal and state governments to be vigilant in enforcement of disposal regulations, to provide start up capital and research for new product and technologies, and to participate in setting standards and purchasing recycled products for use in civil engineering and construction projects. Creating viable products, processes, and markets for scrap tires should be an integral part of federal and state goals, as well as the goals of tire industry participants. Additional research is suggested to assess potential disparity of goals and to examine the potential contributions and from manufacturer, processor, and dealer segments of the industry.

REFERENCES

Aaker, David and R. Bagozzi (1982), "Attitudes Toward Public Policy Alternatives to Reduce Air Pollution," *Journal of Marketing and Public Policy*, 1, 85-94.

Apaiwongse, Tom S. (1991), "Factors Affecting Attitudes Among Buying-Center Members Toward Adoption of an Ecologically-Related Regulatory Alternative: A New Application of Organizational Theory to a Public Policy Issue," *Journal of Public Policy and Marketing*, 10, 2 (Fall), 145-160.

Arndt, Johan (1981), "The Political Economy of Marketing Systems: Reviving the Institutional Approach," *Journal of Macromarketing*, 1 (Fall), 36-47.

_____ (1983), "The Political Economy Paradigm: Foundation for Theory Building in Marketing," *Journal of Marketing*, 47 (Fall), 44-54.

Bagozzi, Richard P. (1975), "Marketing as Exchange," *Journal of Marketing*, 39 (October), 32-39.

Barnes, James (1982), "Recycling: A Problem in Reverse Logistics," *Journal of Macromarketing*, 2 (Fall), 31-37.

Belsley, D. A., E. Kuh, and R. E. Welsch (1980), *Regression Diagnostics: Identifying Influential Data and Sources of Collinearity*, New York, NY: John Wiley & Sons.

Crane, Andrew (2000), "Marketing and the Natural Environment: What Role for Morality?" *Journal of Macromarketing*, 20 (December), 144-154.

Cunningham, Bernard, S. Sethi, and T. Turicchi (1982), "Public Perception of Government Regulation," *Journal of Macromarketing*, 2 (Fall), 43-51.

Diefendorf, Sarah (2000), "Venture Capital & the Environmental Industry," *Corporate Environmental Strategy*, 7 (4), 388-399.

Farrell, Molly (1999), "Building Sustainable Recycled Tire Markets," *Biocycle*, 40 (5), 50-53.

Fedchenko, Vera (2002), "Scrap Tire Plan Being Considered," *Tire Business*, 19 (January 21), 10.

Friedlander, Philip (1990), Testimony Before the Committee on Small Business, United States House of Representatives, *Scrap Tire Management and Recycling Opportunities*, Washington, DC: U. S. Government Printing Office.

Granzin, Kent L. and Janeen E. Olsen (1991), "Characterizing Participants in Activities Protecting the Environment: A Focus on Donating, Recycling, and Conservation Behaviors," *Journal of Public Policy and Marketing*, 10, 2 (Fall), 1-27.

Grether, E. T. (1966), *Marketing and Public Policy*, Englewood Cliffs, NJ: Prentice-Hall.

Grewal, R. and R. Dharwadkar (2002), "The Role of the Institutional Environment in Marketing Channels," *Journal of Marketing*, 66 (July), 82-97.

Harris, Robert and J. Carman (1983), "Public Regulation of Marketing Activity: Part I: Institutional Typologies of Market Failure," *Journal of Macromarketing*, 3 (Spring), 49-58.

_____ and _____ (1984), "Public Regulation of Marketing Activity: Part II: Regulatory Responses to Market Failures," *Journal of Macromarketing*, Vol. 3, Spring, 41-52.

Jupiter Consultancy (2002), "New Report Asks: Is Pyrolysis of Scrap Tires Poised for Widespread Acceptance?" *Scrap Tire News* (February), available online March 16, 2003, *www.scraptirenews.org/archives*.

Marcus, Alfred, D.A. Geffen, and K. Sexton (2002), "Business-Government Cooperation in Environmental Decision Making," *Corporate Environmental Strategy*, 9 (4),
345-355.

Marshall, Kimball P. (1998), "Generalized Exchange and Public Policy: An Illustration from Support for Public Schools," *Journal of Public Policy and Marketing*, 17
(2), 274-286.

_____, W. S. Piper, and L. Micich (2002), "An Assessment of School System and
Mass Media Sources of Information on the Relationships of Generalized Exchange
Perceptions to Voter Support for Public Education," *Journal of Nonprofit and Pub-
lic Sector Marketing*, 10 (1), 23-40.

McCarty, J. and L. Shrum (2001), "The Influence of Individualism, Collectivism and
Locus of Control on Environmental Beliefs and Behavior," *Journal of Public Policy and Marketing*, 20 (Spring), 93-104.

Miller, Chaz (2002), "Profiles in Garbage: Scrap Tires," *Waste Age*, 33 (3), 20-21.

Mills, Michael (1984), "Energy Issues and the Retail Industry: Public Policy/Market-
ing Implications," *Journal of Public Policy and Marketing*, 3, 167-183.

Moore, Miles (1998), "Solutions Vary to Scrap Tire Woes," *Waste News*, 4 (26), 21.

_____ (2001), "Processors Want Voice in Creating Scrap Tire Policy," *Rubber &
Plastics News*, 31 (October 15), 25-26.

_____ (2002), "N.Y. Senate Votes for Scrap Tire Bill," *Tire Business*, 20 (7), 18.

Nason, Robert (1986), "Externality Focus of Macromarketing Theory," in G. Fisk
(ed.), *Marketing Management Technology as a Social Process*, New York, NY:
Praeger.

Noga, Edward (2001), "Scrap Tire Piles Declining, STMC Says," *Tire Business*, 18
(25), 3.

Northwest Product Stewardship Council (2003), "Tires and Product Stewardship,"
Products & Sectors, available online February 11, 2003, *www.productstewardship.
net*.

Onorato-Jackson, Danielle (2001), "Old Process Aims to Convert Tires into New Oil,"
Waste Age, 32 (September), 29-30.

Owen, Kenneth C. (1998), "Scrap Tires: A Pricing Strategy for a Recycling Industry,"
Corporate Environmental Strategy, Vol. 5 (2), p. 42-50.

Premus, Robert (2002), "Moving *Technology* from Labs to Market: A Policy Perspective," *International Journal of Technology Transfer and Commercialisation*, Vol. 1
(1/2), 22-39.

Reidenbach, R. E. and T. Oliva (1983), "Toward a Theory of the Macro Systemic Effects of the Marketing Function," *Journal of Macromarketing*, 3 (Fall), 33-40.

Rowbotham, Keith (1990), "State, National Associations Rally: Don't Look to Government to Solve the Scrap Tire Disposal Problem," *Modern Tire Dealer*, 71
(April), 16.

Rubber Manufacturers Association (2002a), *U.S. Scrap Tire Markets 2001*, available
online December 10, 2002, *www.rma.org/scraptires*.

_____ (2002b), *Scrap Tire Laws*, available online December 10, 2002, *www.rma.org/
resources*.

_____ (2003), "Tire Manufacturers Take a Broad Look at Scrap Tire Management," available online February 27, 2003, *www.rma.org/newsroom*.

Scrap Tire Management Council (1991a), "Survey on Applications of Scrap Tires," *Rubber World*, 205 (Dec.), 10-11.

_____ (1991b), "Scrap Tire Legislation Status, January 1991," *Scrap Tire News*, 5 (1), Leesburg, VA: Recycling Research Institute.

Tire Business (2001), "Industry Scrap Tire Effort Pays Off," *Tire Business*, 18 (25), 8.

Tire Retread Information Bureau (2003), "2002 Fact Sheet," available online February 27, 2003, *www.retread.org*.

Truini, Joe (2002), "Scrap Tire Recycling on the Rise, Study Says," *Waste News*, 8 (18), 4.

United States Environmental Protection Agency (1991), *Markets for Scrap Tires*, Office of Solid Waste, Washington, DC.

_____ (2002), *Municipal Solid Waste: Tires*, available online December 29, 2002, *www.epa.gov/epaoswer/nonhw/muncpl/tires.htm*.

_____ (2003), *The Environmental Technology Commercialization Center*, available online March 16, 2003, *www.etc2.org/index_technology.htm*.

Wallace, Joseph (1990), "All Tired Out," *Across the Board*, 27 (Nov.), 24-29.

APPENDIX

Measures

ITEM	DESCRIPTION	SCALE
1. Preferences for Solutions		
GPREF	Preference for (1) federal and (2) state government leadership in scrap tire disposal and uses	Average of two items (alpha = .74); 5-pt. Likert scale*
MPREF	Preference for manufacturer leadership in scrap tire disposal and uses	5-pt. Likert scale
PREF	Dealer categorization based on comparison of preference scores for GPREF and MPREF	1 = government preferred; 2 = equal preference; 3 = manufacturer preferred
MKT1-5	Preference for each of 5 market uses (Table 5)	5-pt. Likert scale
MFG1-5	Preference for each of 5 manufacturer services (Table 5)	5-pt. Likert scale
GOVT1-9	Preference for each of 9 government remedies (Table 5)	5-pt. Likert scale
2. Environmental and Business Concerns		
ENV1	Scrap tire disposal is a major environmental concern	5-pt. Likert scale
ENV2	Recycling is more of an environmental issue than an economic issue	5-pt. Likert scale
ENV3-6	Each of 4 items (wetlands, ozone layer, landfill capacity, scrap tires in landfills) as a major environmental concern	5-pt. Likert scale
PROB	Scrap tire disposal a serious problem, compared to others the dealer faces	5-pt. Likert scale

APPENDIX (continued)

ITEM	DESCRIPTION	SCALE
2. Environmental and Business Concerns (continued)		
REGS	Scrap tire regulations faced by the dealer are restrictive	5-pt. Likert scale
INFO	Dealer needs more information on disposal alternatives	5-pt. Likert scale
CCONC	Disposal of scrap tires is a growing concern of the dealer's customers	5-pt. Likert scale
CBUY	Dealer's customers would be more likely to buy brand of manufacturer with scrap tire disposal program	5-pt. Likert scale
CHNG	Disposal method changed in past 3 years	0 = no; 1 = yes
INVST	Investigating new disposal method	0 = no; 1 = yes
3. Solution Effectiveness		
GEFF	(1) Federal and (2) state government doing a good job of working on solutions to scrap tire disposal	Average of two items (alpha = .78); 5-pt. Likert scale
MEFF	Manufacturers doing a good job of working on solutions to scrap tire disposal	5-pt. Likert scale
TEFF	Trade associations doing a good job of working on solutions to scrap tire disposal	5-pt. Likert scale

*All 5-point Likert scales ranged from (1) strongly disagree to (5) strongly agree.

The Comparative Influence
of Government Funding
by the U.S. Commerce Department's
Advanced Technology Program
and Private Funding
on the Marketing Strategy
of High-Tech Firms

Conway L. Lackman

SUMMARY. This paper conducts an exploratory study on whether high-tech venture firms funded by a United States Federal Government program–Department of Commerce Advanced Technology Program (ATP)–pursue different types of marketing strategies and different goal levels than firms without such funding. A sample of ATP and non-ATP-funded firms was used. Nineteen hypotheses relating to commercialization processes, i.e., types of strategy (licensing), types of commercial advantage (new, innovative solutions), and types of commercialization applications (product

Conway L. Lackman, PhD, is Associate Professor of Marketing, Duquesne University, Pittsburgh, PA 15282 (E-mail: Lackman@duq.edu).
The author thanks Christine Koenig for her secretarial support.

[Haworth co-indexing entry note]: "The Comparative Influence of Government Funding by the U.S. Commerce Department's Advanced Technology Program and Private Funding on the Marketing Strategy of High-Tech Firms." Lackman, Conway L. Co-published simultaneously in *Journal of Nonprofit & Public Sector Marketing* (Best Business Books, an imprint of The Haworth Press, Inc.) Vol. 13, No. 1/2, 2005, pp. 179-197; and: *Government Policy and Program Impacts on Technology Development, Transfer and Commercialization: International Perspectives* (ed: Kimball P. Marshall, William S. Piper, and Walter W. Wymer, Jr.) Best Business Books, an imprint of The Haworth Press, Inc., 2005, pp. 179-197. Single or multiple copies of this article are available for a fee from The Haworth Document Delivery Service [1-800-HAWORTH, 9:00 a.m. - 5:00 p.m. (EST). E-mail address: getinfo@haworthpressinc.com].

Digital Object Identifier: 10.1300/J054v13n01_10

vs. process) and four hypotheses relating to levels of performance (levels of cost reduction) were tested. Small ATP firms are more likely to pursue best practice marketing policies such as R&D collaboration, and licensing strategies, and to pursue higher cost reduction and performance goals than non-ATP firms. *[Article copies available for a fee from The Haworth Document Delivery Service: 1-800-HAWORTH. E-mail address: <docdelivery@ haworthpress.com> Website: <http://www.HaworthPress.com> © 2005 by The Haworth Press, Inc. All rights reserved.]*

KEYWORDS. Venture funding, government funding, technology transfer

INTRODUCTION

This paper conducts an exploratory study on whether high-tech venture firms funded by a United States Federal Government program–Department of Commerce Advanced Technology Program (ATP)–pursue different types of marketing strategies and different goal levels than firms without such funding. The mission of ATP is to prevent the "the most promising new technologies, with the greatest potential for widespread national benefits, from remaining in the research laboratory" (ATP 2003). ATP steps into the gap between the research lab and the marketplace by funding projects judged too risky or too far outside the mainstream to warrant private funding. ATP's major goal is twofold: to fund the technologies that deliver new, innovative products and services to the marketplace more quickly and to foster new, innovative process or manufacturing technologies (ATP 2003). The goal is to deliver multibillion dollars worth of economic benefits to the U.S. economy. These benefits stand as the justification for the use of U.S. taxpayer dollars to fund a $240 million agency, awarding $55-65 million funds annually to rigorously scrutinized awardee companies. The ATP provides cost-share funding in the high risk early stages of R&D that are prohibatitive for other sources of funding. ATP funding limits are $2,000,000 in direct costs over a period not to exceed three years for a single company and up to half of the total project costs for a maximum of five years for a joint venture involving more than one company (ATP 2003). Between 1990 and May 2003, ATP has awarded $1.97 billion in funding to companies to develop high-risk, enabling technologies. Industry has matched this funding with $1.89 billion in cost sharing. Of the $1.97 billion of ATP funds:

- 195 joint ventures with 882 member firms and organizations have received about 59% of total ATP award funds.
- 454 firms in single company projects have received about 41% of total ATP award funds (ATP 2003).

Justification for use of these funds requires performance by ATP awardees that achieve goals specified by the ATP program: advancement of U.S. technological and commercial advantage and provision of economic benefits to U.S. economy. A barely explored issue of whether government programs, focusing here on ATP-funded firms, foster marketing best practices and by inference more effectively transfer technology than non-ATP-funded firms is explored in this paper. Specifically, the ATP is used to investigate two important issues: (1) whether marketing strategies and levels of outcomes pursued by ATP-funded venture firms are different than types of strategies and levels of outcomes pursued by non-ATP-funded firms, and (2) whether these ATP-funded firm strategies and outcomes are more consistent with marketing best practices than those of non-ATP-funded firms.

Pursuant to investigation of these issues, a random sample of 361 ATP-funded firms was drawn from the U.S. Commerce Department's ATP database. A random sample of 398 non-ATP-funded firms was also drawn from the Department's U.S. Small Business Administration (SBA) database. Twenty-three hypotheses were tested regarding the differences between ATP-funded and non-ATP-funded firms. Fourteen relate to commercialization processes, i.e., types of strategy (licensing vs. in-house production), two to types of commercial advantage (new solutions vs. performance improvement vs. cost reduction), and three types of commercialization applications (product vs. process). Four relate to levels of performance, i.e., levels of cost reductions and performance (productivity). Strategies include R&D collaboration, alliances with customers, distributors, and suppliers, in-house production of products and services, adoption of new processes for in-house use, and licensing. Types of competitive advantage include new, highly innovative solutions, and performance goals. Performance outcomes or goals relate to performance improvement, cost reduction and performance/cost improvement combined.

BACKGROUND LITERATURE

Small business success strategy is of increasing scholarly interest (Hall and Rifkin 1999, Kaplan 2001, Patsula and Nowik 2002). However, little research has been done regarding the impact of government funding programs on marketing strategies and goals of small business, much less high-tech ventures. Most of the literature on venture firm management is limited to descriptions of the process of obtaining government funding. Little attention has been given to assessment of the impact of government programs (Parbotecah 2000, Chawla et al. 1997, Sheahen et al. 1994, Doctors 1969). Some basic small business strategies that have been studied include (1) global marketing niche entry strategies with a focus on low involvement, i.e., licensing, use of export houses (Brown and Eisenhardt 1998), (2) concentration on products with sustainable competitive advantage focusing on products strategy (Levinson 1998, Lussier 1996, Ries and Trout 1993), (3) integration of R&D and marketing strategies using government funding to foster integration (Robinson and Herron 2001, Bernstein 1994, Spann et al. 1993), (4) use of government R&D grants for liquidity and enhanced credibility including use of a second mortgage as a funds source when obtaining (Small Business Investment Research) grants (Brown and Turner 1999), (5) patents for positioning with investors for narrowly focused products (Sink and Easley 1994), and (6) use of horizontal or vertical strategic alliances with little connection to relative performance of the alliances (Ensley et al. 1998, Hoffman and Viswanathan 1996, Gersony 1994, Litvak 1992). Intra-strategy tradeoffs have been examined such as the tradeoff between competitive advantage achieved via innovativeness and incompatibility with existing customer practices (Hall and Rifkin 1999, Kwestel et al. 1998, Ali et al. 1995). The problem of financial constraints on R&D (hence on risk and innovativeness) has just begun to be examined (Gompers and Lerner 1999, Piper and Lund 1996). Finally, the tradeoff between R&D intensity and need for strategic alliances and increasing management complexity has only been superficially as opposed to rigorously examined (Jonash 1996, Maynard 1996).

Five key high-tech marketing best practices emerge from the literature:

1. Low involvement, niche entry strategies such as licensing.
2. Narrow product focus such as product oriented strategy.
3. Integration of R&D and marketing via alliances.
4. Balance of innovation with existing customer practices.
5. Rising to the challenge of increasing management complexity resulting from alliances.

These marketing best practices have proven to be associated with successful high-tech start-ups (Barringer et al. 1997, Harper 1996). Therefore, ATP-funding that fosters best practices is likely to increase the chances of success for ATP-funded high-tech firms.

HYPOTHESES

The objective of this paper is to determine (1) if ATP-funded firms' marketing strategies are different from non-ATP firms, (2) if marketing strategies of ATP-funded are more consistent with marketing best practices than non-ATP funded firms, and therefore, (3) if ATP is an effective program and a good model for government support of high-tech business. To determine if there are differences in strategies and objectives between ATP-funded firm strategies and non-ATP, twenty-three hypotheses are tested. Fourteen relate to commercialization processes, i.e., types of strategy (licensing vs. in-house manufacturing), two to types of commercial advantage (new highly innovative solutions vs. moderately innovative performance improvement vs. low innovation cost reduction solutions), and three to types of commercialization applications (product vs. process vs. service). Four relate to levels of performance, i.e., levels of cost reductions and performance (productivity) targets. Seventeen of the twenty-three hypotheses also can be used to test whether ATP funding encourages any of the five best marketing practices advanced in the literature.

The rationales for selection of these hypotheses are as follows.

H1a, b: If ATP-funded firms tend to pursue strategies consistent with best practices more than non-ATP-funded firms, the inference is that the impact of ATP funding is more consistent with good marketing policy.

H2a, b: If ATP-funded firms pursue higher levels of innovation than the non-ATP-funded firms, the inference is that ATP-funded firms are more likely to deliver more innovative products.

H3a, b, c: If ATP-funded firms are more product (vs. process) oriented than the non-ATP funded-firm, the inference is that ATP funding fosters more product development among ATP funded-firms. As a result, non-ATP-funded firms are more process or service application oriented.

H4a, b: If ATP-funded firms have higher cost reduction goals than non-ATP, the inference is that ATP funding fosters higher cost reduction goals among funded firms.

H5a, b: If ATP-funded firms have higher performance goals than non-ATP, the inference is that ATP funding fosters higher performance goals among funded firms.

Seventeen of the twenty-three hypotheses test whether ATP funding encourages any of the five best marketing practices. H1a, b, positing that ATP-funded firms are strategically more R&D collaboration, alliances and licensing oriented, tests practices 1, 3 and 5. H3a, b, c, positing that ATP-funded firms are more product oriented, tests practice 2. Regarding H1a, b, H3a, b, H4a, b and H5a, b, the inference of a functional relationship between strategies and best practices is only an inference, not a model. No model specifying a functional relationship that would imply a tautology is offered. Regarding H2a, b, only an inference is implied about delivering more innovative products. It does not necessarily follow that more effective technology transfer will result. Technology transfer is out of the scope of the analysis of this relationship between pursuit of high innovation and innovative products. To test the hypotheses, samples of 361 ATP-funded firms and 398 non-ATP-funded firms were drawn. Z-tests for statistical differences in proportions were used to compare characteristics of ATP-funded firms and non-ATP funded firms with size of firms controlled. Firms will fewer than 500 employees were classified as small, and firms with 500 or more employees were classified as large.

The first hypothesis is:

H1a: Among small firms, a greater percent of ATP firms pursue business strategies that are R&D collaboration oriented, alliances oriented, licensing strategy and product oriented than non-ATP firms.

H1b: Among large firms, a greater percent of ATP firms pursue business strategies that are R&D collaboration oriented, alliances oriented, licensing strategy and product oriented than non-ATP firms.

To prevent confounding "double barrel" effects in testing H1a and b, each strategy was tested independently. To accomplish this objective, H1a and b are broken down into fourteen sub-hypotheses, seven each for the seven strategies utilized by the each of the two firm size categories.

H1a-1: Among small firms, a greater percent of ATP firms pursue R&D collaboration than non-ATP firms.

H1a-2: Among small firms, a greater percent of ATP firms pursue alliances with customers than non-ATP firms.

H1a-3: Among small firms, a greater percent of ATP firms pursue alliances with suppliers than non-ATP firms.

H1a-4: Among small firms, a greater percent of ATP firms pursue alliances with distributors than non-ATP firms.

H1a-5: Among small firms, a greater percent of ATP firms pursue licensing than non-ATP firms.

H1a-6: Among small firms, a greater percent of ATP firms pursue process applications than non-ATP firms.

H1a-7: Among small firms, a greater percent of ATP firms pursue product applications than non-ATP firms.

H1b: Among large firms, a greater percent of ATP firms pursue business strategies that are R&D collaboration oriented, alliances oriented, licensing strategy and process oriented than non-ATP-funded firms.

H1b-1: Among large firms, a greater percent of ATP firms pursue R&D collaboration than non-ATP firms.

H1b-2: Among large firms, a greater percent of ATP firms pursue alliances with customers than non-ATP firms.

H1b-3: Among large firms, a greater percent of ATP firms pursue alliances with suppliers than non-ATP firms.

H1b-4: Among large firms, a greater percent of ATP firms pursue alliances with distributors than non-ATP firms.

H1b-5: Among large firms, a greater percent of ATP firms pursue licensing than non-ATP firms.

H1b-6: Among large firms, a greater percent of ATP firms pursue process applications than non-ATP firms.

H1b-7: Among large firms, a greater percent of ATP firms pursue product applications than non-ATP firms.

H2a: Among small firms, a greater percent of ATP firms pursue commercial advantage types that are new (highly innovative, path breaking) as opposed to moderately innovative performance solutions or low innovation, cost reduction solutions than in non-ATP firms.

H2b: Among large firms, a greater percent of ATP large firms pursue commercial advantage types that are new (highly innovative, path breaking) as opposed to moderately innovative performance solutions or low innovation, cost reduction solutions than in non-ATP firms.

H3a: Among small firms, a greater percent of non-ATP commercialization applications are product oriented compared with ATP firms.

H3b: Among large firms, a greater percent of non-ATP commercialization applications are process oriented compared with ATP firms.

H3c: Among large firms, a greater percent of non-ATP commercialization applications are service oriented compared with ATP firms.

H4a: Among small firms, a greater percent of ATP-funded firms pursue higher cost reduction targets than do non-ATP firms.

H4b: Among large firms, a greater percent of ATP-funded firms pursue higher cost reduction targets than do non-ATP firms.

H5a: Among small firms, a greater percent of ATP-funded firms pursue higher performance improvement targets than non-ATP firms.

H5b: Among large firms, a greater percent of ATP-funded firms pursue higher performance improvement targets than non-ATP firms.

DATA AND METHODOLOGY

The ATP maintains a database of high-technology companies that has competed for awards under the ATP program. In this study the sample period chosen is 1993-96, because this period had the only complete data sets desired. A random sample of 361 ATP-funded firms (230 small and 131 large) was collected from the population of ATP-funded firms in the database. The survey instrument was used to collect data from all firms receiving ATP grants (Powell 1997). The survey elicits key information on the firms' technologies, applications, and business strategies used to market their technology as well as performance goals. For comparison purposes, a random sample of 398 non-ATP-funded firms (208 small, 190 large) for the 1993-96 period was drawn from the population of non-ATP-funded firms in the U.S. Department of Commerce Small Business Administration (SBA) company database. The SBA survey instrument used elicits essentially the same information as the ATP survey. All hypotheses were tested using 1-tail Z-tests for differences in percentages with the alpha probability level (P) set at .05.

Profile of Companies in the Sample

Table 1 provides a breakdown of the characteristics of the firms in the sample, including organization type (number of firms that are single companies and joint venture organizations), sector type (number of firms that are small for profit, large for-profit, small non-profit, etc.), technologies (number of firms that are in the seven basic technologies groups), and applications (number of firms whose venture funding is primarily in product development applications, process development applications, etc.). Regarding organization type,

TABLE 1. Profile of Firms (percentage of firms by ATP involvement)

	ATP Funded	Non-ATP Funded
Organization		
Single Firm	42	52
Joint Venture	58	48
Sector		
Small For-Profit	39	21
Large For-Profit	30	41
Medium For-Profit	23	21
Non-Profit	4	10
University	4	7
Technologies		
Information/Computer	28	22
Manufacturing	21	30
Material	17	21
Biotech	11	8
Electronics	10	7
Chemicals	8	10
Energy/Environment	5	2
Application		
Products	65	69
Processes	26	20
Services	9	11
Total Number of Companies	361	398

a greater proportion of the non-ATP-funded sample is single companies (as compared to joint venture companies) than the ATP-funded sample. The ATP-funded sample also has more small for-profit companies than non-ATP-funded sample. The non-ATP-funded sample has more large for-profit companies than the ATP-funded sample. In terms of technologies, the ATP-funded sample has slightly more information technology companies than the non-ATP-funded sample, including more biotech, electronics, and energy/environment firms. The non-ATP-funded sample had more companies in manufacturing, materials and chemicals. Regarding applications, the ATP-funded sample has more process applications, while the non-ATP sample has more product and service applications.

Limitations

Given the research goal to determine the difference between marketing strategy behaviors of government funded (ATP-funded) and non-government (non-ATP-funded firms) it would have been ideal if non-ATP-funded firms were entirely non-government funded. This condition does not entirely hold. Some non-ATP-funded firms also received government funding in the form of U.S. Commerce Department SBIR program. However, less than 20% of the non-ATP funded firms had received SBIRs. On a note of caution, ATP and SBIR may not be the only influences on firms in the sample. Transactions with other fund providers and potential partners (or actual partners in the case of joint ventures) could induce ATP-funded and non-ATP funded firms either to conform more closely with marketing best practices, or, in some cases, less closely, for example to resist alliances due to a desire to protect intellectual property. Furthermore, differences in the ATP-funded sample and the non-ATP funded sample are substantial and could impact current findings, and the survey instruments while very close in nature and order of questions were not identical.

RESULTS

Commercialization Process

Three of seven H1a sub-hypotheses in the Commercialization Processes (i.e., types of strategies) category were supported. In the sections commercialization processes of particular note are reviewed.

R&D Collaboration

H1a-1 is supported. More ATP small firms pursue R&D collaboration (76%) compared with non-ATP small firms (51%–Table 2, row 2) which supports marketing best practice #3–integration of R&D and marketing via alliances. H1b-1 is supported. More ATP large firms pursue R&D collaboration (82%) compared with non-ATP small firms (68%–Table 2, row 1).

Alliances

H1a-2 is supported. More ATP small firms (29%) pursue alliances with customers compared with non-ATP small firms (22%) (Table 2, row 2) and the differences in percentages were statistically significant. This finding only partially supports marketing best practice #3, because the percentage difference between ATP small firms (26%, 21%, respectively) pursuing supplier and distribution alliances, respectively, compared to non-ATP small firms (24%, 22%), were not statistically significant (Table 2, rows 3 and 4).

Licensing

H1a-5 is supported. More ATP small firms (38%) pursue licensing (Table 2, row 5) compared with non-ATP small firms (25%, see Table 2, row 5) and the differences in percentages were statistically significant. This finding sug-

TABLE 2. Usage of Primary Business Strategies by ATP Involvement and Firm Size (percentage of firms by type and size)

| | ATP | | Non-ATP | |
Strategy	Small	Large	Small	Large
R&D Collaboration	76[a]	82[b]	51[a]	68[b]
Alliances-Customers	29[a]	18	22[a]	15
Alliances-Suppliers	26	9	24	7
Alliances-Distributors	21	15	22	14
Licensing	38[a]	17	25[a]	13
Process	22	34	29	28
Product	62	66	55	64
Total Number of Observations	230	131	208	190

[a] Denotes statistically significant higher percentage of ATP small firms compared with non-ATP small firms (P < .05, 1-tail).
[b] Denotes statistically significant higher percentage of ATP large firms compared with non-ATP large firms (P < .05, 1-tail).

gests ATP small firms use a strategy more consistent with best marketing practices #1–licensing. A slightly greater percentage of large non-ATP firms (17%) use licensing than ATP firms (13%), but the difference is not statistically significant. ATP small firm licensing may compensate for lack of scale economies for profitable manufacturing. Many ATP large firms apparently do not have this scale economy deficiency.

Process

H1a-6 is not supported. A smaller (not larger as hypothesized) percentage of ATP small firms (22%) pursue process compared with non-ATP small firms (29%). Regarding large firms, a slightly greater percentage of ATP firms (34%) pursue process in-house compared to non-ATP firms (28%), but the difference in percentage is not statistically significant. This is counter-intuitive because large firms with generally more extensive production capabilities would be more likely to focus on processes to achieve scale efficiencies. This could be a useful area for further research.

Product

H1a-7, that ATP small firms are more product oriented, is not supported. The percentage of ATP small firms (62%) compared to non-ATP small firms (55%) pursuing product strategies was not statistically significant.

In summary, R&D collaboration and licensing are the relatively common strategies for small ATP funded firms. Product strategy is pursued by large ATP firms.

Commercial Advantages

H2a is supported. ATP funding does foster new, highly innovative solutions for small firms. A greater percentage of ATP small firms (37%) compared to non-ATP small firms (29%) pursues new, highly innovative, solutions (Table 3, row 1). H2b is not supported. A greater percentage of ATP large firms (31%) compared to non-ATP large firms (26%) pursues new solutions (Table 3, row 1), but the differences in percentages were not statistically significant. However, a larger percentage of ATP large firms (41%) compared to non-ATP large firms (28%) pursues performance improvement. There is little difference in percentage usage between ATP large and small compared to non-ATP large and small firms pursuing the other strategies (performance improvement, cost reduction, performance and cost, and other–Table 3, rows 2-5) and none of the differences in percentages was statistically significant.

TABLE 3. Commercial Advantage by ATP Involvement and Firm Size (percentage of firms by type and size)

Type	ATP Small	Large	Non-ATP Small	Large
New Solution	37[a]	31	29[a]	26
Performance Improvement	43	41[b]	47	28[b]
Cost Reduction	38	17	32	15
Performance and Cost	33	38[b]	40	27[b]
Other	6	7	7	8
Total Number of Observations	230	131	208	190

[a] Denotes statistically significant higher percentage of ATP small firms compared with non-ATP small firms ($P < .05$, 1-tail).
[b] Denotes statistically significant higher percentage of ATP large firms compared with non-ATP large firms ($P < .05$, 1-tail).

Types of Commercial Application

H3b is supported. A larger percentage of non-ATP large firms (57%) compared with ATP large firms (34%) pursue process oriented commercial applications (Table 4, row 1) with the differences in percentages statistically significant. H3a was not supported. ATP small firms do not pursue more product applications. About the same percentage of ATP small firms (22%), compared to non-ATP small firms (20%), pursue product oriented commercial applications (Table 4, row 1). H3c was not supported. Nearly the same percentage of ATP small firms (9%) compared to non-ATP small firms (8%) pursue service oriented commercial applications (Table 4, row 3) and the differences in percentages were not statistically significant.

Performance Outcomes

Cost Goals

Cost reduction goals break down into four main levels of performance as shown in Table 5. H4a is supported. More ATP-funded small firms (38%) have cost reduction goals of 50% or higher compared with small non-ATP-funded firms (30%–see Table 5, row 4). H4b is not supported. The same percentage of ATP large firms (21%) purse cost reduction targets above 50% compared with non-ATP large firms (21%–Table 5, row 4).

TABLE 4. Commercialization Applications by ATP Involvement and Firm Size (percentage of firms by type and size)

Type	ATP Small	Large	Non-ATP Small	Large
Product	22	34[a]	20	57[a]
Process	69	48	72	31
Service	9	18	8	12
Total Number of Observations	230	131	208	190

[a] Denotes statistically significant higher percentage of ATP large firms compared with non-ATP large firms (P < .05, 1-tail).

TABLE 5. Cost Reduction Goal Levels by ATP Involvement and Firm Size (percentage of firms by type and size)

Levels of Cost Reduction	ATP Small	Large	Non-ATP Small	Large
0%	28	36	29	34
1-24%	25	28	30	30
25%-49%	9	15	11	15
50%-99%	38[a]	21	30[a]	21
Total Number of Observations	230	131	208	190

[a] Denotes statistically significant higher percentage of ATP small firms compared with non-ATP small firms (P < .05, 1-tail).

Performance Goals

H5a is supported. A greater percentage of small ATP firms (43%–Table 6, col. 1, row 5) pursue high performance goals (100% or higher) compared with non-ATP firms (30%). H5b is not supported. Nearly the same percentage of large ATP firms (33%) pursues high performance goals (Table 6, col. 2, row 5) as large non-ATP-funded firms (34%), and the differences in percentages were not statistically significant.

CONCLUSIONS

Three broad areas of implications can be drawn from the current study: (1) ATP goals achieved, (2) ATP goals not achieved and (3) contributions

TABLE 6. Performance Improvement Goal Levels by ATP Involvement and Firm Size (percentage of firms by type and size)

Levels of Performance Improvement	ATP Small	Large	Non-ATP Small	Large
0%	19	21	26	20
1%-24%	9	23	11	21
25%-49%	12	13	15	15
50%-99%	17	10	18	10
100% or more	43[a]	33	30[a]	34
Total number of observations	230	131	208	190

[a] Denotes statistically significant higher percentage of ATP small firms compared with non-ATP small firms (P < .05, 1-tail).

from government funding of venture firms that are realistic. However, the implications of this study must be viewed with caution given the previously noted differences in samples. While a substantial number of hypotheses have been supported (H1a-1, H1a-2, H1a-5, H2a, H3b, H4a, and H5a) that suggest that ATP-funded firms compared with non-ATP-funded firms will more frequently exhibit marketing best practices and high performance goals, further research is needed utilizing carefully stratified samples to verify and extend the findings of this study. Still, this exploratory study has provided grounds for provocative observations of importance to public policy.

In terms of ATP goals achieved, a major subset of ATP goals is to build marketing best practices into the performance standards of the ATP funded firm. These include #3–integration of R&D and marketing via fostering alliances with customers, and #2–niche market entry strategy using licensing. The twenty-five percent difference among small firms (and fourteen percent difference among large firms) between ATP and non-ATP firms with respect to R&D collaboration shows ATP firm superior consistency with best practice marketing policy. Furthermore, the thirty-two percent difference between ATP small firms licensing and non-ATP small firms shows that ATP funded firms are more likely to get their technology into the market sooner than non-ATP-funded firms. ATP has fostered more pursuit of both highly innovative solutions and high performance goals among small firms, which clearly provide a basis for competitive advantage. The inference is that small firm ATP funding fosters higher goal setting, which might lead to actually achieving higher performance. Raising the odds of achieving higher goals arguably increases the odds of sustainable competitive advantage for the ATP firm consistent with good marketing policy. At best, it may follow that small ATP-funded firms are likely to be more successful. At worst, given small ATP funded-firms tend to pursue strategies consistent with best practices more than small non-ATP funded firms, the inference is that the impact of ATP funding on small firms is more consistent with good marketing policy. Furthermore, among large firms, while ATP has fostered performance solutions, it has not fostered highly innovative solutions.

As a result, the odds that ATP will reach the first of its two major goals of funding more innovative, more rapid, and successful product development should be higher. ATP's second major goal to deliver economic benefits to the nation should be met by the revenue creation and cost savings provided by the products developed by ATP-funded firms. Whether these economic benefits are multibillion dollar in dimension depends on the findings of future studies by ATP and others that should provided reliable estimates of the dollar value of the economic benefits delivered by ATP projects.

In terms of goals unachieved, ATP has not fostered more small firm R&D-marketing integration via supplier and distributor alliances, thereby leaving this effort unbalanced. Despite this more marginal propensity for ATP small firms to more fully utilize alliances (with customer only), demonstrable evidence exists that this policy is consistent with success and provides a foundation for a legitimate argument that ATP small firms are more consistently aligned with best practices. ATP has not fostered product orientation among small ATP-funded firms. The evidence suggests that ATP has not met the goal of process orientation among large firms with respect to implementing the seven commercial practices. Although, among large firms, a greater percent of ATP firms (34%) pursue process applications than non-ATP firms (28%), the differences in percentages were not statistically significant. Despite fostering R&D collaboration and performance improvement among large firms, ATP has not fostered process orientation or higher performance goals among large firms, suggesting that ATP funding is, at best, only marginally advancing marketing strategy best practices among large firms.

In terms of realistic contributions of a government funding program, this relative lack of effectiveness in advancing best practices among large firms may be more due to the conventional wisdom that larger firms are more advanced in the use of best practices so that the impact is likely to be more marginal. However, this argument does not seem to hold regarding why ATP does not seem to have advanced large firm process orientation by which large firms normally obtain market leverage from the efficiencies gained, or small firm product orientation by which small firms normally obtain competitive advantage. Finally, licensing is normally a more effective strategy for small business, rather than large. Therefore, lack of influence by ATP here could be argued to be irrelevant. A more realistic argument for the effectiveness of ATP in reaching its goals may be that by fostering some of the best marketing practices and higher goal setting alone, ATP-funded small companies are given better odds for success and, therefore, ATP is a viable model for an effective government program supporting high-tech venture businesses.

From the viewpoint of the benefit-cost ratio performance standard used to evaluate government-supported projects, an economically viable project must yield a benefit-cost ratio exceeding the value of 1 (Arrow and Kurz 1970). The $1.97 billion investment of ATP in 649 ATP projects to date (cited in the Introduction section) represents the total ATP cost of the projects over their funding period. This is a modest public investment by federal government standards. Any profits generated by the 649 ATP-funded firms count as economic benefits (Sen 1972). To meet the benefit-cost standard of greater than 1, the 649 ATP-funded firms would have to produce total economic benefits (profits) of just above $1.97 billion from products/services funded by ATP over their life cycle. To produce $1.97 billion in benefits (profits), the average ATP-funded firm would only have to generate about $3.04 in profits ($1.97 billion divided by 649) from products/services funded by ATP over their life cycle. It is difficult to believe that the average ATP funded project won't perform at this profit level. Furthermore, it is highly likely that additional economic benefits beyond profits would be generated (in the billions of dollars) by ATP-funded firms in the form of cost savings from the new technologies passed on to the firms' supply chain members. In addition, there could be cost savings spillovers (in the billions of dollars) to companies and industries outside the ATP-funded companies' supply chain that adopted the firm's ATP-funded technologies. Therefore, there is reason to believe that ATP would exceed the minimum benefit-cost ratio standard of 1.0 and possibly deliver a benefit-cost ratio well above the average for federal programs.

AUTHOR NOTE

Conway L. Lackman specializes in B2B marketing and marketing research. He holds a PhD from the University of Cincinnati. His publications have appeared in *Psychology and Marketing*, *Industrial Marketing Management*, *Journal of Marketing Intelligence*, *Journal of Database Marketing*, and *Journal of Marketing Theory and Practice*.

REFERENCES

Ali, A. et al. (1995), "Product Innovativeness and Entry Strategy: Impact on Cycle Time and Break-Even time," *Journal of Product Innovation Management*, 12: 54-69.
Advanced Technology Program website, *http://www.atp.nist.gov/atp/brochure.htm*: 5.

Arrow, K.J. and M. Kurz (1970), *Public Investment, the Rate of Return, and Optimal Fiscal Policy*, Baltimore, MD: The John Hopkins Press: 1-271.

Barringer, B.R., F.F. Jones and P.S. Lewis (1997), "A Qualitative Study of the Management Practices of Rapid Growth Entrepreneurial Firms," *Journal of Business & Entrepreneurship*, 9(2): 21-35.

Bernstein, E.H. (1994), "Small Business Performance: A Test of Contrasting Models," *Journal of Business & Entrepreneurship*, 6(3): 9-18.

Brown, G.E. and J. Turner (1999), "Reworking the Federal Role in Small Business, Research, *Issues in Science and Technology*, 15(4): 51-59.

Chawla, S.K., C. Pullig and F.D. Alexander (1997), "Critical Success Factors from an Organizational Life Cycle Perspective: Perceptions of Small Business Owners from Different Business Environments," *Journal of Business Entrepreneurship*, 9(1): 47-58.

Doctors, S.I. (1969), *The Role of Federal Agencies in Technology Transfer*, Cambridge, MA: MIT Press: 1-159.

Ensley, M.D., J.W. Carland and J.C. Carland (1998), "The Effect of Entrepreneurial Team Skill Heterogeneity and Functional Diversity on New Venture Performance," *Journal of Business & Entrepreneurship*, 10(1): 1-14.

Gersony, N. (1994), "New Technology Ventures and the Strategic Alliance Option," *Journal of Business & Entrepreneurship*, 6(3): 29-36.

Gompers, P. and J. Lerner (1999), *Chasing Money Deals? Impact of Fund Inflows on Private Equity Valuations*. Cambridge, MA: Harvard University and NBER.

Hall, S. and G. Rifkin (1999), *Radical Marketing*, New York: Harper Collins.

Harper, S.C. (1996), "An Analysis of the Prerequisites to Growth in Emerging Firms," *Journal of Business & Entrepreneurship*, 8(2): 1-8.

Hoffman, D.L. and R. Viswanathan (1996), "Strategic Alliances and Small Business," *Journal of Business & Entrepreneurship*, 8(2): 133-146.

Jonash, R. (1996), "Strategic Technology Leveraging: Making Outsourcing Work for You," *Research-Technology Management*, Mar.-Apr., 1996: 19-25.

Kaplan, Jerry (2001), *Startup: A Silicon Valley Adventure*. New York: Penguin.

Kwestel, M., M. Preston and G. Plaster (1998), *The Road to Success: How to Manage Growth*, New York: John Wiley.

Levinson, J.C. (1998), *Guerrilla Marketing*, Boston: Houghton Mifflin.

Litvak, I. (1992), "Winning Strategies for Small Technology-Based Companies," *Ivey Business Quarterly*, 57, 47-51.

Lussier, R.N. (1996), "A Business Success versus Failure Prediction Model for Service Industries," *Journal of Business & Entrepreneurship*, 8(2): 23-38.

Maynard, R. (1996), "Striking the Right Match," *Nation's Business*, 84: 18-28.

Malhortra, N. (1999), *Marketing Research: An Applied Orientation*, Upper Saddle River, New Jersey: Prentice-Hall.

Parbotecah, K.P. (2000), "Choice of Type of Entrepreneur: A Process Model," *Academy of Entrepreneurship Journal*, 6(1): 28-43.

Patsula, P.J. and W. Nowik (2002), *Successful Business Planning in 30 Days: A Step-By-Step Guide for Writing a Business Plan and Starting Your Own Business*, New York: Patsula Media.

Piper, A. and M. Lund (1997), "The Financing of Technology Based Small Firms: An Update," *Bank of England Quarterly Bulletin*, May, 210-213.

Powell, J. (1997), "High-Technology Venture Companies: The ATP Experience," *U.S. Department of Commerce, National Institute for Standards and Technology, Advanced Technology Program*: 1-77.

Ries, A. and J. Trout (1993), *Positioning: The Battle for Your Mind*. New York: Warner Books.

Robinson, R. and L. Herron (2001), "The Impact of Strategy and Industry Structure on the Link Between the Entrepreneur and Venture Performance," *Academy of Entrepreneurship Journal*, 7(2): 31-50.

Sen, A.K. (1972), "Accounting prices and control areas: An approach to project evaluation," *Economic Journal*, 11: 33-42.

Sheahen, T., R.E. Rosenthal, R.A. Hawsey, S.W. Freiman and J.G. Daley (1994), "Evaluation of Technology Transfer by Peer Review," *Journal of Technology Transfer*, 19(3/4): 100-109.

Sink, C.H. and K. Easley (1994), "The Basis for U.S. Department of Energy Technology Transfer in the 1990's," *Journal of Technology Transfer*, 9(2): 52-62.

Spann, M.S., M. Adams and W.E. Souder, (1993), "Improving Federal Technology Commercialization: Some Recommendations from a Field Study," *Journal of Technology Transfer*, 18(3/4): 63-74.

U.S. Small Business Administration (1997), Washington, DC.

Food Product Development, Food Regulations and Policies– Compatible or Not?

Alicea A. Glueck-Chaloupka
Louis M. Capella
Patti C. Coggins

SUMMARY. Laws and regulations govern many aspects of life in order to guide and protect individuals. The Food and Drug Administration regu-

Alicea A. Glueck-Chaloupka, PhD, is affiliated with the Department of Food Science & Technology, Box 9805, Herzer Building, Mississippi State University, Mississippi State, MS 39762 (E-mail: yellowrose519@yahoo.com). Her research interests include new product development, shelf life of dairy products, and innovative methods in marketing food products.

Louis M. Capella, DBA, is affiliated with the Department of Marketing, Quantitative Analysis and Business Law, Box 5288, 111 McCool Hall, Mississippi State University, Mississippi State, MS 39762 (E-mail: lcapella@cobilan.msstate.edu). His research interests include services marketing and international marketing.

Patti C. Coggins, PhD, is affiliated with the Department of Food Science & Technology, James E. Garrison Sensory Evaluation Laboratory, Box 9805, Herzer Building, Mississippi State University, Mississippi State, MS 39762 (E-mail: pcoggins@ foodscience. msstate.edu). His research interests include new product development and perception and sensory evaluation of foods.

[Haworth co-indexing entry note]: "Food Product Development, Food Regulations and Policies–Compatible or Not?" Glueck-Chaloupka, Alicea A., Louis M. Capella, and Patti C. Coggins. Co-published simultaneously in *Journal of Nonprofit & Public Sector Marketing* (Best Business Books, an imprint of The Haworth Press, Inc.) Vol. 13, No. 1/2, 2005, pp. 199-212; and: *Government Policy and Program Impacts on Technology Development, Transfer and Commercialization: International Perspectives* (ed: Kimball P. Marshall, William S. Piper, and Walter W. Wymer, Jr.) Best Business Books, an imprint of The Haworth Press, Inc., 2005, pp. 199-212. Single or multiple copies of this article are available for a fee from The Haworth Document Delivery Service [1-800-HAWORTH, 9:00 a.m. - 5:00 p.m. (EST). E-mail address: getinfo@haworthpressinc.com].

lates the development, production and marketing of foods, pharmaceuticals, and biotechnologies to maintain safe and high quality products for consumers. However, in recent years, food manufacturers seem to have forged ahead of the FDA's regulations and entered uncharted, unregulated areas. This paper examines the impact the FDA has on new food product development, particularly the impact on new technological food products and suggests the use of a team-based approach in evaluating and establishing regulations in the food industry. *[Article copies available for a fee from The Haworth Document Delivery Service: 1-800-HAWORTH. E-mail address: <docdelivery@haworthpress.com> Website: <http://www.HaworthPress. com> © 2005 by The Haworth Press, Inc. All rights reserved.]*

KEYWORDS. FDA, food product development, food regulations, food technology, government regulation

INTRODUCTION

Laws and regulations govern numerous arenas of everyday life and industries. They are devised and implemented to guide and protect consumers. With this being said, corporations find themselves grappling with how to develop new products within these legal and regulatory guidelines. Corporations may ask the question, "Do these laws and regulations hinder or aid the safe development and implementation of new products?"

Food products are one of the most highly regulated industries. Food is defined to include (1) articles used for food or drink by man or other animals, (2) chewing gum, and (3) components of any of the above [21 U.S.C.§ 371(F)]. From packaging to the labeling, ingredients, processing, branding and marketing, all facets of the industry are highly regulated in order to provide the safest, highest quality product available to consumers. Changes in the demographic structure and household composition have affected the public's outlook and interest in their food choices. This combined with increasing market globalization has heightened concern that the food system may be more vulnerable to hazards ranging from misbranding to additives to even product tampering.

The government plays a major role in guarding and protecting the safety of the food supply through a variety of laws, regulations, standards, and agencies. The U.S. Food and Drug Administration (FDA) is part of the federal executive branch that oversees most food products with the exception of meat and poultry. It is the goal of the FDA to enforce adulteration and misbranding prohibitions with powers of enforcement, including labeling requirements, powers of seizure, injunction, and criminal prosecution. Processors, manufacturers, distributors, retailers and food service professionals must adhere to and abide by guidelines set out by the FDA in order to conduct business in the food industry. However, these guidelines can enormously increase the cost of products and may deter companies from branching out and developing innovative new products and technologies.

The following discussion will review the FDA's role in safeguarding consumers against hazards associated with the food supply. Through selected issues, the role the FDA has in the development of new food products will be investigated and speculation will be made regarding alternatives to increase the compatibility between the FDA and new food products and new food technological developments.

HISTORY AND PURPOSE
OF THE FOOD AND DRUG ADMINISTRATION

Established by Congress in 1938, the Food and Drug Administration (FDA) directly impacts the development, production and marketing of foods, pharmaceuticals, animals, and biotechnologies. The FDA is directly responsible for categorizing and regulating substances in the food, drug and cosmetic industries that are "generally recognized as safe" (GRAS) and for ensuring these industries do not produce products that are injurious to consumers' health (Anonymous, 1992). This agency is empowered by Congress to promote regulations for the "efficient enforcement of the Act" through their informal rulemakings [21 U.S.C.§ 371(a) (FDCA 701(a))] and to issue binding, substantive rules with respect to matters deemed of special consequence through the formal rulemaking provisions [21 U.S.C. §371(e) (FDCA 701(e))]. Over the years, they have impacted the types and the degree to which colors and additives may be added to food products, as well as the type of packing available for utilization, and what may be displayed on the products' label. Table 1 presents a historical overview of selected events and policies implemented and regulated by the FDA.

The FDA provides scientific guidance documents to assist applicants with petition preparation and for the safety of the proposed material (Greenberg, 2000). They review petitions and add them to the Code of Federal Regulations if approved, describing the approved usage for manufacturers to utilize and implement. Therefore, the acts and regulations put forth through this agency directly limit or promote the development of food products and technological innovations. The model illustrating the FDA's role in new food product devel-

TABLE 1. Selected Historical Events and Policies of the Food and Drug Administration (FDA)

Year Implemented	FDA Event and/or Policy
1785	First general food law in United States enacted in Massachusetts
1902	Appropriations made to establish food standards and study effects of chemicals on digestion and health
1906	Original Food and Drugs Act passed
1927	Food, Drug and Insecticide Administration formed
1938	Federal Food, Drug and Cosmetic Act
1958	Food Additives Amendment
1969	FDA's role expanded to include milk, food service, shellfish, and interstate travel sanitation and poisoning and accident prevention
1973	FDA issues final regulation on nutrition labeling
1977	Saccharin Study and Labeling Act–stopped FDA from banning saccharin, but stated must add warning on label with foods containing stating that it has been found to cause cancer in laboratory animals
1980	Infant Formula Act
1982	Tamper resistant packaging regulations
1984	FDA issues ruling making sodium content a mandatory part of nutritional labeling and creates definitions for sodium claims on labels
1986	FDA issues final ban on use of sulfites for raw fruits and vegetables
1990	FDA approves use of irradiation on fresh and frozen uncooked poultry to control Salmonella and other bacteria
	Congress enacts Nutrition Labeling and Education Act
1994	Dietary Supplement and Health and Education Act
1996	Food Quality Protection Act
1997	FDA Modernization Act of 1997–new alternative food additive approval procedure for indirect food additives in which marketers submit notice of their intended new use of a food contact substance.
2002	Public Health Security and Bioterrorism Preparedness and Response Act

opment is shown in Figure 1 and was based on the NPD models put forth in Greenberg (2000) and Urban and Hauser (1993). The FDA directly influences the NPD process during the initial product definition and then again in the refinement and final launch of the product or technology. In both these stages the regulations outlined and proposed by the FDA influence the ingredients, technological processes, packaging, and labeling involved in the development

FIGURE 1. The FDA's Role in the New Food Product Development Process

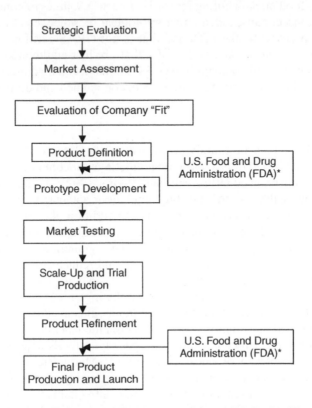

* At these stages in the new food product development process, the U.S. Food and Drug Administration regulates the ingredients utilized, technology process used to produce the new food product, labeling of the product, and the new food product's safety.

Source: Adapted from Greenberg (2000) and Urban and Hauser (1993).

of new food products. Adulteration and misbranding are two general regulatory categories governed by the FDA that have a direct impact on the development process at the product definition stage and refinement stage.

Adulteration

Adulteration is generally summarized as something being awry with a product. If the food is (1) filthy, putrid or decomposed, (2) produced in unsanitary conditions, or (3) containing any substances deleterious to health, it is referred to as adulterated (Ronsivalli and Vieira, 1992). Numerous tests are in place within food manufacturing industries to detect sources of contamination such as dirt, insect parts, rodent, microbiological presence, etc. . . . Any substance not approved by the FDA and that is not GRAS certified and that is found within a food product will render that product adulterated and affect both the product's and company's success. Foods deemed adulterated by the FDA through facility inspections are seized, condemned and destroyed.

Misbranding

If a food (1) fails to comply with written standards of identity, (2) is wrongly labeled, or (3) fails to meet the regulations for container fill, it is considered to be misbranded (Ronsivalli and Vieira, 1992). A misbranded food fails to comply with the FDA's labeling regulations and may be mislabeled according to weight, portions, or ingredients. In addition, the omission or misstatement of ingredients, product attributes, geographic origins, or false claims may render a product misbranded. These guidelines were developed and instituted to prevent fraud and deception and reduce confusion involved with new ingredients and additives used to develop new products.

In 1990, Congress passed the Nutrition Labeling and Nutrition Education Act. This act is a powerful tool, which accurately aids in educating the public through labels and information contained within the label of food products. It makes labeling mandatory for almost all packaged foods and voluntary for raw fruit, vegetables, fish, meat and poultry (Mermelstein, 1992). Even though this act allows consumers to view the origin of growth and processing, the serving sizes in household measures, and health claims that are regulated, it is costly to farmers and manufacturers. In fact, it is estimated that this labeling system, while beneficial to the public, will cost the farmers and food industry approximately $2 billion yearly (Bachrach, 2000; Gersema, 2002).

THE FDA'S IMPACT ON EMERGING FOOD TECHNOLOGIES

New Technological Products

The FDA's broad authority allows it to regulate the introduction of new foods and food additives. This agency holds producers and manufacturers legally responsible for the safety and wholesomeness of the foods they bring to market (Wilkinson, 1997). This, combined with the fact that there are thousands of the FDA regulations, impacts the willingness of manufacturers to branch outside their "safety-net" and develop new ideas, technologies and products. In addition, the food product development arena is now focusing on developing and providing multi-functional food products for consumers. These products not only provide the body with essential nutrients, they also provide the consumer with "healing benefits" and an enjoyable eating experience (Taylor and Feld, 1999). Currently, irradiated products, nutraceuticals, and biotechnological foods are three areas distinctly affected by the FDA's oversight and that directly affect consumers of food products.

Irradiated Products

Foods subjected to irradiation have been marketed and sold in countries other than the United States for a number of years. Irradiated foods had not been marketed and sold in the United States because of the Food Additives Amendment (1958) to the Food, Drug, and Cosmetic Act of 1938 which stated that a food is adulterated if it has been intentionally irradiated, unless that irradiation is carried out in conditions prescribing it safe under regulated conditions. As a result, the only major methods for preserving food products have been canning and drying. In the mid-1980s the FDA began taking steps to amend the Food Additives Amendment and include food irradiation as an approved preservation method for all food products. Many individuals felt this would never occur due to the amount of legal and political action required, as well as the need for consumer acceptance. However, after many years and numerous petitions meat food and poultry products may now be treated with sources of ionizing radiation to lower foodborne pathogens and to extend shelf life of the produce (9 CFR Part 424, 2000). Meat plants began successfully implementing irradiation into their processing methods in the spring of 2000.

Even though the approval of the FDA to irradiate and market food products in the United States was achieved, many associated issues were and still are left unresolved today. The largest issue affecting the implementation of this product development is the labeling issues. Originally, the FDA stated that all irradiated foods must bear the radura symbol, a flower encompassed by a broken circle, as stated in the Code of Federal Regulations (2000). Either "treated with radiation" or "treated by irradiation" in addition to the radura symbol on food products in which the processing treatment is not obvious are needed. However, the regulations did not explain the prominence of the disclosure statement. Therefore, in August of 1998, the FDA cleared up the confusion regarding this issue. They issued a final rule amending the previous labeling requirements stating that a radiation disclosure statement has the same requirements as the "declaration of ingredients" (Code of Federal Regulations, 2000; Littlefield and Hadas, 2000). This means that the disclosure statement and the ingredient statements need to be equally prominent on the label and that one does not need to override the other.

Clearing up the labeling confusion among manufacturers, combined with educational materials, aids the development and implementation of irradiation as a technological food development. A recent survey reported that 51 percent of consumers and 53 percent of restaurant owners claimed their attitudes regarding irradiation are positive (Salvage, 2002a). This is evident in the increasing sales of irradiated meat products to grocery stores, and even more so, sales to foodservice and retail outlets (Salvage, 2002b). Even though a larger percent of today's consumers view irradiated foods in a positive manner, restaurant owners do not want to deal with labeling irradiated food products on their menus (Salvage, 2002a). If the FDA is forcing manufacturers to market their irradiated products in grocery stores with disclosure statements and symbols, why should it be any different for the restaurants if the goal of the FDA is to protect consumers from hazards and keep them abreast of what they are consuming? Therefore, the FDA probably should address the issue of selling and consuming irradiated meat products in foodservice establishments in the near future. Even though great strides have been made to implement this new food product technology, there are still hurdles, which the FDA and manufacturers alike must continue to address.

Nutraceuticals

Currently, numerous food products carry a claim that they are fortified with additional vitamins, minerals or herbs, thus creating havoc for regulatory agencies. One area particularly interested in this labeling issue is the nutraceuticals or functional foods area. There is not a specific category to account for these food products, which are neither dietary supplements nor food products. As a result, many of these products are often marketed as dietary supplements to remove the manufacturer's burden of proof that an additive is safe or a claim misleading. Furthermore, this marketing tactic heightens the legal responsibility of the FDA to educate the food consumer. The director of the Center for Science in the Public Interest charged the FDA with failing to protect the public's safety from questionable ingredients and misleading claims (Allen, 1999). In line with the FDA's regulations and goals, endangering public safety through the use of unapproved ingredients or marketing tactics is against the law and detrimental to consumer safety, and therefore the FDA does everything in their power on a daily basis to protect the safety of consumer's. This is evident in a notice posted in the October 2000 *Federal Register* and with the *Pearson* decision which stated that the FDA could prohibit a health claim if the support of the claim is outweighed by evidence against the claim both qualitatively and quantitatively (U.S. Food and Drug Administration, 2002).

There are several major areas in dispute when discussing society's versus the FDA's stance on nutraceuticals. In today's informational society where many products are bought and sold on the Internet, what regulations, if any, can be placed on those advertising the effects of functional food products? Who regulates content relative to the food products on the Internet, the FDA or the Federal Trade Commission (FTC)? These two questions are pertinent, current issues in the nutraceuticals and functional foods area. In fact, they are having a large impact on the regulations the FDA may or may not design and enforce in the functional food product area.

The FDA and the FTC work together to regulate false or misleading information regarding to labeling of food products (Taylor and Feld, 1999). Food companies are bound by the FDA's regulation on proper labeling and advertising of food or drug products. However, the FDA does not stipulate that manufactures include any conflicting information, only that what they report regarding functional foods is "truthful" and "correct." For example, labeling a product as "Low-Fat" is perfectly legal, yet manufacturers are not required to state "High in Sodium." Obviously there are some conflicting health issues. Kellogg's Ensemble carrot cake is one functional food product that fell into this "loophole." The product boasted heart-healthy benefits due to the addition of psyllium; however, the fat content disqualified the product from bearing the heart-disease prevention claim (Allen, 1999). The question, therefore, that remains is should the FDA require the reporting of all health implications for functional or nutraceuticals food products and if so, how? This issue remains to be solved to the satisfaction of the federal government, the manufacturers and the consuming public. However, on December 18, 2002, the FDA issued a statement trying to provide further guidance on qualified health claims and labeling of food products for manufacturers, distributors and consumer alike. It was stated that the FDA would utilize the "reasonable consumer" standard when evaluating and responding to future health claim petitions in order to assist in formulating truthful and non-misleading messages regarding the health benefits obtainable from food products. This reflects that the FDA believes consumers are active partners in their own health care and can enhance their own health when provided with accurate health information (U.S. Food and Drug Administration, 2002).

Biotechnological Food Products

The federal government has been involved in the field of biotechnology over the last two and a half decades (Wilkinson, 1997). Biotechnology is often referred to as a diverse field that includes research areas involving recombinant DNA/RNA techniques, cellular methods, and supercellular methods dealing with embryo transfer. In fact, it made headlines recently when President Bush commented on the legislation regarding stem cell research and its future in the medical community. Even though all of these applications are pertinent to governmental regulation, we will only consider biotechnology applications in regards to the use of plants and animals as food sources. These include, but are not limited to, foods such as the Flavr Savr™ Tomato, Laurical® Canola Oil, Freedom II™ Squash, Round-Up Ready® Soybeans, and Maximizer™ and NatureGuard™ Corn.

For every "new technology offers opportunity" issue, there are numerous others that report on the detrimental side effects associated with the new technology (Pape, 1989). In regards to utilizing biotechnology with food products, this new development can offer boundless opportunities from producing/modifying enzymes for food production to producing amino acids, and food and color additives, to eliminating undesirable characteristics of foods or even enhancing the desired characteristics in food products. Even though Andrew Pollack reports that this technological advancement is responsible for producing between twenty-five and forty-five percent of the United States major crops, it still remains a controversial issue in the food industry (Winn, 1999). In October of 1999, the FDA searched to find a resolution to this problem between food producers and consumers. They initiated and promoted consumer education regarding the proper usage and benefits of biotechnology by holding public meetings on current policies regarding bioengineered foods (Littlefield and Hadas, 2000; Thompson, 2000).

The FDA has the broad sweeping authority to regulate the introduction of foods derived from new plant varieties produced through genetic engineering. The FDA's statement of policy indicates foods produced through this technological development should be regulated the same as those by conventional means, unless special circumstances apply (Wilkinson, 1997). In November of 1999 Representative Dennis Kucinich introduced the "Genetically Engineered Right to Know Act." This act proposed that foods containing or produced with genetically engineered materials be labeled as such (Littelfield and Hadas, 2000). However, even though this proposed Act was a noble gesture, there is still no definitive regulation set forth by the FDA to date. This is due to the fact that no information has been found to indicate and distinguish genetically engineered foods from those altered through plant breeding methods (Wilkinson, 1997; Thompson, 2000; Anonymous, 2002). As a result, at this time the FDA is encouraging "voluntary labeling" of genetically engineered foods and offered guidelines to aid in this matter in the January 18, 2001, publication of the *Federal Register* (Degnan, 2000; Anonymous, 2002).

Controversy surrounding the application and marketing of biotechnology with food products has prevented the realization of the impact biotechnology can have on new food product development. Producers and manufacturers need to utilize genetic engineering in order to survive in competitive markets, as well as meet the ever changing and growing demands of consumers for higher quality, more nutritious food products. It is definitely a technological development that shows extreme promise for all parties involved provided education of the consumers is included in the framework. Therefore, the FDA needs to take a stance to begin to reinstate consumers' faith and trust not only in the food manufacturer, but in the regulatory agency as well, in order for genetic engineering to reach its most beneficial and useful incorporation in the food industry.

RECOMMENDATIONS TO INCREASE THE COMPATIBILITY BETWEEN REGULATIONS AND NEW FOOD PRODUCT DEVELOPMENT

In 1997, Congress passed the Modernization Act allowing marketers to submit notice of their intended new use of a food contact substance. A food contact substance is defined as "any substance intended for use as a component of materials used in manufacturing, packing, packaging, transporting, or holding food if such use is not intended to have a technical effect in such food" (United States, 1998). Once the FDA receives the submission notice, they have 120 days to question the legitimate use and safety of the product. If no contact or questions are made within this time frame, the food contact substance is automatically approved (Greenberg, 2000). This will aid in adding predictability to the timing of the petition review process of the FDA. It will also aid in speeding up the new product development process for innovative products implementing new uses of food contact substances.

Even though the Modernization Act aids in expediting the food NPD process to an extent, it will still take over a year to finalize any proposed rulemaking on a health or nutrient content claim. Therefore, other alternatives and solutions should be considered. One possible alternative to reduce the time consuming food NPD process is to utilize a team-based approach, similar to ones used in many product development areas. Currently, there are multiple agencies, typically, which must approve new food technologies and products thus taking time and increasing the NPD process. Having representatives, from each agency meet on a quarterly basis to evaluate new foods or processes will reduce redundant evaluations and may promote communication and consensus between agencies and regulations. Adopting a team-based approach in evaluating each product and technology will not only aid in decreasing the time it takes to approve and ensure the process is appropriate for consumers and their safety, but it may also streamline the regulations and policies regarding various foods and technological processes.

CONCLUSIONS AND IMPLICATIONS

Single commodity products or foods are not the norm anymore. Instead, there is a melding of products and technologies to form functional-based food products for consumers. This melding of products and technologies forms a combination of foods and drugs in order to feed the body, and provides basic nutrients and benefits at the same time. The problem is, however, that product development within the food industry is ahead of the FDA's forums and regulations. As a result, many developments and innovations are at a standstill, remaining in laboratories awaiting approval from the FDA. Furthermore, the issue of consumer acceptance based on the current and proposed regulations set forth by the FDA still remains to be seen. Therefore, the question to be examined from here forth is not "do these laws and regulations hinder or aid the safe development and implementation of new products?" Instead we should be asking whether or not the FDA could keep-up with or forge ahead of the food development process in order to allow food manufacturers to supply products needed and demanded by consumers. One way the FDA may be able to forge ahead of the new food development process is to promote a team-based approach among the numerous agencies utilized in food safety research and other areas. This will allow for a melding of ideas and opinions, reduce the number of repetitive evaluations conducted on the food or technological process, and promote creative thinking among top scientists and leaders.

This question will only be determined through time. However, food manufactures can take an active role in promoting this concept by keeping abreast of current regulations and where their products fall within them and constantly questioning and challenging the FDA to think ahead to the future.

REFERENCES

Allen, Andrea, (1999), "Do the Laws Function for Functional Foods?" *Food Processing* (June): 68.
Anonymous, (2002), "Biotechnology and Foods: Letter to the Editor," *FDA Consumer* (July-August).

Anonymous, (1992), "Government Regulation of Food Safety: Interaction of Scientific and Societal Forces," *Food Technology* (January): 73-80.

Bachrach, Eve E., (2000), "The Case for a Substantial Evidence Amendment to the Informal Rulemaking Provision of the Federal Food, Drug, and Cosmetic Act," *Food and Drug Law Journal* 55: 293-299.

Degnan, Fred H., (2000), "Biotechnology and the Food Label: A Legal Perspective," *Food and Drug Law Journal* 55: 301-310.

Code of Federal Regulations, (2000), 21 CFR Part 179 (Irradiation in the Production, Processing, and Handling of Food), 3(170-199), April 1, 2000.

Code of Federal Regulations (2000), Subsection C to the 9 CFR Part 424 (Preparation and Processing Operations), Vol. 2, Parts 200-end, January 1, 2000.

FDA. *Federal Food, Drug and Cosmetic Act.* 21 U.S.C. §301 et seq.

FDA. *Federal Food, Drug, and Cosmetic Act.* 21 U.S.C. §371(a) (FDCA 701(a)).

FDA. *Federal Food, Drug, and Cosmetic Act.* 21 U.S.C. §371(e) (FDCA 701(e)).

Gersema, Emily, (2002), "Food Labeling," FSNET by Doug Powell: November 26, 2002. *www.foodsafetynetwork.ca.*

Greenberg, Eric L., (2000), "Public Policy Issues," *Developing New Food Products For a Changing Marketplace*, editors: Aaron L. Brody and John B. Lord, Technomic Publishing Co., Inc.: Lancaster, Pennsylvania: 409-438, 465-475.

Littlefield, Nick and Nicole R. Hadas, (2000), "A Survey of Developments in Food and Drug Law from July 1998 to November 1999," *Food and Drug Law Journal* 55: 35-56.

Mermelstein, Neil H., (1992), "A Guide to the New Nutrition Labeling Proposals," *Food Technology* (January): 56-60.

Pape, Stuart M., (1989), "Regulation of New Technologies: Is Biotechnology Unique?" *Food Drug Cosmetic Law Journal* 44: 173-179.

Ronsivalli, Louis J. and Ernest R. Vieira, (1992), "Regulatory Agencies," *Elementary Food Science*, 3rd edition, Van Nostrand Reinhold: New York, NY: 50-57.

Rudolph, Marvin J., (2000), "The Food Product Development Process," *Developing New Food Products For a Changing Marketplace*, editors: Aaron L. Brody and John B. Lord, Technomic Publishing Co., Inc.: Lancaster, Pennsylvania: 90.

Salvage, Bryan, (2002a), "KSU Survey Shows Consumers Warming Up–But Still Cautious About–Meat Irradiation," March 1, 2002, *www.meatingplace.com.*

Salvage, Bryan, (2002b), "Food Irradiation Process Is Getting Favorable Press: Survey," March 6, 2002, *www.meatingplace.com.*

Taylor, Sarah E. and Harold J. Feld, (1999), "Promoting Functional Foods and Nutraceuticals on the Internet," *Food and Drug Law Journal* 54: 423-451.

Thompson, Larry, (2000), "Are Bioengineered Foods Safe?," *FDA Consumer* (January-February 2000).

United States, (1998), *Federal Food, Drug, and Cosmetic Act*, FDCA 409(h)(6).

Urban, Glen L. and John R. Hauser, (1993), *Design and Marketing of New Products*, 2nd edition, Prentice Hall: Englewood Cliffs, New Jersey.

U.S. Food and Drug Administration Center for Food Safety and Applied Nutrition Office of Nutritional Products, Labeling and Dietary Supplements, (2002), "Guidance for Industry: Qualified Health Claims in the Labeling of Conventional Foods and Dietary Supplements," December 18, 2002, *www.cfsan.fda.gov.*

Wilkinson, Jack Q., (1997), " Biotech Plants: From Lab Bench to Supermarket Shelf," *Food Technology* (December): 37-42.

Winn, Lara Beth, (1999), "Special Labeling Requirements for Genetically Engineered Food: How Sound Are the Analytical Frameworks Used by FDA and Food Producers," *Food and Drug Law Journal* 54: 667-688.

New Technology Adoption, Business Strategy and Government Involvement: The Case of Mobile Commerce

Michael Stoica
Darryl W. Miller
David Stotlar

SUMMARY. This research focuses on understanding how business organizations are likely to adopt mobile commerce (m-commerce) technology. Mobile commerce adoption represents a complex process that draws in variables external to the firm such as the environment in which the business operates and the government involvement, as well as variables internal to the company such as its business strategy and its organizational culture. A model is formulated and several research propositions are of-

Michael Stoica, PhD, is Professor of Management, School of Business, Washburn University, 1700 SW College Avenue, Topeka, KS 66621 (E-mail: michael.stoica@washburn.edu).

Darryl W. Miller, PhD, is Associate Professor of Marketing, University of Wisconsin-River Falls, River Falls, WI 54022 (E-mail: darryl.w.miller@uwrf.edu).

David Stotlar, MBA, is Owner and CEO, Creative Business Solutions, Inc., 6136 SW 38th Street, Topeka, KS 66610 (E-mail: dstotlar@cjnetworks.com).

[Haworth co-indexing entry note]: "New Technology Adoption, Business Strategy and Government Involvement: The Case of Mobile Commerce." Stoica, Michael, Darryl W. Miller, and David Stotlar. Co-published simultaneously in *Journal of Nonprofit & Public Sector Marketing* (Best Business Books, an imprint of The Haworth Press, Inc.) Vol. 13, No. 1/2, 2005, pp. 213-232; and: *Government Policy and Program Impacts on Technology Development, Transfer and Commercialization: International Perspectives* (ed: Kimball P. Marshall, William S. Piper, and Walter W. Wymer, Jr.) Best Business Books, an imprint of The Haworth Press, Inc., 2005, pp. 213-232. Single or multiple copies of this article are available for a fee from The Haworth Document Delivery Service [1-800-HAWORTH, 9:00 a.m. - 5:00 p.m. (EST). E-mail address: getinfo@haworthpressinc.com].

Available online at http://www.haworthpress.com/web/JNPSM
Digital Object Identifier: 10.1300/J054v13n01_12

fered. They will help understand the mobile commerce applications adoption process. Implications of this model and further research avenues are discussed. *[Article copies available for a fee from The Haworth Document Delivery Service: 1-800-HAWORTH. E-mail address: <docdelivery@ haworthpress.com> Website: <http://www.HaworthPress.com> © 2005 by The Haworth Press, Inc. All rights reserved.]*

KEYWORDS. New technology adoption, mobile commerce, government involvement, business strategy

INTRODUCTION

The mobile Internet is in an evolutionary phase similar to the one the traditional World Wide Web entered almost ten years ago. Currently, limited number of online services are available but many, especially overseas, predict the mobile Internet will very quickly develop into a significant new medium for conducting business (Narduzzi, 2001; Angelos, Shaw, Singh, and Springer, 2001). For example, in Japan DoCoMo had signed up 22 million of its 35 million wireless customers for its I-Mode wireless data services with 40,000 more joining each day. According to the market research firm Strategy Analytics, the global market for m-commerce is expected to reach $200 billion by 2004. As indicated by Mobilocity.com companies are beginning to consider m-commerce as a high priority because they are unwilling to make the same mistakes that led them to underestimate (or wrongly estimate) the potential of the traditional Web (Muller-Veerse, 2000). Pressed by this important technological change, companies are in the process of initiating, developing, and implementing an effective m-commerce strategy (Mobilocity, 2000). The move is critical for the future of the firm's performance and growth. The early evolution of m-commerce has led to important research questions related to its progress, growth and development (van der Poel, 2001).

Researchers have devoted considerable efforts to examine the impact of micro behavioral (i.e., entrepreneurial personality and personal profile), as well as contextual variables (i.e., social network, infrastructure, industry and government support, environmental conditions) on the growth and financial performance of businesses. Clearly, adoption of new technology is critical for the organization's growth and performance (Welsch, Liao, and Stoica, 2001). However, the literature provides no clear or consistent conclusion on the firm's strategy regarding the adoption of new technology and performance. Fraught with many contingencies, this field of research appears to be more divergent than convergent.

Information and communication technology growth is raising productivity, creating jobs, and increasing incomes around the globe. Overall trends in the business climate along with actions taken or not taken by national governments will influence the likelihood of success for companies pursuing m-commerce. Some suggest that countries should make technology a driver for a new national economy (McConnell, 2000). Indeed, governments and business leaders have the responsibility to prepare the national economies for the new technology (McConnell, 2000; Morace and Chrometzka, 2001; Sadeh, 2002, Schneiderman, 1999). Thus, legislative bodies along with regulatory agencies, such as the United States Federal Communications Commission, and the Wireless Communication Bureau, play critical roles in the development of wireless services. The Telecom Act (1996) was supposed to open doors to competition both in the local exchange and in the long distance markets but the results are mixed (Schneiderman 1999). The law provides competitive advantage to operators of wireless systems creating markets for terrestrial microwave systems such as intelligent transportation and high-speed wireless services.

We believe the patterns of technology adoption, such as m-commerce, and its impact on the business strategy in an environment which is highly influenced by government participation has yet to be explored in business research. Investigation in this direction will make a significant contribution in understanding the complex relationships involving performance differences across firms. The aim of this paper is to contribute to the understanding of the new technology adoption process, in particular m-commerce, by developing a model that incorporates the critical contextual as well as internal variables of the firm and explains the relations between business strategy and performance.

CONCEPTUAL FOUNDATIONS

Entrepreneurs and the Adoption of High Technology

Similar to the Internet several years ago, wireless technology represents an important, new innovation that is generating a great deal of enthusiasm (Clarke, 2001). Wireless technology has recently received much attention from entrepreneurs, executives, investors, and business experts. It opens new avenues for businesses (Leung and Antypas, 2001; Seager, 2003). Indeed, adoption of new technology is a form of entrepreneurial endeavor. Entrepreneurs will be the first to explore the business opportunities offered by this quickly developing technology to do things differently and better, as has been the case with many other technological advances (Angelos, Shaw, Singh, and Springer, 2001; Bhide, 2000, Bolton and Thompson, 2000). Some also suggest that firms must adopt new technology in order to acquire the flexibility needed to meet the time and distance challenges created by economic globalization (Gagnon, Sicotte, and Posada, 2000).

Several issues of interest for both entrepreneurs and managers when dealing with the adoption of new technology such as wireless include:

- *Opportunity recognition.* How can opportunities be identified in a changing environment and how can the entrepreneur take advantage of the emerging opportunity (Angelos, Shaw, Singh, and Springer, 2001; Hills, Lumpkin, and Singh, 1997)?
- *Timing.* Is the technology mature enough to offer a significant return on investment for a company in a given industry? Are there applications that can be successfully carried out with the technology in its current stage of development in order to improve a company's business model (Cadeaux, 1997; Glazer and Weiss, 1993)?
- *Trends.* What are the recent developments in the industries related to the new technology and how will they impact the business world? Is the technology fragmented (Harte, Kellog, Dreher, Schaffnit, Campbell, and Gosselin, 1999; Morace and Chrometzka, 2001)?
- *Killer applications.* Are there applications that will capture the market and provide significant revenue in the foreseeable future (Smith, 2000; Wilhelmsson, 2001)?

Entrepreneurs and managers in established firms play important roles in the adoption of new technology. Research indicates that several characteristics influence the success of the adoption process of new technology, such as Internet business and m-commerce, by their respective organizations. Roberts (1991) identifies several variables that could predict the entrepreneur's behavior and explain his/her propensity towards new technology. He also discovers several factors that help technologically-oriented entrepreneurial businesses to succeed. Among these are marketing orientation and team work. Technological-based service firms outperform companies that are primarily selling their personal technical capabilities.

On the other hand entrepreneurs and entrepreneurial companies involved in technology share several transition and growth characteristics that inhibit the successful marketing of their products/services. This includes a tendency to focus on the technology to the exclusion of any demand side or customer focus. Many dot-com entrepreneurs became trapped in this mode of thinking and considered that, being technological wizards, their adeptness with the technology would provide them with a sustainable competitive advantage (Porter,

2001). Indeed, the dot-com crash clearly highlighted the danger of this one-sided approach.

The wireless Internet and its commercial usage, known as mobile commerce, represent opportunities as well as challenges for entrepreneurial companies. Will the findings presented above hold for entrepreneurial companies in a mobile commerce environment? What type of strategy is more likely to succeed? What will be the likely impact of government programs and policies? To investigate these questions further some basic definitions must be outlined.

M-Commerce Defined

So far several alternative definitions are commonly used in the literature (Stoica, 2001):

> *The use of mobile handheld devices to communicate, inform, transact and entertain using text and data via connection to public and private networks;*

> *Any transaction with monetary value that is conducted via a mobile telecommunications network;*

> *The use of wireless technologies to provide convenient, personalized and location-based services to customers, employees and partners.*

One can see that the central elements of m-commerce include offering information, services, providing transactions via devices that allow the user full mobility. It is critical to realize that m-commerce is not merely an extension of e-commerce. Rather, it represents a different business philosophy which necessitates different business models. Operating in m-commerce markets with their speed and location requirements will be very demanding for companies. Characteristics and market drivers are detailed in Table 1.

A significant number of business opportunities will quickly emerge and it is up to a new generation of entrepreneurs to take advantage and fully exploit these opportunities. The question is not if the new technology and whether its mobile commerce applications should be adopted. The question is when and how to adopt m-commerce. What applications are already profitable so that small and medium-sized businesses can successfully implement them? Further, what capabilities are about to emerge from the technology innovation pipeline?

Mobile commerce applications are evolving with a wide variety of procedures and devices in use. Applications have been developed for existing mobile

TABLE 1. Characteristics of the Wireless Technology

Ubiquity	Today's mobile devices fulfill the need for real-time information and communication independent of the user's location (anywhere).
Localization	Technologies like GPS (Global Positioning System), TOA (Time Of Arrival), will enable marketers and consumers to send, receive, and access information, services, and conduct transactions specific to their location. Knowing where the user is physically located at a particular moment will be of particular importance to offering relevant services.
Convenience	Users are not constrained by time and place. Devices are always at hand and are getting easier and easier to use.
Accessibility	Consumers and businesses alike are easy to access and timing and organizational responsiveness will become critical. Responsiveness relates to speed and coordination at the company level; an entrepreneurial organizational culture will become an important differential advantage for companies in the future.
Personalization	The availability of personal information (fed through the mobile phone) will move customization to a higher level.
Capacity (bandwidth)	Capacity represents a main issue today. Technology is changing rapidly and bandwidth will be accessible.
Size and form factors	The size and physical form of devices invokes a different experience from desktop PCs. Limitations due to size and portability are obvious.
Security	The use of the smart card and SSL (Secure Socket Layer) will offer a higher level of security than that available in the fixed Internet environment.

platforms including laptop PCs, PDAs, telephone handsets and specialty pagers (Anckar and D'Incau, 2002). Mobile applications have focused on: (a) delivering existing Internet services to the mobile customer; (b) using location sensing to deliver location based information, (c) using location sensing for tracking and logistics (fleet services, automobiles, pets, etc.), (d) using broadband to deliver mobile entertainment content such as music and games.

Application coverage is not universal and is constrained by the size, location and technology used by the mobile infrastructure provider (Council of Economic Advisers, 2000). A partial list of m-commerce applications available within the United States is presented in Figure 1. Low-End Applications (LEA) include email, web browsing and information services. High-End Applications (HEA) include transactions, inventory management, supplier-buyer relationship, interactivity, etc.

Businesses can choose between adopting LEA and HEA. In time those who adopted LEA will move to the more sophisticated and more profitable HEA. A firm's culture, strategy, management structure and its business environment will determine if and what wireless applications should be adopted. In order to understand the adoption process we must design a model that enables to identify and understand the relationships that govern m-commerce applications adoption.

MODEL DEVELOPMENT

By understanding the determinants of business performance, managers and entrepreneurs can better formulate their business models and comprehend how a technology, such as the Internet, impacts their firms as well as how companies can exploit new technology (Fisher, Reuber, and Carter, 1999). The literature contains several studies designed to determine factors influencing business performance. Afuah and Tucci (2001) for example, have developed a model that explains the relationships among the factors that determine performance. The model includes three major determinants of business performance: (1) company's business model, (2) the internal conditions in which the company operates, and (3) change. Change is represented by the new technology, in this case the Internet. The model does not consider an important variable, namely the broader economic environment. The climate and the circumstances the businesses work in prove to be critical for technology adoption by any business (Williams 1992).

Alternatively, Gagnon, Sicotte and Posada (2000) analyze technology adoption by using the Stevenson Model (Chrisman, Bauerschmidt, and Hofer

FIGURE 1. M-Commerce Applications: Value Creation versus Complexity

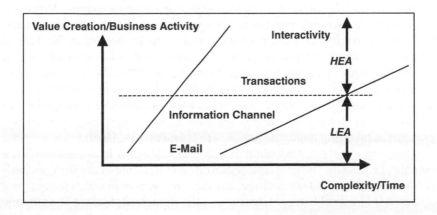

1998). With this model, adoption of a new technology (change) is investigated along dimensions that include: strategic orientation, opportunity seizing, management structure, and control of resources. For Chaston (1997) the interaction between entrepreneurial style (and thus firm's culture) and organizational structure is the main determinant of performance for small firms. Desphande, Farley, and Webster (1993) with their investigation of Japanese firms found a strong relationship between organizational culture and innovativeness. Sadeh (2002) presents several models related to the way firms could adopt m-commerce: the user fee model, shopping model, revenue-sharing model, etc. Srinivasan, Lilien, and Rangaswamy (2002) discuss the adoption process from the resource-based view of the firm. They found that differences in adoption of new technologies among firms can be attributes to a sense-and-respond capability termed technological opportunism. This is dependent upon the structure and culture of the company. Finally, Keen and Mackintosh (2001) discuss m-commerce adoption from the knowledge capital and logistic point of view. All the above models have a common shortcoming; they do not take into consideration the environment, in particular the framework set by legislatures and regulatory agencies.

Addressing this shortcoming, McConnell (2000) has produced a model that accounts for the role of government. The study suggests that evaluating the capacity and readiness of a nation to participate to the global economy is essentially an art not a science because data go rapidly out of date due to rapid changes in technology and markets. The report defines economic readiness, a concept that describes the capacity of an economy to integrate into the digital economy, along five dimensions: connectivity, e-leadership, information security, human capital, and e-business climate. The model was used to analyze 42 countries, and therefore 42 government programs and policies, according to these dimensions. Implicit in the model is the assumption that the government is responsible for the level of readiness acquired by a country. However, the study included only environmental elements controlled by the government and did not take into consideration the way businesses behave in the defined climate, i.e., which policies and type of climate tend to favor which type of business strategy in technology adoption. Maddox (1998, 2002) points to connectivity and security as major components of the business climate that impact a firm's strategy in adopting a new technology. A favorable climate, i.e., an economy ready for the new technology, will allow businesses to be more aggressive in investing in new technology (Schneiderman, 1999).

Thus, one can see that adoption of new technology (or its applications) is a complex process that involves the organizational structure of the firm, its business strategy, organizational culture, and the environment in which the business operates. It is critical that government sets the framework (the socio-economic readiness). The new technology (the change) will impact the environment, will alter, modify, or even revolutionize the business strategy, and in time, will change the firm's structure and its organizational structure. However a comprehensive model that includes all these elements has not been proposed. A major purpose of this study is the formulation of such model (see Figure 2).

The Proposed Comprehensive Model

Business Environment

Strategy development is dependent upon the business environment (Williams, 1992). Thus, the business environment is a major component of the model. Within the business environment, we posit two forces: technological change and government involvement. Both of these factors influence the degree of turbulence within the business environment.

Turbulence is defined (Glazer and Weiss, 1993; Sinkula, 1994) as high levels of interperiod change (in magnitude and direction) of key environmental variables such as market size, introduction of new technology and degree of competition. Turbulence has two dimensions–magnitude and direction. The level of turbulence in a business environment is directly associated with the level of uncertainty faced by business firms. Further, a very turbulent environment might affect negatively the firm's profitability (Williams, 1992). The degree of environmental turbulence, especially that involving many changes happening in a short period of time, will typically produce more or less offensive strategies among existing firms. Offensive strategies (market timing, market share control, speed of response) are used in dynamic industries (intense rivalry, focus on innovation), while defensive strategies (nurture protected market, isolate firm from rivals) will work in relaxed settings (isolated competition according to Williams, 1992). A turbulent environment might affect negatively the firm's profitability (Bolton and Thompson, 2000: Williams, 1992). Strategically aggressive businesses are more likely to adopt advanced m-commerce applications (van Haas, 2001; Weiber and Kollman, 1998).

The firm's business environment is heavily dependent on regulatory framework and the government involvement. As explained in the literature review, the degree of government involvement in promoting technological innovation determines the socio-economic readiness (SER) of a nation for the adoption of the innovation. McConnell (2000) suggested that SER is comprised of five dimensions that include connectivity, government leadership, security, climate, and human capital. Other studies (Maddox 1998; Schneiderman 1999) considered the business climate to incorporate the human capital dimension present

FIGURE 2. The Framework for Technology Adoption

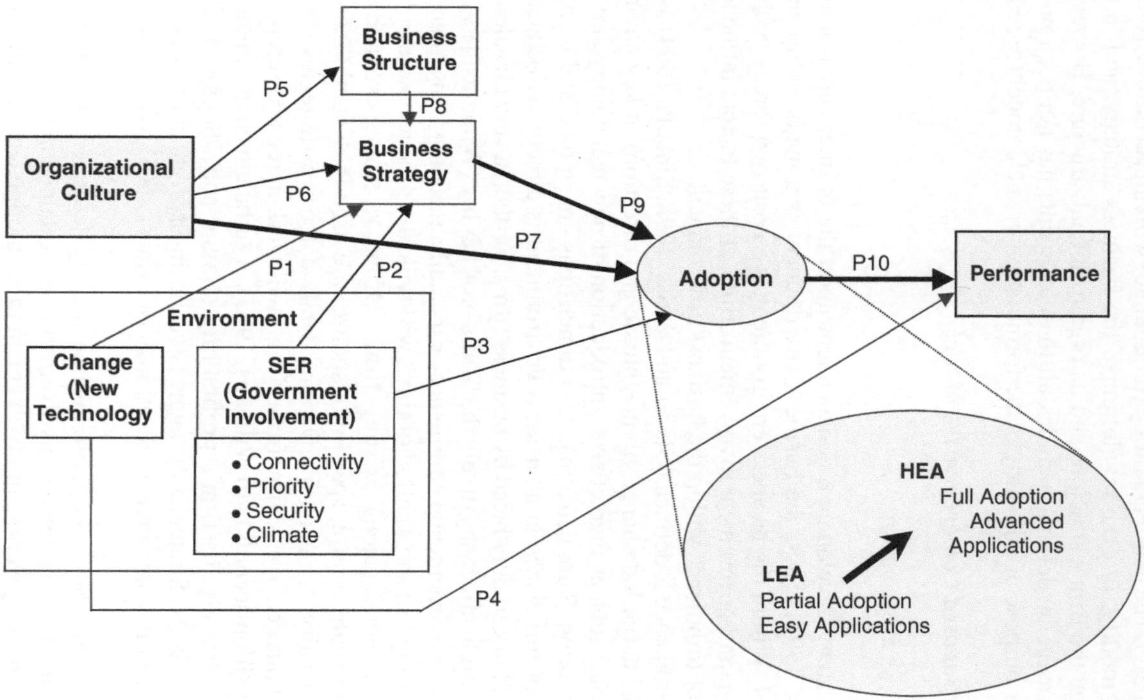

in McConnell (2000). Since education is one of the major tools available for the government to help build a positive climate, through education one can get the kind of human capital needed to generalize the use of the new technology. We will adopt the latter point of view and will discuss SER along four components (see Table 2).

Connectivity relates to the availability, affordability, and reliability of the infrastructure (McConnell, 2000; Maddox, 1998; Schniederman, 1999):

- availability (access centers, networked computers, infrastructure).
- affordability of the services (cost of service).
- reliability of the network access (downtime).

Measurement of the government involvement (McConnell, 2000; Maddox, 1998) can be considered along two dimensions: incentives for private investment in infrastructure and development of wireless government functions (m-government).

Priority relates to the commitment of the government, in partnering with industries, to create favorable climate, i.e., to create an environment that encourages the action of the private sector and to protect consumers (McConnell, 2000; Sadeh, 2002; Smith, 2000; Winge, 2001). Measurement of the government involvement could include: (1) development of wireless government functions (m-government, automate the governmental processes), (2) proactive in the promotion of an m-society, (3) partnerships with the industry leaders, and (4) level of effort to promote access for all businesses and individuals.

TABLE 2. Dimensions of Socio-Economic Readiness and Government Involvement

SER Dimension	Literature Support	Government Involvement
Connectivity	McConnell (2000), Maddox (2002), Schneiderman (1999)	• Incentives for infrastructure investment
Priority	McConnell (2000), Sadeh (2002), Smith (2000), Winge (2001)	• Promotion of wireless networks • Leadership and partnership
Security	Smith (2000), Maddox (1998), Kalakota and Robinson (2002), Sadeh (2002)	• Legal environment (security)
Business climate	Keen and Mackintosh (2001), McConnell (2000), Steve (2003), Peter, Shaw, Sing, and Springer (2001)	• Legal environment (stability) • Competition • Education • Human capital

Security refers to the protection of data and intellectual property. Key elements that could measure government involvement include (Smith, 2000; Maddox, 1998; Kalakota and Robinson, 2002): Strength of legal protection of privacy and protection of intellectual property (particularly for software), effectiveness of the law enforcement, and authorization of digital signature.

Business climate relates to the complex context of institutional arrangements that set and enforce the rules of private action in the marketplace. The climate also refers to a cadre of skilled professionals and a population that is able to use and is interested in the wireless network (Balasubramanian, Peterson, and Jarvenpaa, 2002; Clarke, 2001; McConnell, 2000, Keen and Mackintosh, 2001). Measurement of government involvement appears feasible along the following dimensions: (1) degree of competition among service providers (both communication and application providers), (2) transparency and predictability of regulatory implementation, (3) general business risk (political stability, cultural compatibility, financial soundness), (4) ability to support wireless transactions, (5) quality of the education (emphasis on effort to create and support a knowledge based society), (6) skills and efficiency of the workforce, and (7) information sharing within the society.

Based on the above considerations and on our formulation of the components of the business environment (i.e., turbulence and SER) we offer the following research propositions (see Figure 2):

P1: An environment perceived as competitive, with multiple changes taken place in a short period of time, will determine firms to adopt competitive offensive strategies incorporating m-commerce. An environment perceived as less competitive will determine firms to adopt strategies that are less offensive.

P2: SER will impact the firm's strategy.

P21: High degree of connectivity will encourage companies to be aggressive in adopting the new technology; low connectivity will make companies to adopt defensive strategies.

P22: High priority placed by the government on new technology will determine companies to use offensive strategies in pursuing the adoption of the new technology.

P23: A high security data environment will favor offensive strategies.

P24: The better the business climate the more offensive the companies in pursuing the new technology.

P3: *A SER high on all dimensions will determine the firms to adopt HEA of m-commerce. Accordingly, a SER perceived by businesses as low will determine firms to act more conservatively and adopt m-commerce LEA or not any m-commerce applications at all.*

P4: *A highly competitive environment (many changes in the new technology) will have a negative impact on profits.*

Organizational Culture

One of the key factors that may influence firm's adoption process is its organizational culture and traditions. The field of organizational behavior is rich in theoretical literature on business culture (Deshpande, Farley, and Webster, 1993; Marcoulides and Heck, 1993). Quinn and McGrath (1985) and later Quinn (1988) define organizational culture as the pattern of shared values and beliefs that help individuals understand organizational functioning; thus providing norms for behavior in the organization. Hofstede (1980) defines culture as the interactive aggregate of common characteristics that influence a group's response to its environment.

Two key dimensions have been used to classify organizational culture: the continuum from organic to mechanistic processes and the relative emphasis on internal maintenance versus external positioning (Deshpande, Farley, and Webster, 1993). Modeling these dimensions produce four basic types of organizational culture–adhocracy, clan, hierarchy, and market driven.

Adhocracy culture centers around entrepreneurship, creativity and adaptability. Flexibility and tolerance are given priority. New markets, new technology, new applications, and new sources of growth are important. *Clan* values cohesiveness, participation and teamwork highly. Employee commitment is achieved through participation. Cohesiveness and personal satisfaction are valued more important than financial goals. *Hierarchy* stresses order, rules, regulations, administrative procedures, accountability and predictability. Tracking and control are emphasized. Finally, *Market-Driven* culture focuses on competitiveness and goal achievement. Emphasis is placed on productivity and responsiveness to environmental changes. The four types of culture developed above imply varying degrees of business performance. Indeed, Deshpande, Farley, and Webster (1993) found that competing values of the market culture outperform those of the clan culture. Those of the adhocracy culture outperformed those of the diagonally opposing hierarchy culture. The number of rules and routines (and therefore the management structure) are dependent upon the type of culture (Deshpande, Farley, and Webster, 1993). The speed of response to environmental changes which determine a higher performance is thus culturally dependent. It appears that for technology adoption, culture remains a strong determinant.

Information search and organizational culture are related (Sinkula, 1994). Therefore, the adoption of a new technology being based on information search, will also be related to culture. In traditional bureaucracies and hierarchical cultures, market information need is low. Learning is based on institutionalized experience. In this case, the firms expect to grow and survive, but at a higher level of efficiency and predictability, and continue the same behavior that worked in the past (Sinkula 1994; Menon and Varadarajan, 1992). In other firms with a market or ad-hoc type of culture (thus entrepreneurial behavior), the learning emphasis could be focused more on adaptive behavior. Moreover, companies with market-driven culture may be more systematic in information search and new technology adoption than the ones with adhocracy culture. Therefore, we propose:

P5:   An entrepreneurial type of culture (adhocracy) will determine a more flexible management structure with fewer levels and, therefore, will favor m-commerce adoption.

P6:   An entrepreneurial type of culture will determine an offensive type of strategy for the firm and therefore increase the propensity of m-commerce HEA adoption.

P7:   Overall, the intensity and scope of new technology adoption is greater in companies with an adhocracy type of culture, sequentially followed by market-driven culture, clan and hierarchy cultures. However, the culture-adoption relationship is moderated by the business strategy and management structure of the company.

Business strategy and management structure will determine the adoption of a new technology or an application of a new technology such as m-commerce (van der Poel, 2001). If adoption fits into the organization's strategy, if it was predicted or anticipated in the strategic planning process (Gagnon, Sicotte, and Posada, 2000) then the company will consider one or more m-commerce applications (Wilhelmson, 2001). The better the fit the more probable the adoption of high-end applications will be (Robert and Weiss, 1988; Winge, 2001). How close is the opportunity to the current and/or projected strategy? Opportunity search, assessment, development, and pursuit were the dimensions used by Robert and Weiss (1988) in conducting its firms' analysis on innovation and technology adoption. Offensive business strategies will suit best new technology adoption (O'Keefe, O'Connor, and Kung, 1998). Managerial autonomy will favor innovation and new technology adoption. The same conclusion was reached by Sahay, Gould and Barwise (1998) when analyzing the experts' perceptions of opportunities and threats of the Internet for existing businesses. Prahalad (1995) advocates a relationship between management structure and business strategy. Therefore:

P8: *A flexible management structure will lead to an offensive business strategy.*

P9: *An offensive business strategy will lead to m-commerce HEA adoption. A defensive business strategy will not consider the adoption of m-commerce HEA. At maximum, they will adopt a low end application.*

P10: *Adoption of m-commerce HEA will lead to an increase in profitability. However, the increase will happen in time after the full implementation of the high-end applications takes place.*

DISCUSSION AND IMPLICATIONS

Because of the social and economic value of m-commerce, models leading to an improved understanding of the wireless applications adoption by companies represent significant contributions to the literature (Narduzzi, 2001; Nelson, 2001). Table 3 suggests several m-commerce applications by functional area. The case of m-commerce and its adoption by the mass of firms, especially small and medium-sized ones (they represent almost 98 percent of all businesses), is very promising for empirical research (Sharma, 2000). Only when the new technology is used by the mass of small and medium-sized companies, one can consider the adoption of the new technology to be successful. Therefore, the "middle class" companies should be studied and their adoption process analyzed. The model should be tested on a sample of small and medium-sized companies. In a relatively short period of time almost all companies will adopt some m-commerce applications (Angelos, Shaw, and Springer, 2001; Held, 2001; Rendon, 2001). This process will affect their business model and eventually help them increase profitability (Rendon, 2001; Shih, 1998).

The model presented in this paper makes a contribution towards a better understanding of the adoption of m-commerce and its applications by companies. Firm's culture (van Haas, 2001) and its business environment (Venkatesh 1998; Weiber and Kollman, 1998) are predictors of new technology adoption. The firm's business strategy will have the final say every time a mobile commerce technology/application will be adopted (Winge, 2001).

TABLE 3. M-Commerce Applications (Examples)

Remote Access (LEA)	• Email
	• Web Browsing
Information Services (LEA)	• News
	• Weather
	• Horoscopes
	• Stock Prices
	• Sports Scores
Directory Services (LEA)	• Restaurant Guide
	• Movie Guide
	• Telephone Directory
	• Hotel Guide
Operations and Maintenance (HEA)	• Fleet/Inventory Management • Supplier-Buyer Relationships
Location Based Services (LEA/HEA)	• Where is the nearest (ATM, Restaurant, etc.)?
	• Town/Travel Navigator
Entertainment (LEA/HEA) (can be interactive)	• Games
	• Gambling
	• Chat (ICQ)
Interactive Transactions (HEA)	• Banking
	• Stock Trading
	• Ticket Booking
	• Mobile Internet Shopping
	• Insurance
	• Car Rental

The rate of adoption of new technology across companies will be heavily influenced by the extent of government involvement in promoting SER. The McConnell (2000) study demonstrated that the most successful nations in participating in the digital economy are those that have an increased connectivity, have a government that provides leadership through partnering with industry leaders, stresses education for the new technology, and provides a competitive, transparent business climate. The model presented in this study incorporates the above dimensions into SER. This represent an important predictor of the business strategy and the adoption of the new technology. Besides SER, the organizational culture of the company will have a critical impact on the adoption of the new technology.

The proposed adoption model consolidates existing findings from the literature. It builds on the work of Afuah and Tucci (2001) and can form the basis of empirical studies of what determines businesses to adopt new technology. By identifying the factors that influence the adoption process, this model will help business owners/managers understand the changes they have to make, the factors that need most attention when deciding the future of their firms.

Attempts should be made to develop multiple indicators for each variable in the model, as that would help in improving the construct validity of the measures used. In addition to developing indicators for each variable in the model, data related to some contingency variables should be obtained for control purposes. For example, data on variables such as size, SIC code or age of the business should be collected. Such data would enable testing for any kind of prevalent systemic biases in the data and provide controls needed to ensure that contingencies do not bias the testing of the model. Once data are collected, multiple regression, path analysis, and/or structural equation modeling may be used to test the model and the hypotheses. While multiple regression and path analytical techniques would help to assess the validity of the model and test the hypotheses developed in this article, structural equations modeling allows the test of causal relationships as well as other relationships not hypothesized.

AUTHOR NOTES

Michael Stoica's publications have appeared in *Entrepreneurship Theory and Practice*, *Journal of East-West Business*, *Journal of Developmental Entrepreneurship*, *Energy*, *Foundations of Control Engineering*, *Journal of Vacation Marketing*. Dr. Stoica's current research interest and work is in small business strategy and entrepreneurship.

Darryl W. Miller's publications have appeared in *Journal of Advertising*, *Psychology and Marketing*, *Journal of Marketing Communications*, *International Journal of Aging and Human Development*, *Journal of Vacation Marketing*, and *Services Marketing Quarterly*. Dr. Miller's current work is in services marketing, advertising content analysis, and mental imagery strategies in marketing communications.

David Stotlar's publications appeared in proceedings at FBD Conferences. His research and consulting interests are in wireless communications.

REFERENCES

Afuah, Allan and Christopher Tucci. 2001. *Internet Business Models and Strategies*. Boston, MA: McGraw-Hill.

Anckar, Bill and Davide D'Incau. 2002. "Value Creation in Mobile Commerce: Findings from a Customer Survey." *Journal of Technology Theory & Application*. 4 (1): 43-64.

Angelos Peter, Rupert Shaw, Sunil Singh, and Iris Springer. 2001. "The Mobile Internet in 2006." published by the *Rotterdam School of Management*. January, 2001.

Balasubramanian, Sridhar, Robert Peterson, and Sirkka Jarvenpaa. 2002. "Exploring the implications of M-commerce for markets and marketing." *Journal of the Academy of Marketing Science*. 30 (4): 348-361.

Bhide, Amar. 2000. *The Origin and Evolution of New Business*. Oxford: University Press.

Bolton, Hill and John Thompson. 2000. *Entrepreneurs: Talent, Temperament, Technique*. Oxford: Butterworth-Heineman.

Cadeaux, Jack. 1997. "Counter-revolutionary forces in the information revolution. Entrepreneurial action, information intensity and market transformation," *European Journal of Marketing*. 31 (11/12): 768-785.

Chaston, Ian. 1997. "Small firm performance: Assessing the interaction between entrepreneurial style and organizational structure," *European Journal of Marketing*. 31 (11/12): 814-831.

Chrisman, James, Alan Bauerschmidt, and Charles Hofer. 1998. "The Determinants of New Venture Performance: An Extended Model." *Entrepreneurship Theory and Practice*. 23 (1, Fall): 5-31.

Clarke, Irvine. 2001. "Emerging Value Propositions for M-Commerce." *Journal of Business Strategies*. 18 (2): 133-148.

Council of Economic Advisers. 2000. *The Economic Impact of Third Generation Wireless Technology*. Report, October 2000.

Deshpande, Rohit, John U. Farley and Frederik E. Webster Jr. 1993. "Corporate Culture, Customer Orientation, and Innovativeness in Japanese Firms: A Quadrant Analysis," *Journal of Marketing*, 57 (January): 23-37.

Fisher, Eileen, Rebecca Reuber, and Nancy Carter. 1999. "A Comparison of Multiple Perspectives on Rapid Growth Firms," *Proceedings of the 13th Annual National Conference, USASBE, San Diego*, January 14-17: 233-254.

Gagnon, Yves-C., Helene Sicotte, and Elisabeth Posada. 2000. "Impact of SME Manager's Behavior on the Adoption of Technology." *Entrepreneurship Theory and Practice*. 25 (2 Winter): 43-59.

Glazer, Rashi and Allen M. Weiss. 1993. "Marketing in Turbulent Environments: Decision Processes and the Time-Sensitivity of Information." *Journal of Marketing Research*. 30 (November): 509-21.

Harte, Lawrence, S. Kellogg, R. Dreher, T. Schaffnit, Nancy Campbell, Richard Dreher, Steve Kellogg, Lisa Gosselin, and Judith Rourke-O'Briant. 1999. *The Comprehensive Guide to Wireless Technologies*. Apdg Publishing, Inc.

Held, Gil. 2001. *Data Over Wireless Networks: Bluetooth, WAP, and Wireless Lans*. Osborne McGraw-Hill.

Hills, Gerald, Tom Lumpkin, and Robert Singh. 1997. "Opportunity Recognition: Perceptions and Behavior of Entrepreneurs." *Frontiers of Entrepreneurship Research*. 17: 168-182.

Hofstede, Geert H. 1980. *Culture's Consequences*. Beverly Hills, CA: Sage Publications, Inc.

Kalakota, Ravi and Marcia Robinson. 2002. *M-Business. The Race to Mobility*. New York: McGraw-Hill.

Keen, Peter and Ron Mackintosh. 2001. *The Freedom Economy. Gaining the M-Commerce Edge in the Era of Wireless Internet.* New York: McGraw-Hill.

Leung, Kenneth and John Antypas. 2001. "Improving Returns on M-Commerce Investments." *Journal of Business Strategy.* 22 (5): 12-23.

Maddox, Kate. 1998. *Web Commerce. Building a Digital Business.* New York: Wiley.

Maddox, Kate. 2002. "Wireless b-t-b ads still not finding connection," *B to B.* 87 (9): 3-5.

Marcoulides, George and Ronald Heck. 1993. "Organizational Culture and Performance: Proposing and Testing a Model." *Organization Science.* 4 (May): 209-25.

McConnell International. 2000. *Risk E-Business: Seizing the Opportunity of Global Readiness. www.mcconnellinternational.com*

Menon, Anil and P. Rajan Varadarajan. 1992. "A Model of Marketing Knowledge use Within Firms," *Journal of Marketing.* 56 (October): 53-71.

Mobilocity. 2000. *Mobilocity.net Report on M-commerce,* published by *www.Mobilocity.com*

Morace, Francesco, Lucia Chrometzka. 2001. "The New Dynamics of the Wired Era. Net Flow and Creative Pollination." presentation at the *M-Conference: Seizing the Mobile Advantage,* Rotterdam School of Management, Rotterdam, Holland. January, 19-20.

Muller-Veerse, Falk. 2000. "Mobile Commerce Report." *Durlacher Research. www. durlacer.com*

Narduzzi, Edoardo. 2001. "Is M-business the Same Game as the E-business?" Presentation at the *M-Conference: Seizing the Mobile Advantage,* Rotterdam School of Management, Rotterdam, Holland January 19-20.

Nelson, Eric. 2001. "Wireless Call-A-Cab: The Early Mover Gamble." *M-business.* (start-up section), April: 90-94.

O'Keefe, Robert, Gina O'Connor, and Hsiang-Jui Kung. 1998. "Early adopters of the Web as a retail medium: small company winners and losers," *European Journal of Marketing.* 32 (7/8): 629-641.

Porter, Michael. 2001. "Strategy and Internet." *Harvard Business Review.* Reprint R0103D, (March): 63-78.

Prahalad, C. K. 1995. "Weak Signals versus Strong Paradigms." *Journal of Marketing Research.* 32 (August): 3-6.

Quinn, Robert E. and Michael McGrath. 1985. "Transformation of Organizational Cultures: A Competing Values Perspective." In *Organizational Culture,* Peter Frost ed., Beverly Hills, CA: Sage Publications, Inc.

Quinn, Robert E. 1988. *Beyond Rational Management.* San Francisco: Jossey-Bass Inc., Publishers.

Rendon, Jim. 2001. "Finland's Wireless Fasttrack." *M-business.* (May): 46-51.

Robert, Michael and Alan Weiss. 1988. *The Innovation Formula: How Organizations Turn Change into Opportunity.* Cambridge, Massachusetts: Ballinger.

Roberts, Edward. 1991. *Entrepreneurs in high technology.* Oxford: Oxford University Press.

Sadeh, Norman. 2002. *M-Commerce. Technologies, Services, and Business Models.* New York: Wiley.

Sahay, Arvind, Jane Gold, and Patrick Barwise. 1998. "New interactive media: Experts' perceptions of opportunities and threats for existing businesses," *European Journal of Marketing*. 32 (7/8): 616-628.

Schneiderman, Ron. 1999. *A Manager's Guide to Wireless Telecommunications*. New York: AMACOM.

Seager, Andy. 2003. "M-Commerce: An Integrated Approach." *Telecommunications International*. 37 (2): 36-38.

Sharma, Chetan. 2000. "Wireless Internet Applications." Luminant Worldwide Corp., (March) *www.luminant.com*

Shih, Chuan-Fong. 1998. "Conceptualizing consumer in cyberspace." *European Journal of Marketing*. 32 (7/8): 655-663.

Sinkula, James M. 1994. "Market Information Processing and Organizational Learning." *Journal of Marketing*. 58 (January): 35-45.

Smith, Clint. 2000. *Wireless Telecommunications FAQs*. McGraw-Hill Publishing Company.

Srinivasan, Raji, Gary Lilien, and Arvind Rangaswamy. 2002. "Technological Opportunism and Radical Technology Adoption: An Application to E-Business." *Journal of Marketing*. 66 (3): 47-59.

Stoica, Michael. 2001. "The Impact of Mobile Commerce on Small Business and Entrepreneurship," *Coleman Foundation White Paper*.

Synchrologic. 2001. *The Handheld Applications Guidebook*. Published by Synchrologic.

van der Poel, Adrien. 2001. "Idea Generation: Killer Ideas = Killer Startups?" presentation at the *M-Conference: Seizing the Mobile Advantage*, Rotterdam School of Management, Rotterdam, Holland. January, 19-20.

van Haas, Theo. 2001. "Mobile Banking and Beyond." Presentation at the *M-Conference: Seizing the Mobile Advantage*, Rotterdam School of Management, Rotterdam, Holland. January, 19-20.

Venkatesh, Alladi. 1998. "Cybermarketscapes and consumer freedoms and identities." *European Journal of Marketing*. 32 (7/8): 664-676.

Walsh, Steve and Bruce Kirchhoff. 2001. "Entrepreneurs' Opportunities in Technology Based Markets," *Proceedings of the Second Annual USASBE/SBIDA National Conference*, Orlando, Florida (electronic proceedings).

Weiber, Rolf and Tobias Kollman. 1998. "Competitive advantages in virtual markets–perspective of 'information-based marketing' in cyberspace." *European Journal of Marketing*. 32 (7/8): 603-615.

Welsch, Harold, Jianwen Liao, and Michael Stoica. 2001. "Absorptive Capacity and Firm responsiveness: An Empirical Investigation of Growth-Oriented Firms." *Proceedings of the Second Annual USASBE/SBIDA National Conference*, Orlando, Florida (electronic proceedings).

Wilhelmsson, Jonas. 2001. "Today's Competitive Edge Is Rounded: Creating Successful Value Added Services for the Mobile Phone." *www.gmcforum.com*.

Williams, Jeffrey R. 1992. "How Sustainable Is Your Competitive Advantage?" *California Management Review*. 34 (3): 1-23.

Winge, Kristian. 2001. "Idea Generation, Selection and Exploitation." Presentation at the *M-Conference: Seizing the Mobile Advantage*, Rotterdam School of Management, Rotterdam, Holland. January, 19-20.

Predicting Core Innovation and Wealth from Technological Innovation and Domestic Competitiveness: The Crucial Role of Anti-Trust Policy Effectiveness

Renée J. Fontenot
Paul G. Wilhelm

SUMMARY. This study examines the determinants of GDP per capita and a nations designation as a core economy. Using data from the global

---

Renée J. Fontenot, PhD, is Assistant Professor of Marketing, The University of Texas of the Permian Basin, School of Business, 4901 East University, MB 286, Odessa, TX 79762-0001 (E-mail: fontenot_r@utpb.edu). Her research interests primarily involve issues related to B2B, entrepreneurship, and international marketing.

Paul G. Wilhelm, PhD, is Associate Professor of Management, School of Business, Department of Management and Marketing, Virginia State University, 1 Hayden Drive, Petersburg, VA 23806 (E-mail: pwilhelm@vsu.edu). His research interests primarily explore issues of government policies, entrepreneurship, competitiveness, culture, game theory, and corruption.

The authors gratefully thank Kimball P. Marshall, Alcorn University for his encouragement and support. His comments and advice were much appreciated. Appreciatively, the authors wish to thank the unknown reviewers for their suggestions and comments.

[Haworth co-indexing entry note]: "Predicting Core Innovation and Wealth from Technological Innovation and Domestic Competitiveness: The Crucial Role of Anti-Trust Policy Effectiveness." Fontenot, Renée J., and Paul G. Wilhelm. Co-published simultaneously in *Journal of Nonprofit & Public Sector Marketing* (Best Business Books, an imprint of The Haworth Press, Inc.) Vol. 13, No. 1/2, 2005, pp. 233-253; and: *Government Policy and Program Impacts on Technology Development, Transfer and Commercialization: International Perspectives* (ed: Kimball P. Marshall, William S. Piper, and Walter W. Wymer, Jr.) Best Business Books, an imprint of The Haworth Press, Inc., 2005, pp. 233-253. Single or multiple copies of this article are available for a fee from The Haworth Document Delivery Service [1-800-HAWORTH, 9:00 a.m. - 5:00 p.m. (EST). E-mail address: getinfo@haworthpressinc.com].

Digital Object Identifier: 10.1300/J054v13n01_13

233

competitiveness report and the World Bank, various measures of technological innovation and competitiveness were tested as predictors of GDP per capita and designation as a core economy. Using multiple regression and discriminant analysis, governmental polices that provide effective anti-trust protection and strong protection of intellectual property are indicated as predictors of a nation's GDP per capita and a nation's degree of innovation and technology transfer. *[Article copies available for a fee from The Haworth Document Delivery Service: 1-800-HAWORTH. E-mail address: <docdelivery@haworthpress.com> Website: <http://www.HaworthPress. com> © 2005 by The Haworth Press, Inc. All rights reserved.]*

KEYWORDS. Core innovation, competitiveness, GDP, anti-trust protection, intellectual property, technology transfer

INTRODUCTION

The United States of America is among the most developed nations in the world, with a population of 284.0 million, a GDP of $10.2 trillion, and a Gross Domestic Product (GDP) per capita of roughly $35 thousand (The World Bank Group 2002). One element that differentiates the U.S. from many other countries in the world is its anti-trust legislation. For more than half of its history, since the Sherman Act of 1890, U.S. anti-trust laws have governed large corporations, corporate trusts, and business combinations.

Anti-trust legislation promotes competition and economic efficiency. Competition results from the efficient allocation of resources; this is one of three important mechanisms for economic growth (McArthur & Sachs 2002). Capital accumulation and technological advances are the other two elements that lead to economic growth. Each of these elements is dependent upon a country's governmental approach to business, specifically large businesses.

Three approaches have generally been, and continue to be, taken by governments to deal with large businesses (Mueller 1996). Laissez-faire governments assume a hands-off approach to business. Communist or socialist governments exercise high degrees of supervision over businesses or assume ownership of businesses. Democratic governments adopt anti-trust policies that assure competition among businesses without interfering in other market decisions. In general, the U.S. and other first world economies tend to operate under this third approach to governmental policy.

Governmental policies can affect a nation's ability to give up current consumption of some goods to make resources available for the accumulation of capital. Savings, a by-product of deferring consumption, depends on the ability of a country to first meet the basic needs of its citizens with existing production technology and resource availability (Baumol & Blinder 1982). To have savings, and therefore accumulation of capital, a country must produce a good or service for which value can be derived. Gross Domestic Product (GDP) is a straightforward measure of a country's wealth or total output of goods and services that allows international comparisons between countries (Alexander, Sharpe, & Bailey 1993).

This paper will explore the effects of anti-trust policies on a country's commitment to technological innovation and anti-trust policies as a predictor of a country's GDP per capita. Using data available from the World Bank (Kaufmann, Kraay, & Zoido-Lobaton 1999, 2002) and the Global Competitiveness Report (Porter, Sachs, Cornelius, McArthur, & Schwab 2002), the authors will test several hypotheses that suggest that a country's GDP and core economy are related to anti-trust policies and technological innovations. Seventeen countries are added to the Global Competitiveness Report data (Porter et al. 2002), boosting its regional diversification and reflecting the rising integration of developing countries into the global economy. This allows a more robust test of the hypotheses to be presented.

First, anti-trust policies and their influences on global competitiveness are discussed. Second, the authors discuss elements of national economic efficiencies, specifically the elements of innovation and technology as they relate to a country's economic growth and their sustained economic growth as a core global economy. Based upon the Global Competitiveness Report (Porter et al. 2002) and data from the World Bank, hypotheses designed to test the relationships between economic wealth, technological innovation, and anti-trust policies are presented. Third, the methodology used to test the hypotheses and the results are discussed.

DOMESTIC COMPETITIVENESS

Anti-Trust

Through anti-trust legislation, governments ensure inter-firm competition without otherwise interfering with pricing or output decisions. Effective anti-trust legislation should result in competition sufficient for low-priced yet high-quality products and ensure a level playing field for competitors (Klein 1999). Antitrust legislation includes those acts designed "to protect trade and commerce against unlawful restraints and monopolies" (United States Code 2000). In general, under U.S. anti-trust laws, illegal business practices must be proven to affect competition adversely, such as when competition is limited or impeded by the joint actions of two or more firms, or when competitors act in a suspiciously uniform manner.

The Sherman Act, the first U.S. Federal anti-trust law (Pitofsky 2000), set the foundation for prohibiting anti-trust behaviors that restrain trade or commerce (Anti-Trust Organization 1994; Klein, 1999). This sweeping act is the main source of U.S. anti-trust law; subsequent acts only expanded its scope (Anti-Trust Organization 1994; Legal Information Institute 2000), and several acts have been added and amended over the last 100 years.

Despite the passing of the years and the increased globalization in many markets, the Sherman Act is not obsolete (Friberg and Thomas 1991). The original act is still relevant as the Microsoft Corporation ("Microsoft") was charged with violations of the Sherman Anti-Trust Act (Jackson 2000). According to Jackson (2000), the Judge presiding in the case, Microsoft did indeed have monopoly power over the operating systems commonly installed on most personal computers and was, therefore, guilty of violations of the Sherman Anti-Trust Act.

The Clayton Act was added to the Sherman Act in 1914 and significantly amended in 1950 (Klein 1999). The Clayton Act permitted U.S. courts to address anti-competitive behaviors at inception (Anti-Trust Organization 1994). The Robinson-Patman Act, which amends Section 2 of the Clayton Act, prohibits price discrimination between purchasers of goods of "like grade and quality" (Brobeck Phleger and Harrison LLP 1999). The Foreign Trade Anti-Trust Improvements Act of 1982, section 6a of the Sherman Act, expands the previous anti-trust legislation to apply to foreign nations if their actions have a direct, substantial, and reasonably foreseeable effect on trade (United States Code 2000). Violations of the Sherman Act can result in criminal felonies, whereas violations of the Clayton, Federal Trade Commission (FTC), or Robinson-Patman Acts are civil offenses that carry no criminal penalties (Klein 1999). Nonetheless, firms found guilty of anti-trust violations in the U.S. Federal courts may be assessed treble damages (Gleim 2000; Mueller 1996).

Though the U.S. anti-trust policies were initially drafted due to domestic issues, they have been used in regard to international trade issues. In the U.S. the courts have ruled that the Sherman Act applies to criminal anti-trust allegations relating to actions taken entirely outside the United States, provided the alleged actions have a foreseeable and substantial effect on U.S. commerce (Klein 1997). Courts have ruled that violations of the Foreign Trade Anti-Trust Improvements Act can result in criminal and civil penalties (Applebaum & Barnett 1997).

Other countries have anti-trust governing rules that encompass many of the same ideas as the U.S. Anti-Trust Acts, which are often called competition laws and policies. More than 90 countries have such laws and policies, many of which have been initiated in the last decade (Janow and Shapiro 2001). A good competition policy has the potential to increase the outputs from scarce resources by providing rules contributing to efficient marketplace behavior (Rill, Stern, Janow, Baird, Donilon, Dunlop et al. 2000). The Competition Act of Canada promotes and maintains fair competition so consumers can benefit from lower prices, product choice, and quality services (Competition Bureau 2001). Article 82 of the European Economic Community Treaty, similar to Section 2 of the Sherman Act, prohibits abuse of a dominant market position within the European Common Market (Berry 2001). Australia's Trade Practices Act of 1974 is considered a comprehensive anti-trust regime based upon previous domestic experience and in part based upon U.S. anti-trust laws (Berry 2001). New Zealand's Commerce Act of 1986 is based on Australia's Trade Practices Act of 1974 (Berry 2001).

In Japan, the Fair Trade Commission is responsible for enforcing the Antimonopoly Act. Like anti-trust policies of other nations, the goal of the Antimonopoly Act is to "realize the efficient operation of the Japanese economy by removing obstacles to free and fair competition in the market in ensuring the sound operation of the market mechanism" (Kazuhiko 2001). Japan has entered into several International Agreements with Singapore, Canada, the E.U. and the U.S. in regard to anti-trust/anticompetitive cooperation between the countries. Countries such as Japan, Canada, the E.U. and the U.S. increasingly enter into joint trade agreements that specifically address issues of anti-trust as they realize that their economies are even more interrelated (Stockton 2002; U.S. Department of Justice 1999).

The World Trade Organization is, as of yet, the only international organization that gives guidance on global rules of trade between countries (World Trade Organization 2002). In 1995, the World Trade Organization replaced the General Agreement on Tariffs and Trade (GATT) that had previously given guidance to member nations on issues related to trade.

The WTO consists of 144 member nations and 32 other nations who have applied but must go through the accession process to become members. Thus the WTO represents roughly 92% or 176 of the 192 independent states or countries in the world (Bureau of Intelligence and Research 2002). The WTO regularly monitors member nations to assure that they adhere to the agreed policies that are designed to help trade flow as freely as possible between members (World Trade Organization 2002). The policies set by the WTO do not replace existing anti-trust laws within a particular nation, but are used as a guideline for acceptable business practices between members.

Fundamental to the anti-trust policies set by individual countries, multinational agreements, or those by the WTO, is a global recognition that competitiveness is necessary for economic growth and development. The *Global Competitiveness Report* (Porter et al. 2002) examines the growth prospects of countries and provides comprehensive data allowing comparison of issues related to domestic competition.

ECONOMIC EFFICIENCIES

Measures of Economic Growth

When countries enter into anti-trust agreements, evidence shows that there is improved trade between the nations (World Trade Organization 2002). Increased trade boosts economic growth for the trading nations. Economic growth from trade improves national income and prosperity. There are limits, however, to the degree to which free trade can promote economic growth (Arrighi 2000). Over time the rate of economic growth diminishes unless the accumulation of capital is reinvested into technological advancement. If the technological change accompanies the economic growth, then the process can become cyclic with technology increasing capital accumulation and capital accumulation increasing technological advancements (McArthur & Sachs 2002).

Just as economic issues are related to technological advancements, so too are they related to social concerns such as the level of investment in education and the training of workers (The World Bank Group 2002). Likewise, social issues influence a country's ability to achieve economic growth such as when population growth exceeds the ability of the country's economic system to meet the population's basic needs and provide jobs for those who want to work. The GDP per capita can be used to measure a nation's productivity, for as productivity increases so too does a nation's standard of living (Porter 2002). Many economists focus on the benefits of free trade as helping countries do what they do best while consumers benefit with the best quality and price on goods they desire (Hilsenrath 2002).

Automatic improvement of technology, social issues, and accumulation of capital is not guaranteed by an increase in GDP per capita. Governments take a purposeful role in facilitating reinvestment, protecting new technologies, and limiting actions by businesses that will limit competition (Porter 2002). The FTC has argued that preserving competition in "innovation markets" is the only way to ensure competition (Garland & Reinhardt 1999). Therefore, governments that help to foster competitive thinking will assure sustainable economic growth. The relationship between business environments and the policies of governments must evolve to support and encourage productive ways of competing (Porter 2002). Effective antitrust policies are positively associated with the variation in GDP per capita across countries (Porter 2001).

H$_1$: Countries with higher GDP per capita will have increased competitiveness.

Technological Innovation

Countries that have the largest GDP per capita also have the highest level of technological innovation (McArthur & Sachs 2002). Technological innovation successfully leads to long-term sustained economic growth for countries whose governmental policies protect and encourage innovation. Technological advances lead to the creation of new goods, service jobs, and security that contributes to economic growth (Schacht 2001b). When governments are unwilling to protect innovation creation through patent law, businesses are less likely to put resources into innovation (Lichtenberg 2001). The number of new patents issued each year is representative of the level of innovation, specifically technological innovation, which occurs within a country (McArthur & Sachs 2002).

Technological advancement is a major method of fostering economic growth for the world's nations. Technology is a tool for development and a means of achieving an economic competitive advantage (United Nations Development Programme 2001). Within a country, technological advancement of industries has a strong influence on employment and productivity, which in turn affects the accumulation of wealth (World Trade Organization 2002). Global competition policy changes are necessary to keep up with technological developments and the drive for competitiveness in world markets (Rill et al. 2000).

The U.S. recognizes the importance of technology competitiveness to the extent that the U.S. Department of Commerce has an Office of Technology Policy which overseas policies that affect domestic technological innovation (Laney-Cummings 2002). The U.S. government considered technological advancement such an important element to the nation's economic development that the Clinton Administration defined it as part of the national economic policy (Schacht 2001b). It is the task of the Assistant Secretary for Technology Policy to work closely with industry to identify critical issues related to the development of governmental policies necessary to ensure continued competitiveness of technologies within the U.S.

Technology improvement (both new goods and better ways of producing goods) can be attained by creating a genuinely new technology, or by adopting (and adapting) a technology that has been developed in another country. *Technological innovation* and *technological diffusion*, or transference, are the respective names for these two processes (McArthur & Sachs 2002). In terms of long-term economic growth, perhaps the most significant global division today is the one between nations that achieve high technological innovation and those that do not. In 2000, fully 99 percent of the patents issued for new innovations by the U.S. Patent Office came from 24 nations that accounted for barely 15 percent of world's population (McArthur & Sachs 2002). This is an indication of the inequitable global distribution of patents and the advancement of technology that they represent.

Governments play a critical role in facilitating business' ability to do basic research that leads to technological advancements that are suitable for the application of patents (Schacht 2001b). Cooperative efforts among universities and industries are indicated as key to sustainable advances in technological advancements. To facilitate these efforts, the U.S. government has passed several Acts including the National Cooperative Research Act and the National Cooperative Production Act (Schacht 2001a). Additionally, tax changes allow industry to receive credit for payments to universities for research and development (R&D), as well as allowing deductions for contributions of equipment used in academic research. Governments can facilitate technological advancements in a variety of ways if they are committed to making their nation competitive in a global environment.

H$_2$: Countries with higher GDP per capita will have an increased commitment to technological innovation.

Core Economies

The World Economic Forum (Porter et al. 2002) distinguishes nations by their degree of innovation and technology transfer. A nation that is a technological innovator is called a *core economy*, the rest are called *non-core economies* (McArthur & Sachs 2002). The core group is defined as all economies achieving at least 15 patents, per year, per million in population. Table 1, Core Economies as of 2000, lists the 24 countries meeting this core criterion. They are the world's richest nations and typically have achieved sustained economic growth for a number of years or decades (McArthur & Sachs 2002). The rest of the countries in the world are considered non-innovating economies.

TABLE 1. Core Economies as of 2000

Australia	Denmark	Hong Kong	Italy	New Zealand	Switzerland
Austria	Finland	Iceland	Japan	Norway	Taiwan
Belgium	France	Ireland	Korea	Singapore	United Kingdom
Canada	Germany	Israel	Netherlands	Sweden	United States

The ability of any particular country to be competitive is highly related to its ability to innovate and create global demand for its technologies (Habib & Coombs 2001). The boundaries between core and non-core economies are not set in stone. Technologically lagging countries can become innovators, although this is hard to do. Hence, the vast majority of nations have not achieved this transition. The most recent countries or city states to achieve this breakthrough are Iceland, Ireland, Hong Kong, Korea, Singapore, and Taiwan. Technological prowess and competitiveness has allowed them to grow rapidly (Porter et al. 2002).

H_3: Countries with core economies will have increased competitiveness.

H_4: Countries with core economies will have an increased commitment to technological innovation.

METHODOLOGY

Data

Competitiveness. Using data from surveys of between 40 and 60 business executives in 75 nations, The World Economic Forum (2002) has identified seven dimensions that serve to measure competitiveness. These items measure elements of competition and barriers to competition including: the intensity and extent of local competition, the ability to enter local markets, the number of permits and the number of days needed to start a firm, the degree of administrative burden to start a firm, and the effectiveness of anti-trust policies.

GDP. Gross domestic products (GDP) per capita is considered the best single measure of current competitiveness across all countries (Porter 2002). Countries with efficient infrastructures and distribution systems tend to be those with high GDP per capita (Shane 1993). GDP per capita has been found to have a positive relationship with innovation (Habib and Coombs 2001). Using GDP per capita as a reliable measure to compare across countries is consistent with the use of this measure in other studies done by organizations such as the World Bank and International Monetary Fund (IMF) (Porter 2002). Gross domestic product (GDP) per capita (purchasing power parity) figures were obtained from the United Nations Development Programme (United Nations Development Programme 2001).

Technological Innovation. The *Global Competitiveness Report* (Porter et al. 2002), examines the Technological Innovation and Diffusion of a country based upon twenty-three measures. Using the U.S. Office of Technology Policies as a guide, six measures were selected that closely align with their policies. Measures were chosen based upon their level of analysis, national level versus firm level and specifically their applicability to national policies. Four measures were chosen from the twenty-three Technological Innovation and Diffusion Competitiveness (TIDC) variables for national level of analysis and their relationship to innovation. Two other competitiveness variables were chosen based upon the guidelines set by the U.S. Office of Technology Policies. The other two key factors, according to McArthur and Sachs (2002), that help nations become innovators are Information Technology (IT) Training and Education, and Intellectual Property Protection. In total, these items are Quality of Scientific Research Institutions, Company Spending on Research and Development (R&D), Tax Credits for Firm-Level R & D, and University/Industry Research Collaboration.

Core Economies. Countries are distinguished as core economies through their efforts to encourage and develop technological innovations and technology transfer (McArthur & Sachs 2002). These countries (discussed above and listed in Table 1) have previously been identified by the World Economic Forum (Porter et al. 2002).

Measurement

Of the seven items used to measure competitiveness five were measured on a seven-point Likert scale. These five items were: the intensity and extent of local competition, the ability to enter local markets, the degree of administrative burden to start a firm, and the effectiveness of anti-trust policies. The number of permits and the number of days needed to start a firm were measured on continuous scales. A seven-point Likert scale was used to measure the six items of technological innovation: Quality of Scientific Research Institutions, Company Spending on Research and Development (R&D), Tax Credits for Firm-Level R&D, and University/Industry Research Collaboration. GDP per capita was reported as a continuous variable in dollars for the year 2000 and was available for 56 countries. The measure of core economies is a binomial variable, with 24 countries considered core economies and 51 considered non-core economies totaling 75 in all as taken from the World Economic Forum data.

Analyses

Correlation, multiple regression and discriminant analysis were employed. Simple associations between variables are indicated by the correlation coefficients. The means, standard deviations, and zero-order correlations between variables in the hypotheses are presented in Table 2.

As demonstrated by the correlation matrix all six measures of technological innovation are positively and significantly related to both GDP per capita and core economies at the 0.001 level. Of the seven measures of competitiveness, four are significantly related to both GDP per capita and core economies at the 0.001 level: intensity of competition, extent of local competition, degree of administrative burden for start ups, and effective anti-trust policies. Entry into local markets is positively related at the 0.005 level with core economies but is not significantly related to GDP per capita. Number of days to start a firm is negatively related to GDP per capita and significant at the 0.005 level but not significantly related to core economies. Number of permits required to start a firm is not significantly related to either GDP per capita or core economies. In general, these findings provide support for a relationship between countries considered to be core economies and a country's GDP per capita to a country's commitment to technological innovation. Similarly, though not fully supported by all measures, relationships between competitiveness and GDP per capita and competitiveness and core economies are indicated.

Test of Hypotheses

To further test the hypotheses multiple regression and multiple discriminant analysis are used. Multiple regression is used to predict GDP per capita based upon the measures of competitiveness and those of technological innovation. Discriminant analysis is used to predict whether countries are members of core economies based upon the measures of competitiveness and those of technological innovation.

An initial full model using the enter method of regression analysis was developed for the independent variables of domestic competitiveness and those of technological innovation as regressed on the GDP per capita. A second model was developed for both sets of independent variables using stepwise regression to predict the best model. The results are shown in Tables 3 and 4. For each of the two regression equations, through stepwise regression the best model proved to be determined by a single variable.

As shown in Table 3, Effectiveness of Anti-Trust Policies contributed to 84% of the explained variance ($p < 0.001$) for the competitiveness model, versus a full model where the seven variables together explained 78% of the vari-

TABLE 2. Means, Standard Deviations, and Zero-Order Correlations

	Means	S.D.	1	2	3	4	5	6	7	8	9	10	11	12	13	14
1. Intensity of Local Compet.	5.13	0.67	1.00													
2. Extent Local Competition	4.44	0.71	0.73*	1.00												
3. Entry into Local Markets	5.24	0.44	0.78*	0.52*	1.00											
4. Admin. Burden–Start-Ups	4.47	0.90	0.66*	0.55*	0.55*	1.00										
5. Permits to Start a Firm	4.82	1.92	−0.28	−0.07	−0.32**	−0.57*	1.00									
6. Days to Start a Firm	40.22	22.60	−0.22	−0.10	−0.22	−0.42*	0.62*	1.00								
7. Effective Anti-Trust Policy	4.24	1.05	0.79*	0.64*	0.51*	0.66*	−0.24	−0.29	1.00							
8. Qlty. of Scientific Res. Inst.	4.58	0.99	0.79*	0.60*	0.54*	0.67*	−0.27	−0.36*	0.86*	1.00						
9. Comp. Spending on R&D	3.85	1.01	0.73*	0.63*	0.48*	0.65*	−0.29	−0.36*	0.85*	0.90*	1.00					
10. Tax Credits for Firm R&D	3.22	0.99	0.56*	0.59*	0.38*	0.52*	−0.10	−0.18	0.65*	0.72*	0.69*	1.00				
11. Uni./Ind. Res.Coll.	3.90	0.99	0.79*	0.67*	0.54*	0.71*	−0.28	−0.34*	0.87*	0.91*	0.90*	0.75*	1.00			
12. IT Training and Education	4.32	1.03	0.77*	0.60*	0.51*	0.74*	−0.30	−0.33**	0.85*	0.87*	0.82*	0.71*	0.89*	1.00		
13. Intellectual Prop. Protect.	4.09	1.35	0.72*	0.55*	0.46*	0.63*	−0.33**	−0.29	0.89*	0.84*	0.83*	0.58*	0.83*	0.84*	1.00	
14. Core Economies	0.32	0.46	0.59*	0.48*	0.32**	0.61*	−0.25	−0.39*	0.81*	0.77*	0.83*	0.60*	0.81*	0.76*	0.82*	1.00
15. GDP	14.546	9.558	0.69*	0.45*	0.36	0.58*	−0.35	−0.39**	0.86*	0.80*	0.82*	0.55*	0.83*	0.80*	0.90*	0.90*

*Significant at the 0.001 level.
**Significant at the 0.005 level.

TABLE 3. Multiple Regression Analysis of Competitiveness Variables with GDP Per Capita

Independent Variables	All Variables				Best Fit				
	Unstandardized Coefficients		Standardized Coefficients		Unstandardized Coefficients		Standardized Coefficients		
	b	Std. Err.	Beta	t	b	Std. Err.	Beta	t	Beta ln
Intensity of Local Compet.	4,392.88	2,877.37	0.28	1.53	*** Dropped***				−0.001
Extent Local Competition	−104.19	1,401.78	−0.01	−0.07	*** Dropped***				−0.018
Entry into Local Markets	−5,918.24	2,667.31	−0.27	−2.22*	*** Dropped***				−0.103
Admin. Burden–Start-Ups	106.07	1,139.09	0.01	0.09	*** Dropped***				0.03
Permits to Start a Firm	−116.13	464.02	−0.03	−0.25	*** Dropped***				−0.069
Days to Start a Firm	−47.73	36.56	−0.12	−1.31	*** Dropped***				−0.108
Effective Anti-Trust Policy	6,734.16	1,237.12	0.73	5.44***	8,018.23	635.66	0.864	12.61***	
	$R^2 = 0.78$				$R^2 = 0.84$				
	$F = 24.30^{***}$				$F = 44.23$				
	d/f = 7/48				d/f = 6/49				

**p< 0.01
***p< 0.001
*p< 0.10
F = F-ratio
R^2 = explained variance
d/f = degrees of freedom

245

TABLE 4. Multiple Regression Analysis of Technological Innovation Variables with GDP Per Capita

Independent Variables	All Variables				Best Fit				
	Unstandardized Coefficients		Standardized Coefficients		Unstandardized Coefficients		Standardized Coefficients		
	b	Std. Err.	Beta	t	b	Std. Err.	Beta	t	Beta In
Qlty. of Scientific Res. Inst.	−1,231.09	1,697.64	−0.01	−0.73			*** Dropped***		0.175
Comp. Spending on R&D	1,944.23	1,383.93	0.21	1.41			*** Dropped***		0.252
Tax Credits for Firm R&D	75.7	790.59	0.01	0.1			*** Dropped***		0.102
Uni./Ind. Res. Coll.	1,707.1	1,638.33	0.17	1.04			*** Dropped***		0.257
IT Training and Education	353.46	1,314.45	0.04	0.27			*** Dropped***		0.162
Intellectual Prop. Protect.	4,645.28	838	0.66	5.54***	6,373.89	410.27	0.9	15.54***	
	$R^2 = 0.84$				$R^2 = 0.82$				
	$F = 44.23$***				$F = 241.36$***				
	$d/f = 6/49$				$d/f = 1/54$				

*p< 0.10
**p< 0.01
***p< 0.001
F = F-ratio
R^2 = explained variance
d/f = degrees of freedom

ance (p < 0.001). As shown in Table 4, Intellectual Property Protect contributed to 82% of the explained variance (p < 0.001) for the technological innovation model, versus a full model where the six variables together explained 84% of the variance (p < 0.001). Thus support was found for hypotheses one and two, countries with higher GDP per capita will have increased competitiveness and increased technological innovations.

Discriminant analysis was used to analyze the predictive ability of the independent variables of domestic competitiveness and those of technological innovation in determining membership of countries as core economies. As indicated by the group means for core and non-core economies in Table 5, countries designated as core economies on average reported greater amounts of competition as measured by the intensity and extent of local competition; nearly half as many days needed to start a firm; and most significantly more effective anti-trust policies. This confirms hypothesis three, countries with core economies will have increased competitiveness. The standardized coefficients measure the relative importance of each variable in the model. When the sign is ignored, the coefficients represent the relative contribution of each variable to the discriminant model, as in predicting GDP per capita; effective anti-trust policies is the most significant variable. Effectiveness of anti-trust policies are more than two-and-a-half times as important, as revealed by the standardized coefficients, as the variable number of days to start a firm in terms of its relative contribution to the discriminant model.

By examining Table 6, the group means for core and non-core economies, core economies on average reported greater amounts of technological innova-

TABLE 5. Discriminating Competitiveness Variables and Core Economies

	Means				
Variables	Core	Non-Core	F-Ratio	Std. Coefficients	Relative Importance
Intensity of Local Compet.	5.71	4.86	39.45***	0.06	7
Extent Local Competition	4.94	4.21	22.14***	−0.08	6
Entry into Local Markets	5.45	5.15	8.53*	−0.35	3
Admin. Burden–Start-Ups	5.27	4.1	42.54***	0.33	4
Permits to Start a Firm	4.13	5.15	4.82*	0.21	5
Days to Start a Firm	27.5	46.21	12.99**	−0.37	2
Effective Antitrust Policy	5.48	3.65	142.35***	0.96	1

*significant at the 0.05 level.
* significant at the 0.01 level.
***significant at the 0.001 level.

tion. Unlike in the prediction of GDP per capita, company spending, not Intellectual Property Protection, contributes the most, but just barely to the discriminant model. Looking at the relative contribution of the variables company spending on R&D and intellectual property protection show that the two contribute nearly equally to the discriminant model. Individually, each contributes more than twice the contribution of the third most important variable in the model, university and industry research collaborations. The results of this discriminant analysis indicates support for hypothesis four, countries with core economies will have an increased commitment to technological innovation.

After assessing the discriminant models for core economies, the data used to derive the discriminant function was examined for its ability to correctly assess group membership for core economies. Comparing the predicted against the actual group memberships, as predicting by competitiveness, Table 7, 93.3% of the cases were correctly classified. Similarly, when comparing the predicted against the actual group memberships, as predicting by technological innovation, Table 8, 94.7% of the cases were correctly classified. This further supports hypotheses three and four.

DISCUSSION

The findings in this analysis provide evidence to support the efforts of governments to create public policies that will protect businesses from monopolistic markets. Specifically, the public policies that are most indicated in this research is effective anti-trust legislation and strong protection of intellectual

TABLE 6. Discriminating Technological Innovativeness Variables and Core Economies

Variables	Means Core	Non-Core	F-Ratio	Std. Coefficients	Relative Importance
Qlty. of Scientific Res. Inst.	5.7	4.06	110.17***	−0.18	4
Comp. Spending on R&D	5.08	3.28	163.45***	0.57	1
Tax Credits for Firm R&D	4.08	3.82	40.78***	0.04	6
Uni./Ind. Res.Coll.	5.07	3.36	135.53***	0.24	3
IT Training and Education	5.45	3.8	96.85***	−0.05	5
Intellectual Prop. Protect.	5.71	3.34	151.03***	0.56	2

***Significant at the 0.001 level.

TABLE 7. Classification of Core Economies by Competitiveness

		Predicted	
Actual	Non-Core	Core	Total
Non-Core	49	2	51
Core	3	21	24

*93.3% correctly classified.

TABLE 8. Classification of Core Economies by Technological Innovation

		Predicted	
Actual	Non-Core	Core	Total
Non-Core	49	2	51
Core	2	22	24

*94.7% correctly classified.

property. In this research, these policies were found to be positively related to countries GDP per capita as well as being predictors of countries as core economies.

For nations that have yet to adopt strong anti-trust policies there are many countries to which they may look for a model as well as organizations that are willing to assist a nation in developing effective anti-trust policies. As evidenced by their membership as core economies and the recognition as discussed in the first part of this paper of their anti-trust policies, nations such as the U.S., Japan, Canada, and Australia, and groups of nations such as the European Union, can provide proven models for developing effective anti-trust policies. Organizations such as the WTO will provide guidance to member nations needing to improve their anti-trust policies. As with anything political, it is not necessarily the lack of examples but more often a matter of leadership and organization that prevents rapid changes.

Further, the findings in this analysis provide evidence to support the efforts of governments to create public policies that will provide incentive for business opportunities through increased competition. The positive relationships between technological innovations and GDP per capita (H_2), and technological innovations and core economies (H_4), indicates that countries will be rewarded by the implementation of such policies. Specifically, the public policies that are indicated are those aimed at enticing companies to spend more on research and development. Companies will be more inclined to increase their research and development spending when governments adopt and enforce laws that provide protection of the companies' efforts specifically through intellectual property rights.

Non-innovating economies often achieve very high growth rates by absorbing the advanced technologies and capital of the core economies (McArthur & Sachs 2002). However, their "catch-up" growth has inherent limits, which ultimately widens the income gap between them and the technological innovators (Poter et al. 2002). If they are to catch up, they need to become technological innovators (Porter 2002). To do this most effectively and efficiently, they need to know which domestic policies are most competitive or predictive in gaining the *core economic* status. One way that non-core nations can achieve core innovator status, with the concomitant wealth, appears to be through ensuring effectiveness of anti-trust policies, encouraging companies to spend more on research and development, and protecting intellectual property rights.

This study looked at seventy-five nations, their GDP per capita for the year 2000, and their degree of technological innovation and their designation as a core economy. Future research needs to examine multiple years to assure the findings are not coincidental. Time series analysis is needed to compare across and with-in countries for their adoption of effective anti-trust protection and intelligential property protection and their membership in the group designated as having core economy status.

REFERENCES

Alexander, Gordon J., William F. Sharpe, and Jeffery V. Bailey (1993), *Fundamentals of Investments*. Englewood Cliffs, New Jersey: Prentice Hall.

Anti-Trust Organization (1994), *Anti-Trust Policy–Anti-trust Enforcement Guidelines For International Operations*. Owen Graduate School of Management. Retrieved April 9, 2000, from the World Wide Web: *http://www.anti-trust.org/law/US/intnl-guide.html#intro*.

Applebaum, Harvey M. and Thomas O. Barnett (1997), "Anti-Trust: Sherman Act Can Apply to Criminal Anti-Trust Actions Taken Entirely Outside the Country," *The National Law Journal*, (April 21) B4.

Arrighi, G. (2000), "Globalization, State Sovereignty, and the 'Endless' Accumulation of Capital," in Don Kalb, Marco van der Land, Richard Staring, Bart van Steenbergen, and Nico Wilterdink (Eds.). *The Ends of Globalization: Bringing Society Back In*. New York: Rowman & Littlefield Publishers, Inc.

Baumol, William J. and Alan S. Blinder (1982), *Economics: Principles and Policy* (2nd ed.). New York: Harcourt Brace Jovanovich, Inc.

Berry, Mark N. (2001), "The Uncertainty of Monopolistic Conduct: A Comparative Review of Three Jurisdictions," *Law and Policy in International Business*, *32*(2), 263-330.

Brobeck Phleger and Harrison LLP. (1999), *Price Discrimination (Robinson-Patman Act)*. Retrieved June 11, 1999, from the World Wide Web: *http://www.brobeck. com/docs/sept96pricediscrim.html*.
Bureau of Intelligence and Research (2002), *Independent States in the World*. Office of The Geographer and Global Issues, U.S. Department of State. Retrieved December 7, 2002, from the World Wide Web: *http://www.state.gov/s/inr/rls/4250.htm*.
Cohen, L. and M. Holliday (1982), *Statistics for Social Scientists*, London: Harper & Row.
Competition Bureau, I. C. (2001), *Competition Bureau*. Retrieved December 7, 2002, from the World Wide Web: *http://strategis.ic.gc.ca/SSG/ct01250e.html*
Friberg, Emil and Celia Thomas (1991), "Is Anti-Trust Obsolete?" *Journal of Economic Issues*, Vol. 25, No. 2 (June): 617-623.
Garland, Susan B. and Andy Reinhardt (1999), "Making Anti-Trust Fit High Tech," *Businessweek Online* (March 22): *http://www.businessweek.com/1999/99_12/ b3621076.htm*.
Gleim, Irvin N. (2000), *Anti-Trust*. Gleim Internet, Inc. Retrieved December 5, 2000, from the World Wide Web: *http://www.gleim.com/buslaw/LCE-CH36.html*.
Habib, Mohsin and Joseph E. Coombs (2001), "Cultural and Socioeconomic Determinants of Invention: A Multicountry, Multiperiod Analysis," *New England Journal of Entrepreneurship*, Vol. 4, No. 1 (Spring): 15-29.
Hilsenrath, Jon E. (2002), "Globalization Gets Mixed Report Card in U.S. Universities," *The Wall Street Journal*, (December 2): 13-14.
Hirst, P. (2000), "The Global Economy: Myths or Reality?" In Don Kalb et al. (Eds.). *The Ends of Globalization: Bringing Society Back In*. New York: Rowman & Littlefield Publishers, Inc.
Jackson, Thomas Penfiled (2000), *United States Of America v. Microsoft Corporation*, Civil Action No. 98-1232 (TPJ) United States District Court For The District of Columbia: Washington, DC.
Janow, Merit E. and Andrew J. Shapiro (2001), "A Trusting Approach to Anti-Trust," *The International Economy*, Vol. 15, No. 5 (Sep/Oct): 42-44.
Kaufmann, Daniel, Aart Kraay and Pablo Zoido-Lobaton (1999), *Aggregating Governance Indicators* (WPS2195): The World Bank.
Kaufmann, Daniel, Aart Kraay and Pablo Zoido-Lobaton (2002), *Governance Matters II–Updated Indicators for 2000-01* (WPS2772): The World Bank.
Kazuhiko, Takeshima (2001), *Fair Trade Commission of Japan*. International Affairs Division. Retrieved December 6, 2002, from the World Wide Web: *http://www.jftc. go.jp/e-page/f_home.htm*.
Klein, Joel I. (1997), *United States v. Nippon Paper Industries Co. Ltd*, 109 F. 3d 1, U.S. Court of Appeal for the First Circuit: District of Massachusetts.
Klein, Joel I. (1999), *Anti-Trust Enforcement and the Consumer*. U.S. Department of Justice, Washington, D.C. Retrieved July 5, 2000, from the World Wide Web: *http://www.usdoj.gov/atr/public/div_stats/1638.htm*.
Laney-Cummings, Karen (2002), *Technology Competitiveness*. Technology Administration, Office of Technology Policy, U.S. Department of Commerce. Retrieved December 7, 2002, from the World Wide Web: *http://www.ta.doc.gov/TechComp/ default.htm*.

Legal Information Institute. (2000), *Anti-Trust: An Overview*. Retrieved July 5, 2000, from the World Wide Web: *http://wwwsecure.law.cornell.edu/topics/anti-trust. html*.

Lichtenberg, Frank R. (2001), "Cipro and the Risks of Violating Pharmaceutical Patents," *National Center For Policy Analysis, 380*, 1-2.

McArthur, John W. and Jeffery D. Sachs (2002), "The Growth Competitiveness Index: Measuring Technological Advancement and the Stages of Development," in M. E. Porter & J. D. Sachs & P. K. Cornelius & J. W. McArthur & K. Schwab (Eds.). *The Global Competitiveness Report 2001-2002* (pp. 28-51). New York: Oxford University Press, Inc.

Mueller, Charles E. (1996), *Anti-Trust Law & Economics Review*. Anti-Trust Law & Economics Review, Inc. Retrieved July 5, 1999, from the World Wide Web: *http:// webpages.metrolink.net/~cmueller/i-overvw.html*.

Pitofsky, Robert (2000), *Promoting Competition, Protecting Consumers: A Plain English Guide to Anti-Trust Laws*. Federal Trade Commission. Retrieved November 18, 2000, from the World Wide Web: *http://www.ftc.gov/bc/compguide/intro.htm*.

Porter, Michael E. (2001), "Competition and Anti-Trust: Toward a Productivity-Based Approach to Evaluating Mergers and Joint Ventures," *Anti-Trust Bulletin*, Vol. 46, No. 4 (Winter): 919-959.

Porter, Michael E. (2002), "Enhancing the Microeconomic Foundations of Prosperity: The Current Competitiveness Index," in M. E. Porter & J. D. Sachs & P. K. Cornelius & J. W. McArthur & K. Schwab (Eds.). *The Global Competitiveness Report 2001-2002* (pp. 52-77). New York: Oxford University Press, Inc.

Porter, Michael E., Jeffrey D .Sachs, Peter K. Cornelius, John W. McArthur and Klaus Schwab (2002), *The Global Competitiveness Report 2001-2002*. New York: Oxford University Press, Inc.

Rill, James F., Paula Stern, Merit E. Janow, Zoë Baird, Thomas E. Donilon, John T. Dunlop et al. (2000), *International Competition Policy Advisory Committee Report to the Attorney General and Assistant Attorney General for Antitrust*, Washington, DC: U.S. Department of Justice.

Schacht, Wendy H. (2001a), *Cooperative R&D: Federal Efforts to Promote Industrial Competitiveness* (IB89056): CRS Issue Brief for Congress.

Schacht, Wendy H. (2001b), *Industrial Competitiveness and Technological Advancement: Debate Over Government Policy* (IB91132): CRS Issue Brief for U.S. Congress.

Shane, S. (1993), "Cultural Influences on National Rates of Innovation," *Journal of Business Venturing*, Vol. 7: 59-73.

Stockton, Kilpatrick (2002), *European Union and Japan Create Anti-Trust Alliance*. Legal Alert. Retrieved December 7, 2002, from the World Wide Web: *http://www.kilpatrickstockton.com/site/print/detail?Article_Id=1099*.

U.S. Department of Justice. (1999), "United States and Japan Sign Anti-Trust Agreement," Washington, D.C. Retrieved December 7, 2002, from the World Wide Web: *http://www.usdoj.gov/opa/pr/1999/October/470at.htm*.

United Nations Development Programme. (2001), *Human Development Report 2001*. New York: Oxford University Press, Inc.

United States Code, Title 15 (2002), Commerce and Trade, Chapter 1. Monopolies and Combinations in Restraint of Trade. Washington, DC: Law Revision Counsel, U.S. House of Representatives.

World Bank Group, The (2002), *World Development Indicators Database–United States Data Profile*. Retrieved December 7, 2002, from the World Wide Web: *http://devdata.worldbank.org/external/CPProfile.asp?SelectedCountry=USA& CCODE=USA&CNAME=United+States&PTYPE=CP*.

World Trade Organization (2002), *The World Trade Organization*. World Trade Organization. Retrieved December 7, 2002, from the World Wide Web: *http://www. wto.org/*.

Incorporating
the Dominant Social Paradigm
into Government Technology
Transfer Programs

William E. Kilbourne

SUMMARY. This paper examines problems implicit in transferring technology from developed countries to lesser developed countries. This is done from the perspective of the dominant social paradigm, and argues that the limited success of governmental transfer programs can be attributed to paradigm conflicts between the developer of technology and the recipient. The conclusion suggests that the effectiveness of government technology transfer programs can be improved if a dual approach is adopted. For transfers to lesser developed countries, the initial step should be to understand the differences in paradigms, and this should be followed by the development of appropriate technologies that enhance the development potential of the recipient in terms of the recipients' values and worldviews. *[Article copies available for a fee from The Haworth Document Delivery Service: 1-800-HAWORTH. E-mail address: <docdelivery@haworthpress.com> Website: <http://www.HaworthPress.com> © 2005 by The Haworth Press, Inc. All rights reserved.]*

William E. Kilbourne, PhD, is Professor of Marketing, Department of Marketing, 245 Sirrine Hall, Clemson University, Box 341325, Clemson, SC 29634-1325 (E-mail: kilbour@clemson.edu). His research interests are in globalization and environmental issues in marketing.

[Haworth co-indexing entry note]: "Incorporating the Dominant Social Paradigm into Government Technology Transfer Programs." Kilbourne, William E. Co-published simultaneously in *Journal of Nonprofit & Public Sector Marketing* (Best Business Books, an imprint of The Haworth Press, Inc.) Vol. 13, No. 1/2, 2005, pp. 255-269; and: *Government Policy and Program Impacts on Technology Development, Transfer and Commercialization: International Perspectives* (ed: Kimball P. Marshall, William S. Piper, and Walter W. Wymer, Jr.) Best Business Books, an imprint of The Haworth Press, Inc., 2005, pp. 255-269. Single or multiple copies of this article are available for a fee from The Haworth Document Delivery Service [1-800-HAWORTH, 9:00 a.m. - 5:00 p.m. (EST). E-mail address: getinfo@haworthpressinc.com].

KEYWORDS. Dominant social paradigm, lesser developed countries, technology transfer, government, absorptive capacity

INTRODUCTION

There are a number of government programs that focus on developing, transferring, and commercializing technology. Technology transfer (TT) was enacted as national policy with the Stevenson-Wydler Act of 1980 and then amended to allow federal labs to enter into cooperative R&D agreements in 1986. Interest was further expanded by the perceived need to enhance the competitiveness of small firms through the dissemination of automated technologies because such firms had failed to keep up with foreign rivals. However, the large research labs are not aligned with the needs of small firms and few U. S. states have provided the type of assistance necessary for transferring technologies effectively (Olms 1992). Olk and Xin (1997) also suggest that U. S. policy has attempted to increase technological innovation between countries to develop better relations between them and the U. S. government and U. S. corporations. Thus, the need for successful programs for the transfer of technology is well recognized for both domestic purposes and to enhance international relations.

There are roughly two types of programs, and they are designed either to develop and commercialize technologies, or to commercialize technologies that already exist. Once technologies are developed, they are expected to be employed either by industrial, consumer, institutional, or governmental agencies (Piper and Marshall 2000). Such programs have been relatively unsuccessful, and it is argued that this might be due to the technical character of most organizations involved in such programs and lack of focus on marketing (Piper and Marshall 2000; Piper and Nahshpour 1996). Technology oriented organizations focus more on technology than on marketing, so such programs often fail in the commercialization stage. The process is further complicated by the fact that technologies must first get from the developer to the firm and then be brought to the market, and failure can occur at either of these stages (Linton et al. 2001). Failure could result from excessive costs for some types of technologies that emerging entrepreneurial firms cannot afford or excessively long lead times to profitability. When this occurs, economic growth potential inherent in technology transfer is not realized (Walsh and Kirchhoff 2002).

Along with the push for new technologies, it has also been mandated that agencies demonstrate both the socioeconomic and quality of life benefits of their programs (Piper and Marshall 2000). Among the socioeconomic benefits to the firm are lower costs, higher quality, and an increased ability to produce better products on a more timely basis. Government agencies, on the other hand, are compelled to look for social, economic, and quality of life benefits, although this is usually reduced to traditional economic measures such as new product proliferation, increased competitiveness, trademarks, and jobs because of the difficulty in getting more direct quality of life measures. It is questionable whether these measures actually assess the quality of life changes brought by technology transfers (Piper and Marshall 2000). An examination of the last seventy articles published in the *Journal of Technology Transfer* tends to confirm this conjecture. The majority of articles are comparisons of different countries' approaches to technology transfer, economic implications of transfers, and impediments in public transfers to private organizations. Most appeared to use a highly restricted definition of development that focuses primarily on economic variables.

The forgoing suggests that there are two categories of technology transfer, and the approaches to them can vary. The first is domestic transfers of technologies developed within the country and intended for use in the country of origin. A second type of transfer is that developed within a country and transferred to a different country. Cross-national transfers imply an important dimension for consideration, the similarity of the originating and receiving countries or regions. This provides the framework within which to examine the technology transfer process. It is argued that the nature of the technology and the method of transfer are influenced by the level of development of the country or region and by whether the transfer is international in scope in the sense that it crosses national borders. These dimensions are represented in Table 1.

One of the critical factors in the successful utilization of technology is the absorptive capacity (Cohen and Levinthal 1990; Gann 2001) of the recipient organization. This refers to the recipient's ability to understand and incorporate innovative ideas. The concept is expanded by Glass and Saggi (1998) who relate it to the ability of countries to incorporate new foreign technologies successfully. Referring to lesser developed countries (LDCs) they state, ". . . the limited absorptive capacities of such countries must act as a constraint on the

TABLE 1. Absorptive Capacity and Transfer Conditions

Type of Transfer	Development Level of Recipient	
	Developed	Lesser Developed
Domestic	High	Intermediate
Foreign	Intermediate	Low

ability of foreign firms to transfer state-of-the-art technologies in other situations" (p. 370). Based on this argument, Table 1 suggests that with technologies developed for domestic use that are transferred to highly developed areas, the absorptive capacity would be the highest. For domestic transfer to less developed domestic areas or for foreign transfer to more developed areas, the capacity would be intermediate. For transfers from developed countries to LDCs the absorptive capacity would be low. The analysis that follows from this will focus on the foreign transfer to LDCs, as this is the most problematic type of technology transfer.

If a framework can be developed within which to assess the absorptive capacity of a country in advance, we might significantly reduce impediments to TT. This could speed the process through which technologies are developed within government sponsored programs, accelerate the transfer process, and encourage development within LDCs in both economic and broader quality of life terms. The objective of this paper is to provide a framework to incorporate the concept of the dominant social paradigm into the study of technology transfers to less developed countries.

THE DOMINANT SOCIAL PARADIGM

One approach to examining differences between countries in diverse contexts was initiated by Pirages and Ehrlich (1974) who first used the phrase "dominant social paradigm" (DSP). This paradigm level of analysis was popularized by Kuhn (1996) in the context of scientific progress, but it was expanded to the cultural context suggesting that it contained the shared values and beliefs that constitute the culture's worldview (Cotgrove 1982; Milbrath 1984). Dunlap and Van Liere (1984) further elaborated this perspective and provided the first empirical studies of the DSP. Kilbourne, McDonagh and Prothero (1997) have provided further conceptual development (1997). Their conceptualization will be used in the present analysis as a means for assessing the absorptive capacity of LDCs. A brief description of the DSP will be provided first, and then the implications for the TT process will be examined.

The DSP is described as containing three socio-economic dimensions, political, economic, and technological. The dimensions are interactive and mutually reinforcing, and they form the basis of a culture's worldview. The DSP of Western industrial society, the primary developer and exporter of technology, has been examined empirically and found to be consistent across multiple industrialized countries (Kilbourne et al. 2002). Each of the dimensions will now be briefly characterized.

The Political Dimension

The political dimension of the DSP contains three primary elements all of which were the product of the Enlightenment and were legitimized through the works of Locke (1980). The first is possessive individualism that suggests there is an individual who is separate from society and who is in possession of himself or herself (MacPherson 1962). Because they are in possession of themselves, all they acquire through their labor is rightfully their own possession, and further, that these possessions are held in exclusivity. No one else has a right to them. This then becomes the justification for private property, which is the second part of Locke's theory. Private property here refers to a system of rights that surround possessions and not mere possessions themselves. This leads to the fundamental belief in Western societies that the accumulation of private goods is not only acceptable, but also desirable. Finally, a limited government protects the rights of the individual, including property rights, and enforces contracts regarding such rights. The superordinate goal within the political dimension is the development of democratic institutions conducive to individual freedom and responsive government. This forms the basis for what is now referred to as political liberalism.

The Economic Dimension

The economic dimension was the product of Adam Smith (1937) and the Scottish Enlightenment, and it follows directly from the political dimension. The relevant propositions here are that preferences of free, self-interested, atomistic individuals are best satisfied through the operation of free markets. If market forces are left unfettered, the interests of society will be best served through the functioning of the invisible hand. Here the basic principles are again possessive individuals who seek their own interests and limited government that does not intervene in the operation of markets. Thus the idea of free markets follows from Locke who created the necessary conditions for markets to exist. This is now referred to as the doctrine of *laissez faire*, or economic liberalism, and forms the basis for economic organization within Western industrial societies. The superordinate goal in the economic dimension is efficient economic growth, and the criterion of efficiency is considered to be Pareto optimality in which the absolute gains of one party are not at the expense of another. While Western industrial societies are not necessarily equally committed to all these principles and are certainly not in agreement on the best strategy for achieving them, they do form the basis for those societies. The term neoliberalism is now in popular usage to describe the resurgence of economic and political liberalism since the early 1980s and is frequently attached to the globalization process (Hertz 2001; Stiglitz 2002).

The Technology Dimension

One of the difficulties in the study of technology is the definition itself. The vernacular expression generally refers to any type of mechanical or electronic device, or more generally, machines. On the other end of the scale, technology refers to the systematic application of knowledge to practical tasks (Galbraith 1972). More recently, the definitions have tended to converge in the center of this spectrum. For purposes here, a synthesis of the two extremes will be adopted. Technology is defined as "the application of scientific and other organized knowledge to practical tasks by ordered systems that involve people and organizations, living things and machines" (Pacey 1983, p. 6).

The roots of the technological dimension lie within the Enlightenment and trace back to Bacon (1944). Throughout the history of modern science, the superordinate goal has been the "betterment of man's estate." This suggests that the means through which the inconveniences of life are removed is through scientific advances that lead to technological development. That this project has been successful in the West is subject to little dispute so long as the evaluative criteria are derived from both political and economic liberalism, i.e., the DSP. The West has achieved unparalleled growth in the accumulation of private property, the development of individual freedom, and the lessening of the inconveniences of our estate. Because these goals have been achieved so extensively, the prevailing attitude of Western industrial societies is technological optimism, or the belief that all problems can and will be solved by advancing technology. This includes those problems that are themselves caused by technology. The superordinate goal of technology is material progress through the application of science (Leiss 1990; Winner 1986). This justifies the desire of Western societies to initiate programs, both public and private, for the development of more advanced technologies and for their dissemination on both domestic and global levels.

Thus, the DSP consists of political, economic, and technological dimensions. An essential feature of the DSP is that the three dimensions are interrelated and mutually reinforcing. While they are fairly consistent across Western industrial societies with possible differences in implementation strategies, they vary little on fundamentals. Another characteristic of the DSP, as is true of all paradigms, is that its proponents accept its principles as axiomatic not requiring justification. By providing both the principles and the criteria of evaluation, paradigms become self-justifying. Because of this, paradigm shifts are, as suggested by Kuhn (1996), infrequent and are typically the product of unusual circumstances that cannot be predicted or controlled. It is this aspect of the DSP that can be difficult to understand and incorporate into TT programs whether they are private or public. It is within this context, however, the programs for TT to LDCs need to be examined. This is reflected again in Table 1 if we consider the absorptive capacity of a country to be a function of its DSP, or more specifically, the deviation of the technology developer's DSP from the receiver's. Then it is clearer why some contexts might present a different set of problems than others. It is to this issue that the discussion now turns.

TECHNOLOGY TRANSFER AND THE DSP

Transfer of technology from developed to underdeveloped economies has been examined for a number of years now. It is usually examined from one of two major perspectives, the traditional economic perspective or the cultural perspective. Most studies have focused on the economic aspects (Kedia and Bhagat 1988). This can be seen in a casual review of technology related journals in which the majority of articles in the last five years relate to methods of transfer and the economic implications of such transfers. Characteristic of such approaches is Glass and Saggi (1998) who examine the efficacy of TT between developed nations with high levels of technology and LDCs that are far behind them. In this situation, a technology gap exists that must be bridged. They conclude that strategies resulting in a shrinking of the technology gap facilitate transfers of advanced technology through direct foreign investment that would be too risky when the gap is too large. Unfortunately, they refer to such LDCs as "backward," failing to recognize that there is more to cultural development than technology. Researchers who focus on the economic dimension of the TT process infrequently examine the cultural dimension.

Kedia and Bhagat (1988) suggest that the economic approach is too limited and argue that the efficacy of technology transfers to LDCs is significantly affected by cultural differences. They further argue that the cultural factor plays a smaller role in transfers within or between industrialized cultures, but it is most important in transfers from industrialized nations to LDCs. They expand the study of TT by adding Hofstede's (1980) conceptualization of cultural values into their model. The cultural values that are added to the assessment of technology transfer include such factors as uncertainty avoidance, masculinity-femininity, individualism-collectivism, power distance, and abstractive-associative. Power distance refers to the willingness of members to accept unequal distribution of power and rewards, while the associative factor refers to the importance of context. In associative cultures, for example, the context of communication is important while in abstractive cultures it is less so. The other factors are well characterized by their names.

Kedia and Bhagat (1988) offer propositions regarding the impact of cultural values on TT. They hypothesize that cultures that are high power distance, accepting of uncertainty, individualistic, masculine, and abstractive are more amenable to successful TT. While this broadens the perspective on TT between different cultures, it does not achieve the depth of analysis that is suggested in the study of alternative paradigms. While both economic and cultural variables play a part in the DSP of a society, they are not, as suggested above, a complete picture of it. We can achieve a more complete picture of the TT process if we expand the domain of inquiry to the paradigm level. This is because technological development is not the sum of development; it is only a piece of it.

It has been argued by Sen (1999) that an integral part of the development of any country is the substantive freedoms that it enjoys. The substantive freedoms he refers to are such qualities as political freedom, public choice institutions, labor markets, capabilities, freedom from deprivation, gender equality, and security. He goes even further to argue that such freedoms *are* development and cannot be considered separately from it. Substantive freedoms are prior to technological or economic development and should be considered a starting point rather than the end of the process. This suggests that the starting point for government sponsored TT programs ought not be the technology itself. This echoes Piper and Marshall's (2000) argument that government programs tend to be too technology driven and fail in implementation as a result.

From the perspective of the DSP, this caveat is amplified because the elements of the DSP are mutually reinforcing and attempts to effect changes at one level only diminish the prospects for successful transfer. This perspective is supported by Lado and Vozikis (1996) who state, "Thus, a critical step in the transfer of technology involves the determination of the content of technology necessary to complement existing capabilities in the recipient country" (p. 60). From the perspective of Sen (2002), existing capabilities should be considered when developing technologies for particular applications in particular countries, as this is directly complicit in the absorptive capacity of the country. This again suggests that a country's absorptive capacity is a function of its DSP, or more precisely, the variation between the DSP of the developing country and that of the recipient. This is a case of paradigm conflict, and, because such conflict reduces the likelihood of success in TT, the nature of the conflict should be understood. The discussion now turns to the nature of paradigm conflict and its role in TT.

Paradigm Conflict

Within the industrialized countries, there is a consistency in worldviews. This does not suggest that there is in any sense, a monolithic West that agrees on all aspects of their paradigm. There are often highly contested policies on such things as TT. However, the differences are generally in the area of strategy rather than structure. This suggests a commonality on the question of whether technologies ought to be transferred but differences as how best to do it. The DSP of the industrialized West is based on political and economic liberalism and the prevalent technological attitude is one of optimism leading to what O'Riordan (1995) refers to as the technofix. This is the belief described earlier that all problems, both technical and social, can be solved with technology. This Enlightenment based DSP provides the framework within which Western societies have developed. From the perspective of TT, the essential feature of the paradigm is that it selects which forms of social or technical development will survive. Those forms that are consistent with the paradigm develop and those that are not must change or disappear. So long as the society functions well and in accordance with the evaluative criteria the paradigm provides, it continues on a consistent developmental path. Its political processes, economic processes, and technological development converge and are mutually supportive. This suggests that democratic institutions, economic growth, and technological progress evolve and characterize Western industrial society, but not all societies.

This process works well so long as the context does not change dramatically or quickly. Government programs for the development and transfer of technology exist within the Western industrial context, and, as a result, both their structure and output will be consistent with the elements of the DSP in which they are immersed. An important aspect of DSP theory is that this consistency with the DSP is not recognized as occurring by members of the society. It is axiomatic to the participants that this is the way it ought to be. Any alternative version of reality, politically, economically, or technologically, becomes defined as "backward." Technologies developed within this context will be consistent with democratic political institutions and promote material economic growth. They will be most adaptable to organizations that have evolved within the same paradigm because they will reflect the institutions of that paradigm.

As the receiving country becomes more paradigmatically distinct from the developing country, its absorptive capacity for the new technology will diminish because the characteristics of the technology will be more suitable in the paradigm in which it was developed. The characteristics of the technology include such factors as its infrastructural requirements, its cultural integration, and its fundamental purpose. Technology cannot be considered as disembodied instruments that are equally applicable in any context. Or as Karake (1988) observes:

> Technology of Western origin has different characteristics than that of Eastern origin. Each is created with different perceptions and end-goals according to the economic requirements, objectives, costs, and factor endowments of supplier countries. (p. 1100)

As with many studies, this refers primarily to economic factors, but the passage suggests that the DSP of the origin dictates the character of the technology. To the extent that the DSP of the developing nation deviates from that of the LDC, there is the potential for paradigm conflict, the resolution of which can be difficult because it requires paradigm shifts.

Kilbourne (1999; 2002) has addressed the nature of such paradigm shifts in a social context and suggests the choice between paradigms is not just a transformation of methods as the disembodied conception of technologies suggests. It is a choice between sometimes incompatible models of social life that may be contained within a technology. When such conflicts develop, they cannot be resolved by paternalistic exhortations to join the "modern" world. The standards of one paradigm cannot be judged by the standards of another because failure of consensus results from differences in traditions not questions of strategy. Because the conflict is over premises rather than conclusions, there is no neutral mechanism through which to choose between alternatives. Each of the competing paradigms has their own self-justifying mechanisms. From the standpoint of governmental programs for technology development and transfer, this suggests that there are two approaches to the solution of the problems of paradigm conflict.

MERGING THE DSP AND TT

The first solution is to use persuasion to convince the LDC that a new technology is the solution to a long-standing problem that the indigenous paradigm has failed consistently to solve. This, however, presumes that there is consistency between superordinate goals suggesting that both agree that economic growth or democratic institutions, for example, are desirable within the receiving country. This is precisely where the greatest disagreement is likely to occur, however. What is required in this resolution mode is a translator who is conversant in both paradigms. Then problematic areas can be exposed and examined critically by both sides in an intellectual negotiation that will put the

LDC on a negotiated development path incorporating the appropriate technologies. This is, however, a difficult proposition for two reasons. The first is that there are few such translators who are capable of bridging the DSP gap. The second and more important reason is that, if the DSPs of the two countries are indeed incompatible, then persuading the recipient that this is not the case will not enhance the likelihood of a successful transfer.

This leaves a second approach that incorporates Sen's (1999) suggestion that the political, economic, and technological dimensions be commensurate with each other at the outset. In this second approach it is argued that the process should begin one step removed from the technological development process. This requires that, as suggested by Piper and Marshall (2000), technology programs not be driven by technology alone. This would result in a one-dimensional technology assessment program. Technologies that are developed are the technologies that can be developed in the current development programs. Their commensurability with competing paradigms is not a consideration in such programs because technologies are developed first, and then a potential candidate for transfer is sought. This is where the difficulty lies and where a solution can be found.

Rather than developing a technology and then seeking transfer opportunities, the first step would be to do a thorough assessment of the country, or types of countries, into which the new technology is to be imbedded. An examination of its political, economic and technological structures can be undertaken objectively without paternalistic intentions. The objective must be to develop and transfer technologies that lead to the ultimate objective such programs are designed to achieve. That is the development of long-term relations between receiving countries and U. S. corporations and the U. S. government. Transfer failures do not lead toward this objective and frequently result in what sociologists refer to as "blaming the victim" (Ryan 1976).

To increase the probability of success in this domain, government programs must be geared to the DSP of LDCs, not their own. There are several reasons for this. First, choices of technology should be made directly based on institutional compatibility and not on strictly economic or scale criteria. This suggests, but does not dictate, intermediate technologies that do tend toward labor intensity and smaller scale. The second critical factor is the transfer of technologies that increase production and productivity for domestic consumption of products that are needed within the receiving country (Schumacher 1973). Large-scale technologies for the manufacture of exports have been shown to worsen conditions in LDCs if the proper institutional structures are not in place (Sen 2002). The Russian experience has demonstrated adequately that market like institutions do not spontaneously develop with liberalization. It cannot be assumed that technological convergence will evolve in all cases alike.

Finally, after a thorough examination of the institutional structure of the LDC, focus should be shifted to the development of institutions that facilitate development. As Sen (1999) argues, these institutions are not the result of development as much as they are the preconditions for it. Because development requires substantive freedoms suggested above, appropriate technologies are those that serve to develop the institutions that are deficient within an LDC. Health care, education, courts to resolve contract disputes, public health systems, and labor markets all are necessary conditions for development to occur. Having government programs that develop technologies that are compatible with the DSP of the country and foster institutional growth in these and similar areas serves to strengthen relations as desired and to foster further technology transfers as the success of initial transfers creates the demonstration effect. Programs that develop and transfer advanced technologies to LDCs try to accelerate the development process, but they fail to recognize that development within an LDC must be a slow, evolutionary process because paradigm change is a slow process. It involves changing the superordinate goals of the political and economic processes through technological development. When these goals are inconsistent, the reinforcing nature of the political and economic dimensions will militate against the technological to dampen the developmental process. Once this is understood, the approach to TT can be more conducive to development.

CONCLUSION

Over the last two decades, the transfer of technology has been enacted as national policy. While multiple programs have been developed, many have met with only limited success. This is particularly true for TT to LDCs from developed countries. Explanations for this phenomenon are now being sought. To develop such an explanation, this paper has focused on the concept of absorptive capacity and argues that, for this particular type of transfer, absorptive capacity of LDCs is low when importing advanced technologies. The framework within which this is analyzed is the difference between the DSP, or worldview, of receiving societies and that of the developers. The conceptualization of the DSP used here is that it contains three dimensions. These are the political, economic, and technological dimensions that exist in a mutually reinforcing relationship. The paradigm provides not only the objects of interest in society, but also the evaluative criteria by which its functioning is judged.

When the TT process is between two countries with different paradigms, a situation of paradigm conflict emerges which impedes the transfer process because the absorptive capacity of the receiving country is limited by its paradigm. While technologies developed in Western industrial societies are consistent with their originating paradigm, they may be incompatible with the receiver's paradigm. Resolution to this conflict requires one of two approaches. The first is to try to build an intellectual bridge between the paradigms, but this requires conditions that are not typically present. The receiving culture must be convinced, through persuasion, that its paradigm is deficient and that TT is a way to overcome the deficiency. Only a "translator" who is conversant in the paradigm of both cultures can achieve this. Because the paradigm is self-reinforcing and there is no neutral mechanism to use for choices between alternatives, this approach is problematic.

The second approach is to modify the TT process itself. The absorptive capacity of LDCs is a function of their paradigm, so the nature of their paradigm must be understood before the TT process is begun. This suggests that national policy for government controlled TT institutions should consist of two different approaches, one for domestic transfers and transfers to similarly developed economies and a second for transfers to LDCs. The DSPs in the first instance are more compatible with the paradigm of the developer while, in the second instance, they can be incompatible. Programs such as the Sandia National Laboratories have met with some success in domestic transfers. International transfers to LDCs have been more problematic, however.

The approaches used in TT to LDCs can be improved if a different perspective on development is initiated. It is argued here that Sen (1999; 2002) provides a starting point in the recognition that DSPs of many LDCs frequently lack substantive freedoms that are preconditions for the development process involved in absorbing a technology. This refers to the political, economic, and technological institutions prevalent in the country, i.e., its DSP. Rather than transferring technologies that are incompatible with the existing DSP, it is argued here that government TT should begin with a thorough understanding of the prevailing paradigm into which technologies will be embedded. This becomes a paradigm driven enterprise rather than a technology driven one. Technologies that are appropriate to the DSP of the intended recipient can be developed. This generally, though not necessarily, implies intermediate level technologies because of the infrastructural and perceptual gaps that are a product of different paradigms. It has been demonstrated that inappropriate technology transfers can lead to onerous conditions that were not intended by the developer or the recipient (Sen 1999; Stiglitz 2002). Some of these unintended and unfortunate consequences can be avoided when incompatible DSPs are reconciled in the beginning of the technology development process. When technologies are not considered neutral, disembodied instruments, they can be compatible with any paradigm. Government programs for TT that have as their objective, better relations with LDCs for their mutual benefit, can profit from taking this dual approach in the development and dissemination of technology. It can result in a win-win relationship.

REFERENCES

Bacon, Francis (1944), *Advancement of Learning and Novum Organum*. New York: Willey Book Co.

Cohen, W. M. and D. A. Levinthal (1990), "Absorptive Capacity: A New Perspective on Learning and Innovation," *Administrative Science Quarterly*, 35: 128-52.

Cotgrove, Stephen (1982), *Catastrophe or Cornucopia: The Environment, Politics, and the Future*. New York: Wiley.

Dunlap, Riley E. and Kent D. Van Liere (1984), "Commitment to the Dominant Social Paradigm and Concern for Environmental Quality," *Social Science Quarterly*, 65: 1013-28.

Galbraith, John Kenneth (1972), *The New Industrial State*. London, UK: Andre Deutsch.

Gann, David (2001), "Putting Academic Ideas Into Practice: Technological Progress and the Absorptive Capacity of Construction Organizations," *Construction Management and Economics*, 19 (3): 321-30.

Glass, Amy J. and Kamal Saggi (1998), "International Technology Transfer and the Technology Gap," *Journal of Development Economics*, 55 (2): 369-99.

Hertz, Noreena (2001), *The Silent Takeover: Global Capitalism and the Death of Democracy*. London, UK: The Free Press.

Hofstede, Geert (1980), *Culture's Consequences: International Differences in Work-Related Values*. Beverly Hills, CA: Sage Publications.

Karake, Zeinab A. (1988), "Effects of Eastern European and Western Technologies on LDCs: A Macroeconometric Analysis," *Applied Economics*, 20 (8): 1099-115.

Kedia, Ben L. and Rabi S. Bhagat (1988), "Cultural Constraints on Transfer of Technology Across Nations: Implications for Research in International and Comparative Management," *Academy of Management Review*, 13 (4): 559-71.

Kilbourne, William E. (1999), "Contested Rationalities in the Ecological Crisis," in *Sustainable Consumption and Ecological Challenges: Macromarketing XXIV*, Thomas A. Klein and Pierre McDonagh and Andrea Prothero (Eds.). Nebraska City, NE: University of Nebraska College of Business.

Kilbourne, William E. and Suzanne C. Beckmann (2002), "Rationality and the Reconciliation of the DSP With the NEP," in *Environmental Regulation and Rationality: Multidisciplinary Perspectives*, Suzanne C. Beckmann and Erik K. Madsen, Eds. Aarhus, Denmark: Aarhus University Press.

Kilbourne, William E., Suzanne C. Beckmann, and Eva Thelen (2002), "The Role of the Dominant Social Paradigm in Environmental Attitudes: A Multi-National Examination," *Journal of Business Research*, 55 (3): 193-204.

Kilbourne, William, Pierre McDonagh, and Andrea Prothero (1997), "Sustainable Consumption and the Quality of Life: A Macromarketing Challenge to the Dominant Social Paradigm," *Journal of Macromarketing*, 17 (1): 4-24.

Kuhn, Thomas S. (1996), *The Structure of Scientific Revolutions* (3rd ed.). Chicago: University of Chicago Press.

Lado, Augustine A. and George S. Vozikis (1996), "Transfer of Technology to Promote Entrepreneurship in Developing Countries," *Entrepreneurship: Theory and Practice*, 21 (2): 55-73.

Leiss, William (1990), *Under Technology's Thumb*. Montreal: McGill-Queen's University Press.

Linton, Jonathon D., Cesar A. Lombana, and A. D. Romig (2001), "Accelerating Technology Transfer From Federal Laboratories to the Private Sector–The Business Development Wheel," *Engineering Management Journal*, 13 (3): 15-19.

Locke, John (1980), *Second Treatise of Government*. Cambridge, MA: Hackett Publishing Company, Inc.

MacPherson, Crawford, B. (1962), *The Political Theory of Possessive Individualism*. Oxford: The Clarendon Press.

Milbrath, Lester (1984), *Environmentalists: Vanguards for a New Society*. Albany, NY: University of New York Press.

Olk, Paul and Katherine Xin (1997), "Changing the Policy on Government-Industry Cooperative R&D Arrangements: Lessons From the U. S. Effort," *International Journal of Technology Management*, 13 (7/8): 710-29.

Olms, John M. (1992), "Technology Transfer: Federal Efforts to Enhance the Competitiveness of Small Manufacturers," *Economic Development Review*, 10 (1): 86-88.

O'Riordan, Timothy (1995), "Core Beliefs and the Environment," *Environment*, 37 (8): 4-9.

Pacey, Arnold (1983), *The Culture of Technology*. Cambridge, MA: The MIT Press.

Piper, William S. and Kimball P. Marshall (2000), "Stimulating Government Technology Commercialization: A Marketing Perspective for Technology Transfer," *Journal of Nonprofit & Public Sector Marketing*, 8 (3): 51-63.

Piper, William S. and Shahad Nahshpour (1996), "Government Technology Transfer: The Effective Use of Both Push and Pull Marketing Strategies," *International Journal of Technology Transfer*, 12 (1): 85-94.

Pirages, Dennis C. and Paul R. Ehrlich (1974), *Ark II: Social Response to Environmental Imperatives*. San Francisco: Freeman.

Ryan, William (1976), *Blaming the Victim*. New York, NY: Knopf Publishing Group.

Schumacher, E. F. (1973), *Small Is Beautiful: Economics as if People Mattered*. New York, NY: Harper & Rowe, Publishers.

Sen, Amartya K. (1999), *Development As Freedom*. New York, NY: Knopf.

_____ (2002), "How to Judge Globalism," *The American Prospect*, 13 (1): 1-10.

Smith, Adam (1937), *An Inquiry into the Nature and Causes of the Wealth of Nations*. New York: Random House.

Stiglitz, Joseph E. (2002), *Globalization and Its Discontents*. New York, NY: W. W. Norton & Company.

Walsh, Stephen T. and Bruce A. Kirchhoff (2002), "Technology Transfer From Government Labs to Entrepreneurs," *Journal of Enterprising Culture*, 10 (2): 133-50.

Winner, Langdon (1986), *The Whale and the Reactor: A Search for Limits in an Age of High Technology*. Chicago: University of Chicago Press.

Governmental and Corporate Role
in Diffusing Development Technologies:
Ethical Macromarketing Imperatives

Oswald A. J. Mascarenhas
Ram Kesavan
Michael D. Bernacchi

SUMMARY. Despite the recent growth and globalization of production and trade, global income inequalities are widening, especially among the developing nations and global sustainability is weakening. We suggest a macromarketing approach to resolve this situation. It is in the best interest of the governments and the multinational corporations of the developed world to offer and diffuse their development technologies

Oswald A. J. Mascarenhas, PhD, is Charles H. Kellstadt Professor of Marketing, University of Detroit Mercy, 4001 McNichols Road, Detroit, MI 48221 (E-mail: mascao@udmercy.edu). His research interests are ethics of marketing, macromarketing and e-business.

Ram Kesavan, PhD, is Professor of Marketing, University of Detroit Mercy, 4001 McNichols Road, Detroit, MI 48221 (E-mail: kesavar@udmercy.edu). He has served as the President of the Marketing Management Association. His research interests include the areas of entrepreneurship and global marketing.

Michael D. Bernacchi, PhD, JD, is Professor of Marketing, University of Detroit Mercy, 4001 McNichols Road, Detroit, MI 48221. He is the author of a marketing/advertising newsletter, *Under The Mikeroscope*.

[Haworth co-indexing entry note]: "Governmental and Corporate Role in Diffusing Development Technologies: Ethical Macromarketing Imperatives." Mascarenhas, Oswald A. J., Ram Kesavan, and Michael D. Bernacchi. Co-published simultaneously in *Journal of Nonprofit & Public Sector Marketing* (Best Business Books, an imprint of The Haworth Press, Inc.) Vol. 13, No. 1/2, 2005, pp. 271-291; and: *Government Policy and Program Impacts on Technology Development, Transfer and Commercialization: International Perspectives* (ed: Kimball P. Marshall, William S. Piper, and Walter W. Wymer, Jr.) Best Business Books, an imprint of The Haworth Press, Inc., 2005, pp. 271-291. Single or multiple copies of this article are available for a fee from The Haworth Document Delivery Service [1-800-HAWORTH, 9:00 a.m. - 5:00 p.m. (EST). E-mail address: getinfo@haworthpressinc.com].

Available online at http://www.haworthpress.com/web/JNPSM
Digital Object Identifier: 10.1300/J054v13n01_15

among the developing nations. We apply the theories of economic resources, development technologies, global sustainability and distributive justice to characterize the nature, structure and ethics of such developmental interventions. *[Article copies available for a fee from The Haworth Document Delivery Service: 1-800-HAWORTH. E-mail address: <docdelivery@haworthpress.com> Website: <http://www.HaworthPress.com> © 2005 by The Haworth Press, Inc. All rights reserved.]*

KEYWORDS. Global inequality, global sustainability, macromarketing, technology transfer

INTRODUCTION

A major economic development over the past three decades is the increased globalization of production and distribution (Acemoglu 2003). The world of self-contained national markets has continually progressed to one of linked global markets fueled by the homogenization of customer needs, the proliferation of mass communication, gradual trade liberation and the competitive advantage of a global presence (Ohmae 1991). World trade has expanded from $200 billion to more than $4 trillion in the past two decades–an annual compound growth of more than 17 percent (Jeannet and Hennessey 2000). For many, a global economy has provided the opportunity for substantial income growth, and the availability of more products and services that are of better quality and increasingly differentiated (Kaplinsky 2000). The modern corporation is undergoing a fundamental transition from a business serving the home market to a firm that is responding to a global market. In this connection, even the smallest firm is subject to the powerful indirect influences of the global marketplace (Kotler 2003).

Notwithstanding its benefits, globalization has also created much disparity between and within countries in terms of income and opportunity. It has bypassed much of the world's population, particularly those living in poor countries. There has been little correspondence between the global spread of economic activity and the spreading of the gains from those participating in global product markets (Acemoglu 1999, 2002; Andersen 2002). One would even suspect a causal link between globalization and global inequality (Kaplinsky 2000). Hence, the *first critical question:* can *unchecked globalization* simultaneously assure the global spread of income and opportunity and the reduction of global inequalities? Evidence indicates it cannot (Acemoglu 1999, 2002, 2003; Andersen 2002; Kaplinsky 2000; Magretta 1997).

Hence, the cooperation and the strategic alliances between the public and the private sectors are critical. Today, businesses and governments are two of the most powerful institutions in the world. The public sector influences private sector decision making and vice versa. Most analyses of the interactions between business and government focus on the costs and benefits of public policy actions to domestic stakeholders (Eagelton 1991; Weidenbaum 2004), and rarely deal with their global constituencies (Kopfensteiner 1993; Mascarenhas 1980; Vann and Kumcu 1995). In this context, the *second critical question:* can the tremendous power and resources of the developed world's governments and multinational corporations be streamlined to reduce global income inequalities and structural injustices and to generate economic self-sufficiency and competitive global advantage among the developing nations? Indeed, it can, and this is the thesis of our paper.

One of the main catalysts of globalization is technological change that enables a global coordination of widely dispersed activities, while the pressure for globalization drives the extensions of technology into new applications (Day and Montgomery 1999). Thus, the problem of great global inequality regarding income and opportunity can be significantly lessened if the developed countries share and help diffuse their development technologies among the developing nations–the result being increased world consumption that leads to increased worldwide production and hence a stronger global economy. By *developing nations* we include those previously classified as less developed countries (LDCs), especially the largest low-income countries of the world, but exclude the newly industrialized countries (NICs). We specifically propose that the governments (e.g., the G-7 or now the G-8 countries) and the multinational corporations (e.g., *Fortune Global 500* corporations)–a powerful resource group that we designate hereafter as the *Government-Corporation Consortium* (GCC)–take the lead in this equalizing and developmental venture.

Why Should the Developed World Be Concerned with Global Inequalities?

The reduction of global income inequalities directs our attention to the importance of consumption as the main driver of production. That is, the reduction of global income inequalities will result in: (a) an increase in global consumption, (b) an increase in universalizing (non-luxurious) consumption, (c) an increase in universalizing global demand, and therefore (d) an increase in universalized (global) GNP. In turn, these increases will absorb the current excess supply as well as slack production capacities in the saturated developed markets. Moreover, reduced global inequalities and increased global demand among the developing nations will automatically stimulate foreign direct investments from the developed to the developing world markets.

Hence, our specific approach to the problem of global inequalities is *macromarketing* as Fisk (1974, 1982, 1999) understood and expanded it. The concept of macromarketing is multidimensional. It studies: (a) aggregate marketing systems such as institutional and global structures of marketing, (b) their impact on society such as economic development, global inequalities and sustainability, and (c) the impact of society on marketing systems such as legal, political and social value systems (Fisk 1982; Hunt 2002). Fisk (1974) established connections between consumption, quality of life, sustainable economic development and environmental protection. Following this lead, we believe that the global production-consumption system should include the well being of all the citizens of the world. "Radical and postmodern marketing thought seeks to expand social justice by recognizing the interests of inadequately represented stakeholders in the marketing process" (Fisk 1999, p. 120). We need holistic, humanistic and historical approaches to macro-manage our earth and the universe (*Cosmomarketing*) in order to be able to respond to global economic and distribution problems.

Given the aforementioned problems, (1) we invoke the theories of economic resources and development technologies to characterize the nature, structure and purpose for GCC interventions in diffusing such technologies among developing nations; and (2) we rely on the theories of global poverty, sustainability and distributive justice to ethically ground such technology transactions. Part One deals with the "what" and Part Two with the "why" of GCC interventions in commercializing development technologies among developing nations of the world.

PART ONE:
CHARACTERIZING GCC INTERVENTIONS
IN DIFFUSING DEVELOPMENT TECHNOLOGIES

The theory of resources and sustainable competitive advantage is necessary for characterizing the role of the GCC in sharing new technologies for stimulating progress among developing nations.

Theory of Resource and Sustained Competitive Advantage

A firm gains *competitive advantage* (CA) "when its actions in an industry or market *create economic value* and when a few competing firms are engaged in similar actions" (Barney 2001, p. 9). A firm's actions create market or economic value either because of its own internal resources (this is the resource-based view (RBV) of CA) and/or because of external market conditions (this is the Industrial Organization (IO) based view of CA) (Porter 1990). The RBV theory uses the term "resources" to include all assets, capabilities, organizational processes, firm attributes and information that are controlled by a firm and that enable it to conceive and implement strategies to improve a firm's efficiency and effectiveness (Barney 2001). RBV suggests resources are valuable when they are: (1) *Convertible:* the firm should be able to use the resources or assets to convert an opportunity or neutralize a threat, (2) *Rare* (possessed by very few rivals), (3) *Non-imitable* by rivals and (4) *Non-substitutable* (with rivals not having strategically equivalent convertible assets) (Barney 2001). Of these four tests, the first is the most important–the assets must be *convertible into use* (i.e., customer value). Without the first test verified, the other three remaining tests are irrelevant even if verified (Barney 2001; Srivastava, Shervani, and Fahy 1998). Rarity and value are necessary conditions for CA, whereas nonimitability, non-transferability and non-substitutability are necessary but not sufficient conditions for *sustainable competitive advantage* (SCA) (Priem and Butler, 2001). What gives SCA to resources is the *relative difference* in value generated by firms (Srivastava, Shervani, and Fahy 1998).

Thus, invoking the combined RBV/IO framework we propose that an effective macro-marketing strategy may require the GCC to diffuse various development technologies to developing nations such that they create a sustained competitive advantage (SCA) for their markets. Porter (1990) has expanded the theory of corporate CA and SCA to public corporations such as sovereign nations. Thus, the technologies shared by GCC with developing nations should eventually be resources that are convertible, rare, nonimitable and nonsubstitutable for the host nations (Porter 1990).

Development Technologies

Development deals with developing people in the developing nations. In the short term, development seeks to alleviate hunger, poverty and disease among the suffering masses by providing basic necessities such as, clean water, nutritional food, energy, shelter, health, education, job, and transportation. In the long run, development makes a nation economically able to provide for its own basic necessities of life (*World Development Report 2000/2001*). Development can either be economic or community development.

Most economists define *economic development* as an outcome, such as the growth in per capita income. Others define it as the process of creating wealth. Economic development deals with the factors that expand production and service frontiers of local, regional, national or global economies–where people live and work. *Community development* is a much broader concept than economic development. The former primarily relates to the quality of life brought about by "good" consumption or the *demand* side (e.g., health, housing, environmental quality). Good jobs and skills development are the *supply* side (e.g., education, retraining skills). Both support the strength of the social fabric of the community (e.g., values, bonding, religion, cooperation, solidarity) (Hill 1998). The economic value of a community is the total value of its demand and supply sides.

The diffusion of development technologies should bring about both economic and community development of the various constituencies of developing nations. The GCC could connect with the developing nations, mentor them and cooperate to bring about their economic and community development (Nowak 1997). The link between sustained economic development and sustained community development (both of which constitute SCA for the developing nations) is primarily employment, the quality of the work force, affordable housing and modern healthcare. The GCC-spearheaded diffusion of development technologies could strengthen this link (Hill 1998; Kumcu and Vann 1991).

Technology is both the techniques and machines available to firms, and the organization of production, distribution, labor markets, consumer lifestyles and quality of life (Acemoglu 2003). Some technologies are *labor-intensive* while others are *capital-intensive*. In developing nations where labor is abundant and unemployment is high, labor-intensive technologies are preferred to the capital intensive ones. In essence, labor-intensive technologies absorb large labor, spread wage earnings across a large labor market, and thus tend to reduce income inequalities. Capital-intensive technologies are sophisticated technologies that require higher technical skills and hence, tend to be *skills-biased* (i.e., higher skills are in greater demand and enjoy income premium) and increase income inequalities within the labor force (Acemoglu 2002).

According to the skill-biased theory of technology, capital-intensive technology seems to favor skilled (educated) workers to replace tasks previously performed by the unskilled and to increase the demand (and wage-price) for skills (Caselli 1999; Galor and Moav 2000). Such technology virtually neglects the unskilled, thus increasing income inequalities within the nation. Hence, we define *development technologies* to include a healthy combination of both capital- and labor-intensive technologies, as long as this combination ensures economic self-sufficiency and inequality-reduction for the developing nations and global sustainability for the developed and developing world.

Innovation and technology interweave. Innovations in general provide unique and meaningful benefits to peoples, products and services. Innovation is a "new way of doing things" that can be commercialized (Afuah 1998). It is "new knowledge" to be used in the process or production of goods and services. It may be a *breakthrough* knowledge that generates radical innovations or *incremental* knowledge that stimulates incremental innovations (Chandy and Tellis 1998). The latter involve relatively minor changes and relatively low customer benefits while radical innovations are technological discontinuities that substantially advance the technological state-of-the-art that characterizes an industry. Hence, when development technologies are being macromarketed to developing nations, we advocate a well-balanced mix of radical and incremental innovations that are capital as well as labor intensive. These innovations will accelerate global sustainability and progressively eliminate global inequalities (Li and Atuahene-Gima 2001).

Technology-diffusion has several meanings. We define technology-diffusion as the commercialization of development technologies for the long-term growth and progress of the developing nations. Commercialization implies that the technology should make it to the market. It implies down-stream activities of production, production-cycle compression, marketing and distribution (Afuah 1998). It also implies that the GCC would hear and insert the voice of the developing nations as end-users throughout the commercialization process (Andel 1998). The involvement of the developing nations to the success of GCC interventions is critical.

PART TWO:
ETHICAL GROUNDS
FOR DEVELOPMENT-TECHNOLOGY DIFFUSION

In discussing the "why" of governmental and corporate interventions, we need to understand the nature of several fundamental phenomena such as global poverty and inequalities, global sustainability and distributive justice as a foundation for justifying GCC interventions.

Global Poverty and Inequality

To add perspective to the problem, we cite some glaring global income inequalities of the most heavily populated countries in the world, based on *World Bank* data for the year 2000. The five *largest* high-income countries of the world (namely, U. S., Japan, Germany, France, and U. K.) shared only 10 percent of the world's population of 6 billion in 2000 while enjoying 60 percent of world's buying power. As a contrast, the bottom fifteen *largest* low-income countries (China, India, Indonesia, Pakistan, Bangladesh, Nigeria, Vietnam, Ethiopia, Congo, Tanzania, Kenya, Nepal, Uganda, Sri Lanka, and Ghana, in that order) shared 54 percent of the world's population but had access to only 6 percent of the world's buying power. This gross income-inequality situation has worsened since 1990, and chances are this disparity will continue to grow into the near future.

Global food supplies per person are greater today than at any time. If global distribution of food supplies was according to needs, then one could easily meet the calorie requirements of the world's population (Andersen 2002). The problem, therefore, is not one of global food shortage, but rather one of global mal-distribution. Widespread hunger and malnutrition, associated with significant population growth, have reached astronomic proportions. Global income inequalities have reached all-time highs. The richest 1 percent of the world's population now earns and enjoys more than 57 percent of world-income (UNDP 2001). About 1.2 billion people (20% of the world's population) earn less than a dollar a day and 800 million are food-insecure (UNDP 2002).

Without GCC interventions, we believe the rich would become richer and the poor would become poorer. We can break this cycle of global poverty. For instance, the U. S., as the world's economic power, can lead the way. The U. S. has less than 5 percent of the world's population but controls a remarkable 30 percent of the world's buying power. The GCC could voluntarily accept responsibility for sharing some of its development technologies with the developing world. A productive approach to GCC interventions would be to start the process with the largest low-income countries of the world (listed above). Given the resources of global trade (e.g., WTO), global communication networks (e.g., the Internet) and global transportation and distribution systems, a global diffusion of development technologies becomes an economic and ethical imperative.

Global Sustainability

Global sustainability implies an economy under which our "planet is capable of supporting indefinitely" (Hart 1997, p. 66). With the steady depletion of farmland, fisheries and forests, the urban population of the developing nations is choked, impoverished and overrun with infectious diseases (Hart 1997). *Sustainable global development* enriches all peoples without impoverishing the earth. Sustainable development (SD) relates to a means of unlocking the human potential through economic development based on sound economic policy. SD is a continuous global effort by many players working together over a long period. It requires institutions, policies, people and effective partnerships to carry out this common endeavor (*U. S. Government Report 2002*). Hence, in this paper, we redefine *development technology* as any knowledge (agricultural, industrial or informational technology) that increases global sustainability and reduces global inequalities (in income and opportunity), especially in relation to the developing nations. Abject poverty makes the globe unsustainable. For instance, of the 6.2 billion people in the world today, almost 2 billion live in conditions of destitution (UNDP 2001, 2002). Poverty as understood by the *World Development Report 2001/2001* (p. 15) encompasses not only material deprivation (i.e., low levels of income and consumption) but also low achievements in education and health. Poverty and squalor continue to prevent global sustainability (Magretta 1997) by such things as incubating and helping spread infectious diseases (e.g., AIDS, SARS). Current epidemiological studies causally link the genesis and spread of rapidly infecting diseases to conditions of squalor and malnutrition (e.g., Epstein and Roy 2003; Hart 1977; Hart and Milstein 1999; Kras 2000). Hence, GCC interventions that target global sustainability and the reduction of global inequality and poverty can effectively preempt world spread of disease, a benefit that the developed world would appreciate given current global threats of SARS and AIDS.

Macromarketing Development Technologies for Global Sustainability

Reasons for the inequalities of global income and quality-of-life in general, and among developing nations in particular, have been varied, and at times controversial (Acemoglu 1999, 2000; Caselli 1999; Galor and Moav 2000). Most economists have agreed on the following: (a) developing nations primarily concentrate on the commodity (agricultural and extractive industry) sectors; and (b) world prices or terms of trade of the agricultural and extract industry commodities have been declining or barely increasing for the last 30 to 40 years as compared to the industrial, service and the information technology sectors (Kaplinsky 2000). Hence, transfer of development technologies to the developing nations could progressively aim at *structural changes* that gradually bring about a transition of those societies from over-reliance on commodity sectors (agricultural, extract products) to that based on manufacturing of industrial products and services. Such transfers should result in economic sustainability for those developing nations. Economic sustainability among the developing nations implies increased economic activities (i.e., more industrial output, more employment, more skills-training) with higher economic returns (i.e., higher economic "rents" along all the major points of the value-added chain) (Kaplinsky 2000). It also implies better distribution of these returns (increases in real wages across large labor markets) (see Vann and Kumcu 1995), and consequently, a greater share in the world output, a greater participation in the global trade, and most importantly, an increased buying power for the developing nations.

Recent theoretical and empirical literature on international trade has emphasized the contribution of technology and skills to the relative competitiveness of the developing nations (Kumar and Siddharthan 1994). However, besides adopting and adapting development technology transfers, developing nations should also develop capacities to be *creative and innovative* in the use and further development of such technologies (Prahalad and Hamel 1990). We have already seen such results in China as it has welcomed the production facilities of multinational corporations. China has an abundance of educated labor and their participation in the global market has been increasing significantly. The share of manufacturing in Chinese exports has risen from 49.4 in 1985 to 85.6 percent in 1995 (Khan 1999).

Global development, accordingly, must have a dual imperative of economic growth and environmental sustainability (Epstein and Roy 2003). Global development should go "beyond greening" to develop a sustainable global economy (Hart 1997). That is, the GCC could look beyond its greening efforts to prevent further environmental damage to a sustainable global economy via pollution prevention, product stewardship and clean technology (Hart 1997). Today, many companies and governments are accepting the social responsibility to do no harm to the environment (e.g., CERES, discussed later).

Distributive Justice and GCC Interventions

Justice is giving others what rightfully belongs to them (Rawls 1971); it deals with the aspects of one's rights and duties in society. *Justice* and *fairness* are interchangeable terms, even though some may consider fairness as more fundamental. Distributive justice affirms that the morality of an action is also dependent upon how it distributes its benefits and burdens, and rights and duties among all its stakeholders. According to distributive justice principles, for instance, the GCC should ensure that the rights and duties, benefits and burdens resulting from any of its interventions be equitably spread across all stakeholders impacted by them.

Scholars of ethics have developed the theory of distributive justice from Aristotle (1985) who proposed it around 350 B.C. Of the several sub-theories of distributive justice propounded since then, we select ten that apply best to GCC interventions. The first three sub-theories are from Aristotle (1985); the next two from Rawls (1971) and the last five from Frankena (1973) are current expansions of Aristotle's (1985) principles of distributive justice. Table 1 proposes a program of distributive justice-based rights and duties for GCC interventions. Currently, there is an increasing use of distributive justice principles in marketing (e.g., Laczniak 1999; Mascarenhas 1990, 1991; Robin and Reidenbach 1993; Vann and Kumku 1995).

TABLE 1. A Distributive Justice Mandate for GCC Interventions

Distributive Justice Sub-Theory (Set A)	Moral Principles from Set A (Set B)	Moral Rules Derived form Set B (Set C)	Moral Judgments Derived from Set C (Set D)
Compensatory Justice	*Compensate the wronged persons for the wrong done by restoring them to their original position.*	Compensating the wronged persons for the wrong done by restoring them to their original position is a correct moral rule that all can apply.	If GCC interventions were to result in unjust distributions of rights and duties, and benefits and burdens among the citizens of developing nations, GCC could compensate the victims.
Retributive (Remedial) Justice	*Punish the person for the wrong done so that the person learns to refrain from that wrong.*	Punishing the person remedially is a correct moral rule that all can apply.	GCC could remedially punish those who hamper the just distribution of rights and duties, and benefits and burdens of GCC interventions.
Aristotle's Minimum Principle of Distributive Justice	*Treat all equals within a class equally, and unequals between classes, unequally.*	Treating all one's stake-holders within a given class equally is a correct moral rule that all can apply.	GCC interventions could treat all developing nations equally, especially the poorest and the most indigenous.
Rawls' (1971) Equality Principle:	*Offer all stakeholders fair opportunity for benefits.*	Offering fair opportunity to developing nations for self-development is a correct moral rule that all can apply.	GCC interventions could offer developing nations fair opportunity for economic and community development.
Rawls' (1971) Difference Principle:	*Nullify differences that arise from natural disadvantages.*	Nullifying differences that arise from natural disadvantages of the developing nations is a correct moral rule that all can apply.	GCC interventions could attempt to nullify differences that arise from natural or social disadvantages of the developing nations they work with.
Malfeasance or Strict Liability Justice	*Do not inflict evil or harm on any one by any action.*	Not harming the developing nations, directly or indirectly, is a correct moral rule that all can apply.	GCC should avoid all possible harm to the developing nations from their technological interventions.
Preventive Justice	*Prevent people from harm.*	Prevent developing nations from harm is a correct moral rule that all can apply.	Whenever foreseeable, GCC should prevent all harm to developing nations from their interventions.
Protective Justice	*Protect people from harm.*	Protecting developing nations from harm is a correct moral rule that all can apply.	Whenever possible, GCC should protect developing nations from any harm from their interventions.
Procedural Justice	*Set up just procedures to prevent and protect people from harm.*	Establishing just procedures to avoid all harm to developing nations is a correct moral rule that all can apply.	GCC could establish just procedures to prevent and protect the developing nations from any harm accruing from its technological interventions.
Beneficent Justice	*Promote good among all stakeholders.*	Promoting good to all developing nations is a correct moral rule that all can apply.	GCC interventions can promote net good to all developing nations, especially, the most vulnerable ones.

In general, an ethical or moral justification of any judgment, decision or institution involves four sets of beliefs (De George 1999; Velasquez 2002). A: a set of normative ethical theories; B: a set of moral principles derived from set A; C: a set of moral rules derived from set B, and D: a set of considered moral judgments resulting from applying sets A, B, and C to concrete actions. *Normative ethical theory* is the reasoning process that one uses to justify the morality or ethicality of judgments, actions or institutions. Distributive justice is a normative ethical theory. *Moral principles* are general moral standards derived from moral theories and are characterized as *supreme* (they override other considerations such as self-interest or politics). These are *universal* and apply under all conditions with no exceptions based on any socio-biological factors. *Moral rules* are concrete applications of moral principles to a society, corporation, government, or individuals, given the context of technology, markets, culture and other socio-economic factors. *Considered moral judgments* are specific moral standards or norms derived from sets A, B and C, but are less general than moral principles or moral rules. This ethical framework underlies Table 1.

The classic theory of distributive justice, traditionally attributed to Aristotle, is based on the *minimum principle of distributive justice* that states that equals must be treated equally and un-equals must be treated unequally. This principle is called "formal" justice (Feinberg 1973) since it states no criteria for judging what can constitute equality or inequality in a given society, nor does it furnish criteria by which people can be classified as equals versus un-equals. It merely asserts that, regardless of what aspects or criteria we consider for determining equal classes, no person should be treated unequally among equals.

All wrongs may be corrected using the distributive justice rule. Compensatory justice corrects "involuntary" wrongs such as those that result from accidents or harmful products and compensation should at least restore wronged persons to their original position before the harm occurred. Geographical position, for instance, imposes several "involuntary" wrongs on developing countries and their businesses. Compensatory principles of justice apply to them. Retributive justice corrects "voluntary" wrongs such as those resulting from assaults, thefts, and profiteering from defective products. Besides compensating the victims, retributive justice prescribes adequate punishment for the evil-doer. Often developed countries may victimize or exploit developing nations in various ways (e.g., sweatshops, ecological waste, ownership and use of local scarce resources). Retributive justice corrects the voluntary and involuntary wrongs of a social system whereby citizens or equals among developing nations are treated unequally or unjustly (Boatright 2003). The moral right to be treated as a free and equal person is the foundation of distributive justice (Vlastos 1962).

Rawls' Theory of Distributive Justice

A just society is not necessarily one in which all are equal but one in which inequalities are justifiable. In this context, Rawls (1971) proposed two principles of distributive justice to defend equality and inequality. *The Equality Principle* (Libertarian Fair Opportunism) states that each person engaged in an institution or affected by it has an equal right to the most extensive liberty compatible with a like liberty for all. Equality is the impartial and equitable administration and application of rules that defines a practice. *The Difference Principle* (Libertarian Egalitarianism) states that inequalities as defined by the institutional structure or fostered by it are arbitrary unless they work to everyone's advantage and provided that the positions and offices are open to all. The first principle requires basic equal liberty for all. The second principle admits existing inequalities and differences, if (a) they work to the advantage of all, and (b) if the social system offers equal opportunity for all to combat or compensate for these differences. Equality of opportunity does not entail equality of expectations since unequal expectations are inevitable in a social structure. Inequalities are just, only if the social structure allowing or generating them works to the advantage of all engaged in it, especially the least advantaged.

We can apply both Rawlsian principles to GCC interventions. By the first Equality principle, it would follow that all developing nations impacted by the developed nations have an equal right to the most extensive liberty compatible with a like liberty for all. Developed nations may unwittingly impose restrictions on their liberty by imposing more costs than benefits (e.g., economic embargoes, export quotas and duties). Moreover, developed nations may not offer an equal right to the most extensive benefits available to all developing nations, especially the largest low-income countries of the world. Secondly, we hope that GCC interventions will nullify the disadvantages among the developing nations stemming from their accidents of biology, geography and history. Rawls' (1971) central thesis is that a social arrangement should be a communal effort to advance the good of all who are part of the society. Inequalities of birth, gender, age, ethnicity, color, geography, natural endowment and other discriminating circumstances are "undeserved" as they cause naturally disadvantaged members of the society (Karpatkin 1999). The GCC interventions could progressively help in the eradication of such inequalities.

The Principle of Non-Malfeasance and Beneficence

The principle of non-malfeasance as applied to any act can imply four elements (Frankena 1973, p. 47): the act should *not inflict evil* or harm (strict liability), it should *prevent evil* or harm (preventive justice), it should *remove evil* or harm (protective justice) and it should *promote good* (beneficent justice). The fourth element may not amount to a moral obligation, and constitutes the *principle of beneficence*. The principle of non-malfeasance is primarily incorporated in the first element (strict liability). The remaining three elements are more principles of beneficence than of non-malfeasance. Preventing harm and removing harm are alternate forms of promoting good (Boatright 2003; Frankena 1973). The first three rules of the principle of non-malfeasance directly apply to GCC interventions. That is, GCC interventions are ethical in terms of distributive justice as long as they: (a) do not harm (strict liability), (b) prevent harm (preventive justice) and (c) remove harm from any of the stakeholders (protective justice). Duties of non-malfeasance include not inflicting actual harm and not imposing "risks of harm" (Frankena 1973). Besides directly inflicting harm on developing nations, developed countries may also fault by imposing "risks of harm."

Pre-emptive and protective justices are really subsets of *procedural justice*. Procedural justice demands that structures and procedures be set in society which are just and which produce just outcomes. Structures and procedures are relative to each group, society, state or country. Hence, procedural justice is another instance of distributive justice (Boatright 2003). A GCC strategy or an intervention is *ethical* from a procedural justice viewpoint if at a minimum it establishes just procedures to prevent all harm to the developing nations and to treat all of them fairly. If just procedures prevent or remove all harm, then such procedures ensure both pre-emptive and protective justice. GCC could help all the nations of the world to establish rightful procedures to bring about global development and peace. This is one way of achieving "good" economic results and promoting good (beneficent justice).

DISCUSSION AND IMPLICATIONS

The current debate on the federal role in economic development is not about the fundamental legitimacy as much as the degree and type of federal involvement. The debate is about the proper mix of public and private sector roles, the appropriate level of government for different activities, and the types of policy that are critical to economic growth and prosperity (McGahey 1997).

This discussion relates to the "how" of government roles in diffusing development technologies to the developing nations.

According to the National Academy of Public Administration (NAPA), federal governments engage in three types of activities that influence economic and community development: (1) broad monetary and fiscal policies for macroeconomic demand management; (2) programs that influence the economy such as highway and infrastructure spending; and (3) programs to improve the performance of specific industries, regions and communities (NAPA 1996). The same three areas of federal activities could now assume a global dimension, especially supporting the developing nations in their efforts of economic and community development. The GCC should have both efficiency and equity goals for development. The efficiency goal is to promote the productivity and wealth of the developing nations via innovative technology diffusion. The equity goal is to evenly distribute both productivity and wealth so that all lagging regions and communities benefit (NAPA 1996). The actual choice by the GCC for helping developing nations could be based on the size of the population as indicated previously. It could also consider the intensity of structural or cyclical economic distress, a goal of equity, and the likelihood that technology sharing would alleviate their economic distress in the short run, as well as help them achieve economic self-sufficiency in the long run, a goal of efficiency (Hart and Milstein 1999).

NAPA (1996) stressed four Ls, *learning, leveraging, linking and leadership*, to help with the economic and community development of nations. We propose an expanded version of these four principles. The *learning* principle is based on the role of the GCC in solving the development problems of the developing nations by providing relevant data, knowledge, high-quality statistics, national and international standards of accountability and program performance, information and evaluation on what has worked thus far in other developing nations, and requiring program reviews from the grantees. The *leveraging* principle endorses the use of GCC funds as an organizing goal for local economic development efforts. The *linking* goal is a recommendation for program flexibility, coordination, and consolidation across federal, state, corporate and local resources of the developing nations. The *leadership* principle is the development of local leaders who act as "spark plugs" and who represent and span a cross section of the local, regional and national communities of the developing nations.

How GCC Can Benefit from Its Developmental Interventions

If we propose to ethically mandate GCC to diffuse development technologies to the developing nations, what will GCC gain from such transactions?

While the question and its answer are complex, we respond by invoking the theory of generalized exchange as applied to public systems that involve public policies and not-for-profit organizations (Marshall 1998). A generalized system of exchange has at least three actors, say A, B and C, such that benefits are transferred from A to B, B then transfers benefits to C, who then provides benefits to A. This system of exchange is built on trust and is extended over time with self-interest being a motivator for the exchange. Applied to our context, actor A is the GCC: the consortium of governments and multinational corporations of the developed world. Actor A transfers benefits B* (e.g., development technologies for economic and community development) to B, the developing nations of the world. B (as governments and corporations of the developing world) in turn transfers benefits C* (derived from A and B*) to C, its local and regional communities, in the form of reduced income inequalities, self-sufficiency, improved labor skills, and/or improved income levels and buying power. All of these benefits (B* and C*), in turn, spell further benefits A* to A in terms of increased demand and loyalty for the products and services of A, better labor skills available for A, and in addition, other universally common benefits such as global sustainability, global health and prosperity, and global goodwill and peace. This, in essence, is the basic argument we made in the beginning of the paper when discussing why the GCC should reduce global inequalities. The benefits A* may be "deferred benefits" (Houston and Gassenheimer 1987), but nevertheless, real and desired benefits.

Good Precedents for GCC in Its Interventions

Quasi-governmental organizations and institutions such as the World Bank, OECD, UNIDO, and UNCTAD are currently participating in global sustainability. However, we cite two leading non-governmental models, the CERES and the CAUX Round Table, as good practices for GCC to emulate. CERES began its work in 1988 as a *Coalition for Environmentally Responsible Economies* (Ceres is the Roman goddess of fertility and agriculture). In the fall of 1989, CERES announced the creation of the Valdez Principles, a ten-point code of environmental responsibility principles. Later, these were renamed the CERES Principles. Despite much corporate resistance to these principles, Sunoco was the first Fortune 500 Company to endorse them in 1993. Other companies followed such as GMC, Ford, American Airlines, Bethlehem Steel and Polaroid. Today, over 50 companies endorse the CERES, including 13 Fortune 500 firms (Visit *http://www.ceres.org*). The CERES principles advocate a ten-point global sustainability plan: protection of the biosphere, sustainable use of natural resources, reduction and disposal of wastes, energy conversation, risk reduction, safe products and services, environmental restoration, informing the public, management commitment, and audits and reports. By adopting these principles, member firms affirm their belief that the corporations have a responsibility for the environment, should conduct all aspects of business as responsible stewards of the environment in a manner that protects the earth. Moreover, they should not compromise the ability of future generations to sustain themselves.

The *CAUX Round Table* (CRT) is an international group of 28 senior business leaders formed in 1986. Participants from Japan, Western Europe and the United States meet at regular intervals in Caux, Switzerland for discussions aimed at reducing international trade tensions and fostering development of an improved international business climate. The CAUX principles for business (published in 1994) are the first worldwide standards for ethical and responsible business practice. Translated into eleven global languages they are currently distributed to over 150,000 business leaders throughout the world (see *http://www.cauxroundtable.org*). The CAUX principles represent the first international code for business. The goal of the *CAUX Round Table* principles is to "set a world standard against which business behavior can be measured," a yardstick for individual companies to formulate their code of ethics. The base principles are rooted in two basic ethical ideals: *kyosei* (Japanese for living and working together for the common good, enabling cooperation and mutual prosperity with healthy and fair competition) and human dignity (sacredness or value of each person as an end, not simply as a means for others). The seven CAUX principles involve: (1) the responsibilities of business: Beyond shareholders to stakeholders; (2) the economic and social impact of business: Toward innovation, justice and world community; (3) business behavior: beyond the letter of law toward a spirit of trust; (4) respect for the rules; (5) support for multilateral trade; (6) respect for the environment, and (7) avoidance of illicit operations. The theory of Stewardship Responsibility (Donaldson and Davis 1991; Davis, Schoorman and Donaldson 1997) places a higher utility on collective behaviors that help the organization than on opportunistic self-serving behaviors. CERES and CAUX principles are based on the theory of global stewardship.

CONCLUDING REMARKS

Although CERES, CAUX and other intergovernmental organizations have made great progress during the last twenty years, the development has not been even (Acemoglu 2003; Andersen 2002; Kaplinsky 2000; Magretta 1997). Our proposed GCC is unique. It will be a great economic and political power that can witness equally great developmental contributions. The production and equity goals for GCC exchanges are that global income inequalities resulting in global poverty, hunger, malnutrition and disease are immediately alleviated. In the long run, the goals of self-sufficiency, global sustainability, global prosperity and peace could become realities.

Developmental interventions should benefit all developed and developing nations alike. While they are feasible, viable, and achievable, they call for much good will, trust and global solidarity. Both the developed and the developing worlds, together, are in the struggle for global sustainability. In this sense, there is no short-run, only the long run, and no isolation, only cooperation. With organizations such as GCC the world will be a better place to live for all of its citizens with global sustainability assured, global inequalities reduced to the barest minimum, and, above all, human dignity restored to its original status (Mieth 1997).

REFERENCES

Acemoglu, Daron. 1999. "Changes in Unemployment and Wage Inequality: An Alternative Theory and Some Evidence," *American Economic Review*, 89: 1259-78.

_____ 2002. "Technical Change, Inequality and the Labor Market," *Journal of Economic Literature*, 40: 7-72.

_____ 2003. "Technology and Inequality," *NBER Research Summary*, Winter, 2003, accessed at *http://www.nber.org/reporter/winter03/technologyandinequality.html*.

Afuah, Allan N. 1998. *Innovation Management*, Oxford: Oxford University Press.

Andel, Tom. 1998. "Are You a Mother of Invention? Get Funded," *Transportation and Distribution*, 39:3 (March), 86-7.

Andersen, Per Pinstrup. 2002. "Food and Agricultural Policy for a Globalizing World: Preparing for the Future," *American Journal of Agricultural Economics*, 84: 5 (December), 201-214.

Aristotle. 1985. *Nicomachean Ethics*, translated by Terrence Irwin, Indianapolis: Hacket Publishing Company.

Barney, Jay B. 2001. *Gaining and Sustaining Competitive Advantage*, Englewood Cliffs, NJ: Prentice-Hall.

Boatright, John R. 2003. *Ethics and the Conduct of Business*, 4th edition, Englewood Cliffs, NJ: Prentice-Hall Inc.

Caselli, F. 1999. "Technical Revolutions," *American Economic Review*, 87: 78-102.

Chandy, Rajesh K. and Gerard J. Tellis. 1998. "Organizing for Radical Product Innovation: The Overlooked Role of Willingness to Cannibalize," *Journal of Marketing Research*, 35: (November), 474-87.

Davis, James H., David F. Schoorman, and Lex Donaldson. 1997. "Toward a Stewardship Theory of Management," The *Academy of Management Review*, 22 (January), 20-48.

Day, George S. and David B. Montgomery. 1999. "Charting New Directions for Marketing," *Journal of Marketing*, 63 (Special Issue), 146-63.

De George, Richard T. 1999. *Business Ethics*, 5th edition, Prentice Hall.
Donaldson, Lex and James H. Davis. 1991. "Stewardship Theory or Agency Theory: CEO Governance and Shareholder Returns," *Australian Journal of Management*, 16: 49-64.
Eagelton, Thomas F. 1991. *Issues in Business and Government*, Englewood Cliffs, NJ: Prentice-Hall Inc.
Epstein, Marc J. and Marie-Josee Roy. 2003. "Making the Business Case for Sustainability: Linking Social and Environmental Actions to Financial Performance," *The Journal of Corporate Citizenship*, (Spring), 79-96.
Feinberg, Joel. 1973. *Social Philosophy*, Englewood Cliffs, NJ: Prentice Hall.
Fisk, George. 1974. *Marketing and the Ecological Crisis*. New York: Harper and Row.
_____ 1982. "Editor's Working Definition of Macromarketing," *Journal of Macromarketing*, 2: (Spring), 3-4.
_____ 1999. "Reflection and Retrospection: Searching for Visions in Marketing," *Journal of Marketing*, 63 (January), 115-21.
Frankena, William. 1973. *Ethics*. 2nd ed., Englewood Cliffs, NJ: Prentice-Hall.
Galor, O. and O. Moav. 2000. "Ability Biased Technological Transition, Wage Inequality and Economic Growth," *Quarterly Journal of Economics*, 115: 469-98.
Hart, Stuart. 1997. "Beyond Greening: Strategies for a Sustainable World," *Harvard Business Review*, 75:1 (January-February), pp. 66-76.
_____ and Mark B. Milstein. 1999. "Global Sustainability and the Creative Destruction of Industries," *Sloan Management Review*, 41:1 (Fall), 23-39.
Hill, Edward W. 1998. "Principles for Rethinking the Federal Government's Role in Economic Development," *Economic Development Quarterly*, 11:4 (November), 299-312.
Houston, Franklin S. and Jule B. Gassenheimer. 1987. "Marketing and Exchange," *Journal of Marketing*, 51: (October), 3-18.
Hunt, Shelby D. 2002. *Foundations of Marketing Theory*, Armonk, New York: M. E. Sharpe.
Jeannet, Jean-Pierre and David H. Hennessey. 2000. *Global Marketing Strategies*, fifth edition, Boston: Houghton Mifflin Company.
Kaplinsky, Raphael. 2000. "Globalization and Unequalization: What Can Be Learned from Value-Chain Analysis?" *Journal of Developmental Studies*, 37:12, (December), 117-30.
Karpatkin, Rhoda H. 1999. "Toward a Fair and Just Marketplace for All Consumers: The Responsibilities of Marketing Professionals," *Journal of Public Policy and Marketing*, 18:1 (Spring), 118-122.
Khan, A. R. 1999. "Poverty in China in the Period of Globalization: New Evidence on Trend and Patterns," *Issues in Development Discussion Paper 22*, Geneva: International Labor Office.
Kopfensteiner, Thomas R. 1993. "Globalization and the Autonomy of Moral Reasoning: An Essay in Fundamental Moral Theology," *Theological Studies*, 54:3, 485-511.
Kotler, Philip. 2003. *Marketing Management, 11th edition*, Upper Saddle River, NJ: Prentice Hall.

Kumar, Nagesh and N. S. Siddharthan. 1994. "Technology, Firm Size and Export Behavior in Developing Countries: The Case of Indian Enterprises," *Journal of Developmental Studies*, 31:2 (December), 289-310.

Kumcu, Erdogan and Hohn W. Vann. 1991. "Public Empowerment in Managing Local Economic Development: Achieving a Desired Quality of Life Profile," *Journal of Business Research*, 23 (1), 51-65.

Laczniak, Gene R. 1999. "Distributive Justice, Catholic Social Teaching, and the Moral Responsibility of Marketers," *Journal of Public Policy and Marketing*, 18:1 (Spring), 125-29.

Li, Haiyang and Kwaku Atuahene-Gima. 2001. "Product Innovation Strategy and the Performance of New Technology Ventures in China," *The Academy of Management Journal*, 44:6 (December), 1123-1135.

Magretta, Joan. 1997. "Growth Through Global Sustainability," *Harvard Business Review*, 75:1 (January-February), 78-89.

Marshall, Kimball P. 1998. "Generalized Exchange and Public Policy: An Illustration of Support for Public Schools," *Journal of Public Policy and Marketing*, 17:2 (Fall), 274-86.

Mascarenhas, Oswald A. J. 1980. *Towards Measuring the Technological Impact of Multinational Corporations in Less Developed Countries*, (Ph.D. Thesis, Wharton School, University of Pennsylvania, 1976), New York: Arno Publications.

_____ 1990. "An Empirical Methodology for the Ethical Assessment of Marketing Phenomena such as Casino Gambling," *Journal of the Academy of Marketing Science*, 18 (Summer), 209- 220.

_____ 1991. "Spousal Ethical Justifications of Casino Gambling," *Journal of Consumer Affairs*, 25:1 (Summer), 122-143.

McGahey, R. 1997. "Economic Development, Devolution, and Public Policy: Reforming Federal and State Policy in an Era of Global Competition," in *Rethinking National Economic Development Policy*, Harrison, B. and M. Weiss (eds.), Springfield, VA: U. S. Department of Commerce, National Technical Information Service, pp. 103-138.

Mieth, Dietmar. 1997. "The Market and the Inviolability of Human Dignity in the Perspective of Biomedical Technology," in Dietmar Mieth and Marciano Vidal, eds., *Outside the Market no Salvation? Concilium*: SCM Press London, 119-24.

National Academy of Public Administration (NAPA). 1996. *A Path to Smarter Economic Development: Reassessing the Federal Role*. Washington, DC.

Nowak, Jeremy. 1997. "Neighborhood Initiative and the Regional Economy," *Economic Development Quarterly*, 11:1, 3-10.

Ohmae, Kenichi. 1991. *The Borderless World: Power and Strategy in the Interlinked Economy*, Harper Perennial.

Porter, Michael E. 1990. *Competitive Advantage of Nations*, New York: Free Press.

Prahalad, C. K. and Gary Hamel. 1990. "The Core Competence of the Corporation," *Harvard Business Review*, 68:3, 71-91.

Priem, Richard L. and John E. Butler. 2001. "Is the Resource-Based 'View' a Useful Perspective for Strategic Management Research?" *The Academy of Management Review*, 26:1 (January), 22-41.

Rawls, John. 1971. *A Theory of Justice*, Cambridge, MA: Harvard University Press.

Robin, Donald P. and Eric R. Reidenbach. 1993. "Searching for a Place to Stand: Toward a Workable Ethical Philosophy for Marketing," *Journal of Public Policy and Marketing*, 12 (Spring), 97-105.

Srivastava, Rajendra, Tasadduq A. Shervani, and Liam Fahey. 1998. "Market-Based Assets and Shareholder Value: A Framework for Analysis," *Journal of Marketing*, 62 (January), 2-18.

United Nations Development Program (UNDP). 2001. *Human Development Report 2000.*

United Nations Development Program (UNDP). 2002. *Human Development Report 2001.*

U. S. Government Report. 2002. *For the World Summit on Sustainable Development.* Embassy of the USA Public Affairs, Pretoria; accessed at *Africa News Service*, August 23, 2002.

Vann, John W. and Erdogan Kumcu. 1995. "Achieving Efficiency and Distributive Justice in Marketing Programs for Economic Development," *Journal of Macromarketing*, 15 (Fall), 5-22.

Velasquez, Manuel G. 2002. *Business Ethics: Concepts and Cases*, 5th edition, Englewood Cliffs, NJ: Prentice Hall.

Vlastos, Gregory. 1962. "Justice and Equality," in *Social Justice*, Richard Brandt, ed., Englewood Cliffs, NJ: Prentice Hall, 31-72.

Weidenbaum, Murray L. 2004. *Business and Government in the Global Marketplace*, 7th edition, Upper Saddle River, NJ: Pearson Prentice Hall.

World Development Report. 2000/2001. Attacking Poverty, Published for the World Bank by Oxford University Press.

Reflections on Ethical Concerns in Technology Transfer and Macromarketing

Nicholas Capaldi

SUMMARY. The following theses are defended. Technology transfer is (a) a good thing; (b) reflects the universalizing and globalizing impulse of western technologically advanced culture; (c) is based on the fundamental truth of human freedom; (d) requires not only the transfer of technology but of institutions and norms such as the technological project itself, free market economies, limited government, the rule of law, individual rights, religious toleration and personal autonomy; (e) may require the radical transformation of cultures receiving the technology; (f) is impeded largely by the difficulties of the receiving culture to adjust to the full spectrum of change; (g) challenges those in macromarketing to understand the processes of cultural transformation; (h) and, finally, faces a grave threat from those who are in an adversarial relationship to the "grand narrative of western technologically advanced culture." *[Article copies available for a fee from The Haworth Document Delivery Service: 1-800-HAWORTH. E-mail address: <docdelivery@haworthpress.com> Website: <http://www.HaworthPress.com> © 2005 by The Haworth Press, Inc. All rights reserved.]*

Nicholas Capaldi, PhD, is Legendre-Soulé Distinguished Chair in Business Ethics, College of Business Administration, Loyola University New Orleans, 6363 St. Charles Avenue, Campus Box 15, New Orleans, LA 70118 (E-mail: capaldi@loyno.edu).

[Haworth co-indexing entry note]: "Reflections on Ethical Concerns in Technology Transfer and Macromarketing." Capaldi, Nicholas. Co-published simultaneously in *Journal of Nonprofit & Public Sector Marketing* (Best Business Books, an imprint of The Haworth Press, Inc.) Vol. 13, No. 1/2, 2005, pp. 293-309; and: *Government Policy and Program Impacts on Technology Development, Transfer and Commercialization: International Perspectives* (ed: Kimball P. Marshall, William S. Piper, and Walter W. Wymer, Jr.) Best Business Books, an imprint of The Haworth Press, Inc., 2005, pp. 293-309. Single or multiple copies of this article are available for a fee from The Haworth Document Delivery Service [1-800-HAWORTH, 9:00 a.m. - 5:00 p.m. (EST). E-mail address: getinfo@haworthpressinc.com].

KEYWORDS. Distributive justice, dominant social paradigm, technology transfer

INTRODUCTION

Scholars working in the area of macro-marketing are focused on a wide variety of normative issues. One such issue is technology transfer from highly developed countries to developing countries. One paper (Paper A), "Governmental and Corporate Roles in Diffusing Development Technologies: Ethical Macromarketing Imperatives," presents an ethical argument for why such transfers should take place. A second paper (Paper B), "Incorporating the Dominant Social Paradigm into Government Technology Transfer Programs," discusses some of the cultural conflicts and ethical difficulties in carrying out such transfers. Together these papers raise many of the fundamental ethical issues for macro-marketing, and, may be useful counter-points to one other.

DISTRIBUTIVE JUSTICE

Paper A argues in favor of a public policy in which the governments of highly developed countries and multi-national corporations should actively and jointly participate in and encourage technology transfer. They should do so in order to solve the problems of (a) growing inequality and (b) global sustainability. The justification for helping developing countries is by appeal to the normative concept of distributive justice, as articulated primarily by John Rawls.

This paper implicitly raises a number of important issues:

1. Why is economic inequality bad, and, more specifically, why is an increase in "inequality" bad?
2. What credence can be given to "global sustainability?"
3. If we agree that there is a problem in either (a) or (b), what is the correct diagnosis of the problem?
4. If we agree on the diagnosis, can we agree on what constitutes the best solution?
5. Would the solution suggested in Paper A, namely government intervention, even if it were effective, interfere with other legitimate public policies?

Paper A does not explain why inequality is bad. Reading between the lines, it is easy to infer that Paper A presumes that a certain minimal quality of life is the basic ethical standard. It is important to stress the word "minimal" otherwise we would run into the paradox of advocating equality itself as the basic ethical standard. Insistence on a radical equality could lead to a system wide decrease in living standards. This is the paradox which so many advocates of equality fail to address (Capaldi, 2002). Furthermore, it is easy to infer that a growing inequality is presumably bad because it threatens the minimum standard. Moreover, while "global sustainability" is a highly ambiguous and controversial notion, in the context of Paper A it seems to mean economic policies that both maintain the minimum standard and allow for the continued growth of the world economy. Although Paper A recognizes that a good life is more than meeting minimum economic standards it does not elaborate either on the content of the good life or how economic factors are related to the non-economic factors.

When it comes to diagnosing the source of the problem of inequality (understood now as meeting and maintaining the minimum good life for all) Paper A acknowledges that there is controversy. Clearly if there is controversy over the diagnosis there will be controversy over proposed solutions. The only consensus feature in the diagnosis to which Paper A alludes is the view of economists that "developing nations primarily concentrate on the commodity (agricultural and extractive industry) sectors; and (b) world prices or terms of trade of the agricultural and extract industry commodities have been declining or barely increasing for the last 30 to 40 years."[1] The difficulty with this consensus element is that it is more a restatement of the problem than an explanation of the problem.

Paper A, finally, proposes to solve the problem by appeal to the Rawlsian notion of distributive justice.[2] According to Rawls, inequalities are justified only to the extent that such inequalities are "to the greatest benefit of the least advantaged."[3] Moreover, Rawls advocates government policies of redistributing wealth and resources in order to achieve the requisite degree of equality.

Rawls' position, despite its academic preeminence, is based on dubious philosophical premises and has been attacked by both those on the political left as well as those on the political right. Part of Rawls' appeal is that he tries to hold together currently fashionable academic moral and political prejudices, so that while we would like his theory to be somehow correct it just is not. Rawls' notion of distributive justice does not solve the normative problem; rather, criticism of it reveals pervasive fundamental disagreement about moral issues and the human condition. Finally, there is the long-standing criticism made by classical liberals (i.e., libertarians) that trying to solve social and economic problems through government redistribution eventually undermines the free market on which the success of the highly developed countries depends. Paper A fails to grasp that the real world problem is not to fashion a view with which one's academic colleagues feel satisfied on theoretical grounds but actually to transform the world for the better.

DOMINANT SOCIAL PARADIGMS[4]

We turn our attention now to Paper B. What makes Paper B so useful is that it is nuanced in precisely those areas where Paper A is not. This is true in two important respects. First, rather than appeal to dubious and controversial normative premises to ground its public policy recommendations, Paper B offers a sophisticated diagnosis of the differences between highly developed economies and developing economies, by appeal to "dominant social paradigm" theory. Second, Paper B "argues that the limited success of cross-national governmental technology transfer programs can be attributed to paradigm conflicts between the developer of technology and the recipient" country. [5]

According to Paper B, the "dominant social paradigm" in Western technological societies has three interrelated features: political, economic, and technological. The political dimension is expressed by Locke's notion of individual rights (or, as Paper B prefers, "possessive individualism"–to use the words of the Canadian Marxist MacPherson) (MacPherson, 1962). The economic dimension is succinctly described by reference to Adam Smith "and the Scottish Enlightenment, and it follows directly from the political dimension. The relevant propositions here are that preferences of free, self-interested, atomistic individuals are best satisfied through the operation of free markets. If market forces are left unfettered, the interests of society will be best served through the functioning of the invisible hand. Here the basic principles are again possessive individuals who seek their own interests and limited government that does not intervene in the operation of markets."[6] The third dimension is the *technological project*[7] as originally articulated in the seventeenth century by Bacon (1980). Paper B also makes an important observation, namely, that "the prevailing attitude of Western industrial societies is technological optimism," and this optimism justifies "the desire of Western societies to initiate programs, both public and private, for the development of more advanced technologies and for their dissemination on both domestic and global levels."[8]

Paper B recognizes that different cultures have different paradigms (world views). The term "paradigm" is borrowed from Kuhn (1962).[9] Moreover, the paradigm structures the way in which members in these cultures interpret their experience. There is no neutral position from which to challenge these paradigms. If this is the case, then some developing countries with different cultural paradigms will experience difficulty in absorbing technology transfers.

In order to improve the absorption of the technologies, those countries within the dominant social paradigm must either convince developing countries that they have a deficient paradigm or, perhaps better, adjust the technology transfer to the present conditions of the receiving countries paradigm.

Paper B raises a number of important issues:

1. What is the ultimate rationale for technology transfer?
2. Is there a dominant social paradigm, and has it been correctly characterized?
3. Is there a "clash" (reminiscent of Samuel Huntington's thesis) (1993)[10] of cultural paradigms?
4. How should we respond to this potential conflict?

With regard to the rationale for technology transfer, Paper B seems clearly superior to Paper A. Instead of moralizing technology transfer, Paper B recognizes, quite rightly in my opinion, that technological optimism within the framework of the dominant paradigm leads naturally to the desire to implement such transfers. This will become clearer once we look more deeply into the dominant paradigm.

"Western" Paradigm

Paper B is correct to point out that there are political, economic, and technological dimensions to the dominant paradigm, and it is correct to see that these are interrelated. However, it fails to tell us just how. Allow me to offer a fuller version of this paradigm with the purpose of bolstering Paper B's case: The dominant feature is technology, or more exactly, the technological project. Instead of conforming to a given and independent "natural" world order, we seek to transform nature in the service of humanity. Technological advance (Descartes)[11] requires both constant innovation (which, by definition cannot be planned) and inner-directed individuals (Weber, 1930);[12] free-markets are the most effective way of providing for innovation through competition (David Hume, 1987 and Adam Smith, 1981);[13] limited government (Locke, Madison)[14] in the service of markets and the rule of law are necessary for continued advances (Hayek);[15] limited government can only be maintained by a larger culture which sustains and promotes autonomous individuals (Kant, Hegel, and Mill[16] as opposed to MacPherson's possessive individualism); autonomous individuals by their very nature seek to promote autonomy in others (Hegel).[17] Personal autonomy is the driving modern norm behind the technological project. Globalization is nothing but the latest expression of this norm.

Technology transfer is the natural expression of this globalization. All of the foregoing needs some expansion.

Technological Project

In his *Advancement of Learning*, the first great manifesto of the scientific movement, Sir Francis Bacon remarks, "We are much beholden to Machiavelli and others, that write what men do, and not what they ought to do." As this statement suggests, there is a link between Machiavelli's enterprise and that undertaken by Bacon, René Descartes, and their successors, for both endeavors presuppose a debunking of the aspirations underlying classical and Christian idealism. Both, in fact, take as their end the increase of human power; both concern themselves with human welfare in this world as opposed to the next: if they differ, they do so in that the proponents of scientific revolution redirect the ferocity animating the Machiavellian prince from the conquest of men to the conquest of nature.

Market Economy (i.e., Capitalism)

By capitalism we understand an economic system in which privately owned businesses interact within a free market. As an economic system it has transformed entire nations and continues to shape the global economy. The original defense of capitalism centered on the claim that the emphasis on economic interests would contribute to social order by counterbalancing the political passion for glory (Hirschman). It has also been argued that capitalism is much more than an economic system; it is also a kind of culture. This culture, it has been alleged (Weber) was largely a reflection of the Protestant search for individual salvation. Others (Hayek) have pointed to the rule of law as a prerequisite for a free market economy. All of the foregoing theses raise questions about the cultural precedents of capitalism and cause us to want to understand the extent to which a free market economy reflects and is representative of fundamental features of Western civilization.

It will be useful to begin with Smith's own analysis of what he calls the "system of natural liberty and perfect justice" as expounded in the early pages of the *Wealth of Nations*. Smith identified the key mechanisms as the increased productivity resulting from the division of labor, the accumulation of investment capital which result from thrift and political stability, and the expansion of markets. Smith raised questions not only about the cultural preconditions but also about the social effects of this system.

The classical political economists as early as Hume but including Smith and others have maintained the positive thesis that as an economic system, capitalism is efficient, sustains both individual freedom and political freedom, encourages personal autonomy, and promotes world peace. The promise of a free market economy is that it will be the vehicle for extricating all the people of the world from poverty. The notion of individual moral autonomy as expressed in Kant's categorical imperative entails free market economies and limited government as found in commercial republics. Kant went so far as to claim the history of the modern world can be understood as culminating in an alliance of commercial republics. In short, a free market economy seems to have important ethical implications, the most important of which is to teach ethical behavior.

Again starting as early as Smith's own writings but expounded largely by socialists and Marxists, the positive thesis has been challenged by a negative thesis which denies all of the above and faults capitalism for almost all contemporary ills.

Although all such statements are risky, it would appear that the contemporary consensus is that the positive thesis is in the ascendancy. See especially the indexes on economic freedom for some quantificational support of this thesis. Renewed interest by almost everyone in Hayek's *Road to Serfdom* and his argument about the impossibility of large scale economic planning seems to confirm this. What is perhaps more interesting is the literature that claims that despite its apparent and undeniable success capitalism might contain within itself the seeds of its own destruction (Schumpeter, certain aspects of Public Choice, Daniel Bell, etc.). Given the extent to which the global economy is now a reality, these debates have taken on a new urgency for developing economies concerned about the preconditions and the consequences of a free market economy.

Clash of Cultures

Is there a clash of cultural paradigms? There certainly is, and it would be difficult to understand the anguish within Islamic cultures, for example, without recognizing such conflicts (Scruton, 2002). Recognizing this clash of cultural paradigms, explains both why some cultures resist the whole idea of a technological project and why some of them absorb only parts of it. This is a far superior explanation or diagnosis of economic and social and political inequality than any idea about exploitation or western imperialism. Moreover, if there are radically different cultural paradigms, then it is difficult to see how technology transfer can be implemented and improved without potentially delegitimating or radically altering some of these cultures. Paper B is going to have to recognize the full implications of its own thesis. I think we have to be intellectually honest about this prospect.

Three qualifications need to be added. First, the model I have drawn need not encompass the whole of what we normally identify as a culture. What we have singled out and interrelated are technology, markets, government, law, and certain pervasive features or dimensions of culture, such as the notion of an autonomous individual. It is conceivable that there are other aspects of culture that do not come into play. Surely some empirical work needs to be done in this area. This is something that might be a further dimension of macromarketing.

Second, this account permits *diversity within unity*. Short of rejecting the technological project, which is highly unlikely, I would further maintain that all cultures will soon find themselves committed to some version of personal autonomy. *This commitment to personal autonomy, however, is not a commitment to a specific set of guidelines*. The precise content of one's autonomous decisions is not dictated by the principle of autonomy itself. For example, in any culture decisions may be made in a variety of ways including collective ways as long as the individuals involved have voluntarily chosen to be part of that cultural-matrix. The wonderful thing about autonomy is that it is compatible with a great variety of cultural practices.

Third, this is *not* an exercise in *"triumphalism" or a denigration of the cultures of developing nations*.[18] It is an attempt to explicate and capture what I take to be the dominant norms of evolving practices in a world increasingly defined by the technological project. However, those who participate in this discussion are going to have to be more forthcoming about the following. Some would maintain that the dominant social paradigm is the best thing the world has to offer even though it is not perfect and is surely subject to constructive critique. Others, including some western intellectual elites, are in fact in an adversarial relation to that paradigm and are actively working to replace it. The latter, not willing to surrender technological advance, look (perhaps sentimentality or unrealistically) to the cultures of developing countries for alternative models. From their point of view, the worst thing that could happen is the universal and voluntary adoption of what we have identified as the dominant social paradigm in the west. The debate about technology transfer and cultural differences is merely the head of the intellectual iceberg that lies beneath it.

Are There Cultural "Paradigms"?

Paper B, by using the concept of "paradigm," invokes the views of Thomas Kuhn. Kuhn used the notion of "paradigm" to explain the larger theoretical context within which physical scientists evaluated the relationship between specific scientific hypotheses and experimental results. He further argued that there was no common rational ground on the basis of which one could choose among competing scientific paradigms precisely because those paradigms were the ultimate court of rational appeal. Kuhn then went on to speculate, with less success, about those non-rational factors that led to paradigm revision. Social scientists and others have extended Kuhn's notion of paradigm from one area, namely scientific research, to the whole of culture. That is why they can speak of incommensurable cultures. Many writers are happy with this since if they cannot bash the dominant social paradigm they can at least relish the implied relativism. Many writers welcome the notion of a "cultural paradigm" since it seems to imply that individuals are creatures of their culture or that culture is a constitutive part of an individual (communitarianism), and therefore there is something fundamentally flawed about the individualism and personal autonomy of the dominant western social paradigm.

A proper response to this situation would be as follows. The existence of paradigms in physical science in no way entails the existence of rigid cultural paradigms. There is a real social science issue here, namely whether social entities should be construed as having hidden substructures in the way that physical objects are really composed of molecules we do not see with the naked eye. In fact, some would argue that there are no rigid cultural paradigms, and we can see this fact both (a) internally where there is intra-cultural conflict, and (b) externally where cultures are constantly modified by contact with other cultures. Moreover, paradigms within science change over time. Is there an analogous feature in cultures? Any one culture is itself an historical entity that reflects conflict, change, redefinition, absorption, bifurcation, or outright extinction and disappearance. Cultures change. I would contend that cultures are inheritances; that experience never comes pre-packaged but is always open to different imaginative construction (something post-modernism has stressed); that the individual within any culture is free to accept, reject, or reconstruct whatever is inherited; that western cultures were the first to recognize this *universal truth* about human epistemological and imaginative freedom. The existence of "paradigms" does not lead to relativism but to the recognition of (a) the fundamental truth of human freedom and (b) the important sense in which western cultures (i.e., under the dominant social paradigm) are truly capable of being universalized. Some sense of this can be gleaned from Nobel Prize winning novelist V. S. Naipaul in his 1990 lecture "Our Universal Civilization": "The universal civilization has been a long time in the making. It wasn't always universal; it wasn't always as attractive as it is today. The expansion of Europe gave it for at least three centuries a racial tint, which still causes pain. . . . I don't imagine my father's Hindu parents would have been able to understand the idea. So much is contained within it: the idea of the individual, responsibility, choice, the life of the intellect, the idea of vocation and perfectibility and achievement. . . . It cannot be reduced to a fixed system. It cannot generate fanaticism. But it is known to exist, and because of that, other more rigid systems in the end blow away."[19]

The technological project and globalization are expressions of that universalization of personal autonomy. Finally, since cultures do change and adjust it behooves us to try to understand that process. That being the case, then what Paper B should be arguing is a thesis of gradually and responsibly transforming other cultures, in part, through technology transfer. Understanding this process is not something for which social scientists are trained because too often they are looking for permanent structures behind the change when in fact the change ultimately reflects free human decisions. Understanding that process is something that, ideally speaking, artists, philosophers and historians are best equipped to do.

POLICY IMPLICATIONS OF ETHICAL CONCERNS

What policies follow from these observations? First, if the very essence of the technological project requires free markets and limited government then we must be vigilant about the dangers of government-sponsored transfer programs. This does not mean that government has no role; far from it. But its role will be circumscribed so that it does not interfere with the vital link between technological innovation and free markets. The object is not just to get others up to our level but to keep advancing technological innovation. What needs to be sustained is technological innovation. On the macro-level the problem is still one of production not distribution.[20] And, in the case of technology, there is no final solution to the production problem. Second, the ultimate source of the problems of inequality and absorption are cultural.[21] Some serious consideration will have to be given to the problems of cultural transformation. I seriously doubt that politicians, CEOs, and marketing professionals are the ones best equipped to understand and address this problem. Nor, for other reasons already mentioned, are most social scientists up to the job. A whole new discipline or cluster of disciplines needs to address the explanatory and normative issues of cultural transformation. Perhaps macromarketing can be this discipline or part of it. Third, and more important, in order to have a policy with regard to technology transfer one would need a big picture, that is, a larger grand narrative which would at least legitimate such transfers in general. The revised version of the dominant social paradigm that I outlined above is just such a grand narrative.

"Liberalism" as a Grand Narrative

"Liberalism" is a term that has many meanings. We need to distinguish between liberal culture (an interrelated set of historically identifiable institutional practices) and liberal social theory (various attempts to explain, justify, and/or critique liberal culture).

What is liberal culture? By liberal culture, we understand the kind of culture that emerged in Western Europe in the post-Renaissance and post-Reformation period and eventually spread to the United States. The most distinctive institutions of liberal culture are individual rights, personal autonomy, the rule of law, a form of limited government, toleration, and a free market economy.

To begin with, liberal culture is the greatest force in the modern world; it has transformed and continues to transform the moral landscape by improving the material conditions of life and by institutionalizing individual freedom. One would think, therefore, that such a phenomenon deserves special attention.

Liberal culture is not understood even by those of us who are surrounded by it, and that is why we are engaged in an act of retrieval. One explanation for why liberal culture is not understood is, ironically, that it has been defined largely by its critics; so much so that even the defenders of liberal culture have unwittingly adopted the framework of their critics. *At present there exists no positive, internal, comprehensive framework for understanding liberal culture as a whole*.

There is an additional reason for why liberal culture is not understood. The institutions and practices of liberal culture do not exist in a vacuum. Too little attention has been given to understanding the relation between liberalism and the totality of our culture. What is not usually made clear even in very illuminating discussions of specific institutions is that a liberal culture depends upon and presupposes a framework of moral presuppositions. Conflicts within our own culture often reflect ignorance, misunderstanding, or deep disagreement over what the moral presuppositions are. To provide a comprehensive framework that would identify the moral presuppositions of liberal culture would be to fill a great lacuna in the contemporary intellectual environment.

The third reason for embarking upon this explication is that given the attempts on the part of others around the world to emulate liberal culture we are concerned that they might fail by copying the form without the spirit.

There are three important questions we need to ask about liberalism. *What is the source of its authority* (legitimacy)? Can it sustain itself without appeal to some form of transcendence? Is it universalizable or is it a culturally limited and historically transient phenomenon? [22]

With regard to its legitimacy, there have been three historically distinct forms of legitimation: pre-Enlightenment, Enlightenment, and post-Enlightenment. During its pre-Enlightenment period, liberalism was justified by appeal either to transcendence, nature, or reason. During the Enlightenment, liberalism was justified by an appeal to science. The starting point (ontologically, axiologically, and epistemologically) is individualism. From this individualism we *deduce* conclusions about the social world: in its Lockean formulation, individualism reflected a Protestant moral-religious conception of the relation between the individual and God. In its Enlightenment Project form, the atomism reflects the methodological individualism of a wholly scientific naturalism. Secondly, each individual is alleged to have a built-in end or set of such consistent ends: in its original Lockean formulation, these ends (e.g., life, liberty, property, etc.) are designated as *"rights"* (qualified as "natural," "human," etc.); these ends are teleological; rights, so understood, are absolute, do not conflict, and are possessed only by individual human beings; rights are morally absolute or fundamental because they are derived from human nature and God (or later the categorical imperative), and as such cannot be overridden; the role of these rights is to protect the human capacity to choose. Finally, such rights impose only duties of non-interference. The pre-Enlightenment version is the one that animates the American Founding (Thomas West). In its Enlightenment Project form, the ends are not rights; rather, rights are means to the achievement of the ends. As such, rights are only *prima facie*, may be overridden, and may be possessed by any entity, not just individual human beings. Such rights can be welfare rights, i.e., they may be such that others have a positive obligation to provide such goods, benefits or means. What distinguishes one "liberal" social philosopher from another is (a) whether rights are understood to be absolute (libertarians and classical liberals) or *prima facie* (modern liberals such as Rawls), (b) the content of the rights, and (c) the lexical ordering of those rights.

The most significant intellectual development from the point of view of justifying liberalism has been the collapse of the Enlightenment Project. Specifically, the Enlightenment Project is the claim that physical science and its derivative social science are self-legitimating. It is now generally recognized (Rorty, MacIntyre, Gray, Capaldi) that science is not self-legitimating. This has led to what Arendt has called the legitimation crisis at least among secular liberals. The secular response has been to try to justify liberalism simply in terms of conventional assent (Rorty, Habermas, Gray). The other contemporary response is to retrieve the sense of the transcendent for liberalism (Maritain, Novak, and Neuhaus).

This leads to the following questions. Can the issue of universalizability, that is, whether liberal culture is a potentially useful model for the rest of the world, even be posed except within Western culture? That is, are there even resources in non-Western cultures for dealing with this issue? (See Voegelin.) Can liberal culture survive as just one among many if it loses the belief in its own universalizability? Can transcendence be retrieved? Would the belief in transcendence allow for both universalizability and convergence among diverse cultures?

Finally, consider the following argument: *The most significant feature of modern humanity is that it has been and is irreversibly engaged in the technological project, namely, the conquest of the physical environment. Free market economies are the most efficient in carrying out that project. Free market economies can only operate in a political system of limited government (distinguish between a republic and a democracy). Limited government requires some sense of the rule of law (not to be confused with the rule of laws). The rule of law presupposes a conception of personal autonomy (distinguish self-rule from self-definition). Personal autonomy requires some conception of transcendence.*

The over-arching difficulty is that many who live within advanced technological societies committed to the dominant western social paradigm, as revised, neither understand nor accept that grand narrative. For them, discussions of technology transfer become ways of bashing the dominant view or using the global stage to transform the domestic stage. So, for example, being enamored of government economic planning and being unable to generate such planning domestically, the world stage becomes the new locus for the pursuit of the socialist ideals. Discussions of technology transfer are not merely academic or practical but covert discussions of private political agendas. Finally, to the extent that many western intellectuals and academics are themselves ignorant of, hostile to, or alienated from the dominant paradigm, I am not optimistic about the job getting done at all in academe. My best guess is that technology transfer and accompanying cultural change will be done in a costly and painful manner through markets. Finally, I hope this article will be taken as a challenge.

NOTES

1. Paper A, original manuscript, p. 13.
2. Rawls (1971). In this work Rawls presented his theory as a universal truth. However, in 1993 he published *Political Liberalism* in which he recast his earlier view as a reflection only of western liberal democracies. In that same year, he also published "Law of Peoples," in which he specifically denied that the theoretical framework of *A Theory of Justice* could be directly transferred to the international level. For further elaboration of the international implications see Barry (1973) as well as Beitz (1979)

and P. T. Bauer (1981). For an equally prominent view in general opposition to Rawls see Robert Nozick (1974).

3. Rawls (1971), pp. 150 and 302.

4. Paper B refers to the work of Pirages and Ehrlich (1974) who first used the phrase "dominant social paradigm." The expansion to the cultural level can be seen in Cotgrove (1982).

5. See Piper, William S. and Kimball P. Marshall (2000) for a discussion of how the problems of technology transfer are not merely technical problems but go much further.

6. Paper B, p. 6 of original manuscript.

7. The expression "technological project" (TP for short) is my own and is explained below where I give a different account of the dominant social paradigm. The TP means the control of nature for human benefit. This is a new idea introduced in the seventeenth century by Bacon and Descartes. In the ancient and medieval world (and in much of the world today) people thought in terms of conformity to nature, not transforming nature for human benefit. The TP requires inner-directed individuals (autonomy) cooperating to produce innovative ideas (scientific and technical thinking) for understanding and controlling natural processes.

8. Paper B, Page 7 of the original manuscript.

9. A postscript was added in the 1970 edition. For a review of the controversies and implications surrounding Kuhn's views see N. Capaldi (1998), Chapter Two.

10. In a seminal article and controversy in *Foreign Affairs*, Samuel Huntington of Harvard has extended into the foreign policy arena some of the domestic debates in the U.S. that have gone on under the name of multi-culturalism. He points out that what we are witnessing is not simply a contrast between a romantic belief in national roots and a more encompassing cosmopolitanism which espouses universal themes. Given the levels of prosperity in new parts of the world, accompanied by the perception, real or imagined, of Western decadence, and the interest in recovering an historical past, there are now *new transnational sentiments* which are neither nationalistic nor cosmopolitan, but informed by "Civilizational" differences. The idea is that Serbia, Russia and Greece are in sympathy, not for reasons of power politics, but because of a shared Eastern Orthodox Christianity. Similarly, Hindu South Asia, Muslim Middle East and Asia, Catholic and Indian Latin America, Confucian China, and Japan (interestingly standing alone) are the bases for a "Clash of Civilizations."

11. Rene Descartes, *Discourse on Method*, originally published in French in 1637, now a philosophic classic available in many editions. See also Francis Bacon (1995), nos. 13, 16-17, and (1980). See also Hiram Caton (1988).

12. Max Weber originally published in 1904-05.

13. Adam Smith, Bk. I, chapters 1-4; See also Peter L. Berger (1986), the Annual Indexes of Economic Freedom published by the Fraser Institute and Index of Economic Freedom, Interactive Edition of the Heritage Foundation. See also Hirschman (1997), Buchanan (1975), Hayek (1976). For some sense of the religious dimension see Novak (1993). See also Schumpeter (1984) and Fukuyama (1995).

14. See James Madison, Federalist Paper #10 (1961).

15. Hayek (1976), originally published in 1944 as a critique of Keynes, especially Chapter Six, "Planning and the Rule of Law."

16. Mill (1977), especially Chapter Three entitled "Individuality." For a full discussion of the importance of personal autonomy in Mill and how he derived it from Kant

through Humboldt see the biography *John Stuart Mill* by N. Capaldi (Cambridge University Press, 2004).

17. Hegel (1991). See also the discussion of the relationship between master and slave in Hegel (1977).

18. There is actually a U.S. Federal court case in which the issue of *cultural genocide* is invoked but dismissed. See *Beanal v. Freeport-McMoran, Inc.*, 197. F. 3d 161 (5th Cir 1999).

19. Lecture at the Manhattan Institute.

20. The classical economists (Smith, Ricardo, etc.) were focused on production. The publication in 1958 of John Kenneth Galbraith's *The Affluent Society* began a trend among economists of the left (i.e., those who favored massive government intervention in to the economy) to focus on distribution by claiming that the production problem had been solved.

21. Oakeshott (1991), introduces and develops the idea of the anti-individual. Briefly, the anti-individual is the person who has not made the transition to individuality or personal autonomy. Anti-individuals can be found in the west as well as in the remainder of the world. In his (1996) book, Oakeshott further elaborates on the historical development and meaning of autonomy and why to some it appears as burdensome and as running the risk of self-estrangement or self-destruction (pp. 234-39).

22. See Arendt (1958), Capaldi (1998), ch. 10, Sandel (1984), Manent (1998), Neuhaus (1986), Habermas (1973), Maritain (1951), Novak (1993), Oakeshott (1991), Rorty (1990), Voegelin (1987), and West (1997).

REFERENCES

Arendt, H. (1958) "What was Authority?" in *Nomos I: Authority*. Cambridge, MA: Harvard University Press.

Bacon, Sir Francis (1980) *The Great Instauration and New Atlantis*, ed. Jerry Weinberger. Arlington Heights, IL.

_____ (1995) *Essays*. Amherst, NY.

Bauer, Lord P. T. (1981) *Equality, the Third World, and Economic Delusion*. Cambridge: Harvard University Press.

Barry, Brian (1973) *The Liberal Theory of Justice*. Oxford: Oxford University Press.

Beanal v. Freeport-McMoran, Inc., 197. F. 3d 161 (5th Cir 1999).

Beitz, Charles R. (1979) *Political Theory and International Relations*. Princeton: Princeton University Press.

Berger, Peter L. (1986) *The Capitalist Revolution*. New York: Basic Books.

Buchanan, James (1975) *The Limits of Liberty*. Chicago.

Capaldi, N. (1998) *The Enlightenment Project in the Analytic Conversation*. Boston: Kluwer Academic Publishers.

_____ (2002) "The Meaning of Equality," in *Liberty and Equality*, ed. T. Machan. Stanford, CA: Hoover Institution Press. pp. 1-33.

Caton, Hiran (1988) *The Politics of Progress: The Origins and Development of the Commercial Republic, 1600-1835*. Gainesville, FL.

Cotgrove, Stephen (1982), *Catastrophe or Cornucopia: The Environment, Politics, and the Future*. New York: Wiley.

Descartes, René (1989) *Discourse on Method*. Buffalo: Prometheus.

Fraser Institute, (1997-2002) *Annual Indexes of Economic Freedom*. (www.fraserinstitute. ca).

Fukuyama, Francis (1995) *Trust: The Social Virtues and the Creation of Prosperity*. New York: Free Press.

Galbraith, John Kenneth (1998) *The Affluent Society*. Boston: Houghton Mifflin.

Habermas, J. (1973) *Legitimation Crisis*. Boston.

Hayek, F. A. (1976) *The Road to Serfdom*. Chicago: University of Chicago Press.

Hegel, G. W. F. (1977) *Phenomenology of Spirit*. Oxford: Clarendon Press.

_____ (1991) *Elements of the Philosophy of Right*, ed. By Allen W. Wood. Cambridge: Cambridge University Press.

Hume, David (1987) *Essays: Moral, Political and Literary*. Indianapolis: Liberty Classics, 1987. "On Commerce" (pp. 253-267) and "Refinement in the Arts" (pp. 268-280).

Heritage Foundation, *2001 Index of Economic Freedom*, Interactive Edition. (www.heritage.org).

Hirschman, Albert O. (1997) *The Passions and the Interests, Political Arguments for Capitalism Before Its Triumph*. Princeton.

Huntington, S. P. (1993) "If not civilization, what? Paradigms of the post-cold war world," *Foreign Affairs*, 72, pp. 186-194.

Klein, Daniel B. (ed.) (1997) *Reputation: Studies in the Voluntary Elicitation of Good Conduct: Economics, Cognition, and Society*. Lansing, MI: University of Michigan Press.

MacPherson, C. B. (1962), *The Political Theory of Possessive Individualism*. Oxford: The Clarendon Press.

Madison, James (1961) Federalist Paper #10 in C. Rossiter (ed.), Hamilton, A., Jay, J., and Madison, J.: *The Federalist Papers*. New York: New American Library.

Manent, Pierre (1998) *City of Man*. Princeton: Princeton University Press.

Maritain, Jacques (1951) *Man and State*. Chicago: University of Chicago Press.

Mill, John Stuart (1977) *On Liberty*, vol. XVIII of the *Collected Works*, edited by John Robson. Toronto: University of Toronto Press.

Neuhaus, Richard J. (1986) *Naked Public Square*. Grand Rapids: Wm. B. Eerdmans Publishing Co.

Nozick, Robert (1974) *Anarchy, State, and Utopia*. New York: Basic Books.

Novak, Michael (1993) *Spirit of Democratic Capitalism*. Lanham, MD: Rowman and Littlefield Publishers, Inc.

Oakeshott, Michael (1991) "The Masses in Representative Democracy," in *Rationalism in Politics and Other Essays*, ed. T. Fuller. Indianapolis: Liberty Press. pp. 363-383.

_____ (1996) *On Human Conduct*. Oxford: Clarendon Press.

Piper, William S. and Kimball P. Marshall (2000) "Stimulating Government Technology Commercialization: A Marketing Perspective for Technology Transfer," *Journal of Nonprofit & Public Sector Marketing*, 8 (3): 51-63.

Pirages, Dennis C. and Paul R. Ehrlich (1974) *Ark II: Social Response to Environmental Imperatives*. San Francisco: Freeman.

Rawls, John (1971) *A Theory of Justice*. Cambridge, MA: Harvard University Press.

_____ (1993) *Political Liberalism*. New York: Columbia University Press.

_____ (1993) "Law of Peoples," *Critical Inquiry*, 20 (1): 36-68.

Rorty, Richard (1990) "The Priority of Democracy to Philosophy," in Alan Malachowski (ed.), *Reading Rorty*. Basil Blackwell. pp. 279-302.

Sandel, Michael (ed.), (1984) *Liberalism and Its Critics*. New York: New York University Press.

Schumpeter, J. (1984) Capitalism, Socialism, and Democracy. New York: Harper Collins.

Scruton, Roger (2002) *The West and the Rest*. Wilmington: Intercollegiate Studies Institute.

Smith, Adam (1981) *An Inquiry Into the Nature and Causes of the Wealth of Nations*. Indianapolis: The Liberty Fund.

Voegelin, Eric (1987) *New Science of Politics*. Chicago: University of Chicago Press.

Weber, Max (1930) *The Protestant Ethic and the Spirit of Capitalism*. London: Allen and Unwin.

West, Thomas G. (1997) *Vindicating the Founders*. Lanham: Rowman & Littlefield.

Response to Dr. Capaldi

William E. Kilbourne

SUMMARY. This a response to Professor Nicholas Capaldi's reflections on Professor William E. Kilbourne's discussion of dominant social paradigms and government technology transfer programs in this volume. Dr. Kilbourne notes his agreement with Dr. Capaldi on most points, and reiterates his position, in his view consistent with Dr. Capaldi, that difficulties of cross-cultural technology transfers cannot be overcome by "persuasion," and that cultural adaptation must involve a gradual process of successive technology transfers and the development of institutions that further develop technologies with consideration of the needs of users. *[Article copies available for a fee from The Haworth Document Delivery Service: 1-800-HAWORTH. E-mail address: <docdelivery@haworthpress.com> Website: <http://www.HaworthPress. com> © 2005 by The Haworth Press, Inc. All rights reserved.]*

KEYWORDS. Dominant social paradigm, technology transfer, culture

THE RESPONSE

Dr. Capaldi's response to my paper is a set of well-developed suggestions for the improvement of what has been presented. I must say that, with only

William E. Kilbourne, PhD, is Professor of Marketing, Department of Marketing, 245 Sirrine Hall, Clemson University, Box 341325, Clemson, SC 29634-1325 (E-mail: kilbour@clemson.edu). His research interests are in globalization and environmental issues in marketing.

[Haworth co-indexing entry note]: "Response to Dr. Capaldi." Kilbourne, William E. Co-published simultaneously in *Journal of Nonprofit & Public Sector Marketing* (Best Business Books, an imprint of The Haworth Press, Inc.) Vol. 13, No. 1/2, 2005, pp. 311-312; and: *Government Policy and Program Impacts on Technology Development, Transfer and Commercialization: International Perspectives* (ed: Kimball P. Marshall, William S. Piper, and Walter W. Wymer, Jr.) Best Business Books, an imprint of The Haworth Press, Inc., 2005, pp. 311-312. Single or multiple copies of this article are available for a fee from The Haworth Document Delivery Service [1-800-HAWORTH, 9:00 a.m. - 5:00 p.m. (EST). E-mail address: getinfo@haworthpressinc.com].

Digital Object Identifier: 10.1300/J054v13n01_17

very minor differences, I agree with everything he has suggested. The only difficulty is how to incorporate it all into a single journal length paper that has a necessarily limited scope. I have, in fact, addressed all of the areas suggested in other papers on the related issues of globalization and the environment. However, a complete analysis of liberalism with its history and ramifications in contemporary global development is beyond the scope of most any paper. The limitations suggested are not an oversight on my part, nor do they suggest academic dishonesty in not discussing the full implications of the thesis presented. I fully agree that the technological transfer process across cultures is burdened with almost insurmountable difficulties. The purpose of this paper was to open the discussion regarding the technological project and its global ramifications, not to complete it and lay it to rest. Bringing the dominant social paradigm into the marketing literature in this particular context completed my particular agenda, and I hope that it will succeed in that.

It is also suggested that the paper should be arguing ". . . a thesis of gradually and responsibly transforming other cultures . . ." through the technology transfer process. This is what is argued in the section "Merging the DSP and TT." The first approach using persuasion is dismissed because incommensurable paradigms cannot be made commensurable through persuasion. It also makes the assumption that other paradigms are inferior to one's own. The second method suggests gradual change through successive transfers relying on the demonstration effect. The focus should be on technologies that develop the institutions that then lead to further development in a rational process considering the needs of the receiver. While this may sound easy if you say it fast, as Dr. Capaldi suggests, social scientists are ill equipped for this task. With this, I agree whole-heartedly, and among the least equipped for such a multi-disciplinary assessment are traditionally trained micromarketing scholars.

The objective here was to expand the domain of inquiry within the marketing discipline. My only desire is that the marketing audience of the volume be stimulated enough to expand the inquiry in precisely the directions outlined by Dr. Capaldi. This would, indeed, be intellectual progress as no meaningful assessment of technology transfer can be made without a full understanding of its role in the dominant social paradigm of Western industrial society. And for this to occur, a full understanding of the meta-narrative of liberalism and its implications for global development is necessary. Traditional marketing analyses are not up to the task as they fail to examine their own assumptions and position within the liberal paradigm. Failing such an examination increases the likelihood that Capaldi's best guess will be correct. ". . . Cultural change will be done in a costly and painful manner through markets."

Progressive Reduction
of Economic Inequality
as a Macromarketing Task:
A Rejoinder

Oswald A. J. Mascarenhas
Ram Kesavan
Michael D. Bernacchi

SUMMARY. This is a response to Capaldi's reflections on our article in the same volume. Since reduction of global inequality is a necessary condition for global sustainability, our rejoinder focuses on global in-

---

Oswald A. J. Mascarenhas, PhD, is Charles H. Kellstadt Professor of Marketing, University of Detroit Mercy, 4001 McNichols Road, Detroit, MI 48221 (E-mail: mascao@udmercy.edu). His research interests are ethics of marketing, macromarketing and e-business.

Ram Kesavan, PhD, is Professor of Marketing, University of Detroit Mercy, 4001 McNichols Road, Detroit, MI 48221 (E-mail: kesavar@udmercy.edu). He has served as the President of the Marketing Management Association. His research interests include the areas of entrepreneurship and global marketing.

Michael D. Bernacchi, PhD, JD, is Professor of Marketing, University of Detroit Mercy, 4001 McNichols Road, Detroit, MI 48221. He is the author of a marketing/advertising newsletter, *Under The Mikeroscope*.

[Haworth co-indexing entry note]: "Progressive Reduction of Economic Inequality as a Macromarketing Task: A Rejoinder." Mascarenhas, Oswald A. J., Ram Kesavan, and Michael D. Bernacchi. Co-published simultaneously in *Journal of Nonprofit & Public Sector Marketing* (Best Business Books, an imprint of The Haworth Press, Inc.) Vol. 13, No. 1/2, 2005, pp. 313-317; and: *Government Policy and Program Impacts on Technology Development, Transfer and Commercialization: International Perspectives* (ed: Kimball P. Marshall, William S. Piper, and Walter W. Wymer, Jr.) Best Business Books, an imprint of The Haworth Press, Inc., 2005, pp. 313-317. Single or multiple copies of this article are available for a fee from The Haworth Document Delivery Service [1-800-HAWORTH, 9:00 a.m. - 5:00 p.m. (EST). E-mail address: getinfo@haworthpressinc.com].

Available online at http://www.haworthpress.com/web/JNPSM
Digital Object Identifier: 10.1300/J054v13n01_18

313

equality. Accordingly, our response: (a) defines inequality generally and specifically, (b) determines that the macromarketing effects of global inequality are the inadequate consumption by many and an unacceptable quality of human life for many, and hence (c) argues that the progressive reduction of global inequality is a challenge for macromarketing. *[Article copies available for a fee from The Haworth Document Delivery Service: 1-800-HAWORTH. E-mail address: <docdelivery@haworthpress. com> Website: <http://www.HaworthPress.com>* © *2005 by The Haworth Press, Inc. All rights reserved.]*

KEYWORDS. Global inequality, distributive justice, technology transfer

WHAT IS INEQUALITY?

Inequality is a difference between human groups. Some inequality is *natural* and undeserved such as those associated with one's color, gender, age, personality, looks and innate ability. Some inequality is man-made and *sociopolitical* and it is based upon one's race, place of birth, religion, language and culture. Inequality is also *economic* and often, results from one's natural and sociopolitical inequalities and is determined by income, opportunity and resource inequalities (Sen 1992). In its worst form, economic inequality is a "systematic division of the members of a society into separate groups for the benefit of one group at the harm of the other" (Dugger 1998, p. 286). According to Neale (1991, p.1166), it is the natural "consequence of the way our system distributes property, the power to spend, and the power to acquire assets."

ECONOMIC INEQUALITY AND MACROMARKETING CONCERNS

We do not advocate economic egalitarianism such as the one proposed by Frankfurt (1987). Income equality is not a viable solution to problems. No two families or individuals are symmetrical in their basic needs, nor derive the same set of utilities from equal incomes. Furthermore, any reasonable measure of inequality should include an overall social objective function (Atkinson 1970) such that the social losses of equal income distributions are taken into account. Economic equality takes into account the heterogeneity of individuals and their respective non-income circumstances (Sen 1992).

Rawls (1971, pp. 60-65) defined economic equality in the broader terms of *primary goods* such as rights, liberties and opportunities, income and wealth, and the social bases of self-respect. Broadening the Rawlsian coverage of primary goods, Dworkin (1981) argued for the *equality of resources* to include insurance opportunities to guard against the vagaries of brute luck. Sen (1997) further argued that other non-income moderator variables such as personal heterogeneity, infrastructure, environment and social diversities seriously impact choice behavior and the consequent quality of life. For instance, one's conditions of living, level of unemployment, especially persistent unemployment, and freedom to live a relatively unimpeded life (Cohen 1995) are some of the important factors that define and determine one's quality of life in the midst of equal commodity or resource spaces. Hence, in judging economic equality the appropriate space is not of income, not of utilities (as claimed by welfare economists), not that of primary goods, nor of resources, it is rather of "functioning space" (Sen 1997).

INEQUALITY AS A MACROMARKETING CHALLENGE

The above discussion reveals that even with the barest review of the inequality literature, the concept of inequality is complex, dynamic and growing. The "technological project" or the dominant social paradigm of the West invokes inequality as a means to an end. With too much equality, the economy stagnates into a poverty trap. With too much inequality, its development stops prematurely (Matsuyama 2002). This is another version of the insidious theory of the survival of the fittest. Systems of inequality such as racism, sexism, nationalism, color, age and religion are inherently bad unless they work for the greatest advantage of all (Rawls 1971). A strong argument has been made against all inequality by Dugger (1998) and specifically against economic inequality by Acermoglu (2002) and the Noble Laureate Amartya Sen (1997). They all invoke Rawls (1971) as part of their argument. Based on these and other representative writers, we assume that inequality, especially economic inequality (whose subset is income inequality) is bad, that it is pathological, anti-social, and cumulatively dysfunctional. Inequality breeds more inequalities and greater inequality over time as evidenced in our earlier paper.

Capaldi incorrectly believes that income inequality is a form of social investment. This theory implies that eventually, the rich and the powerful become a social system that controls wealth, resources and opportunity, and very soon they begin to believe that any of the privileges they have are richly deserved. This is an enabling self-serving myth.

We are suspicious of this myth because the gaps between peoples and nations that are rich and poor are widening. Inequality has risen over the past 30 years. Sub-Saharan African countries portray a dismal picture. In 1970, only

11 percent of earned incomes were below $1-a-day and 38 percent earned as much as $2-a-day. By the end of the 1990s, those who earned $1-a-day wage dramatically rose to 66 percent and the $2-a-day wage declined to 11 percent. Thanks to India's and China's vast population numbers and their economic growth over the past 30 years, the overall world poverty level has decreased, but nonetheless, worldwide income inequality has worsened. Income inequality is cumulative, and never self-correcting. "Capitalism may be the most effective way of turning assets into wealth, but does not guarantee that wealth will be spread equally; left to itself, wealth does not trickle down" (Handy 2003, p. 83). Wealth accumulates in the hands of a few. Further, it is "a folly to expect that our current systems of inequality will generate much in the way of new productive technology and new capital accumulation" (Dugger 1998, p. 290). Inequality is not a socially advantageous way to accumulate capital nor to stimulate the development of new technology. Economic inequality of opportunity is a diseased state that presently disables over 70 percent of the world's population with malnutrition, poor health, ignorance, destitution and squalor. Thus, inequality reduces consumption and diminishes the quality of life for the poor. This is a serious macromarketing concern for all of us (Fisk 1982, 1999).

THE PROPOSED MACROMARKETING SOLUTION

The Government-Corporations Consortium (GCC) solution that we proposed in our original paper calls for a progressive reduction of economic inequality. The formula is not the market versus the state; it is the market and the state. It is a collective action to diffuse development technologies among the most indigent nations of the world so that poverty and ignorance are immediately alleviated and so that the developing nations are made self-sufficient in this struggle. The GCC solution does not mandate the redistribution of money from the rich to the poor; this may succumb to and become the "leaky bucket analogy" of Okun (1975). Neither do we pretentiously advocate that the technology transfer from the West should precede a "cultural transformation" in the developing nations. Rather, we propose a diffusion of development technology–technologies that produce employment, buying power, increased consumption, and a better quality of life for the more than 70 percent of the world's population who are economically impaired and deprived of the normal "functionings" (Sen 1997). These functionings are the entitlements of all humans. Capitalism, as Adam Smith visualized it, was for the promotion of universal consumption among all and not for the increased consumption among the few. Reduction of economic inequality universally raises demand, consumption and the quality of life for both developed and developing markets. We have not detailed the GCC solution in terms of all conditions of market entry and exit. Neither have we detailed what labor laws, corporate laws, partnership laws, and anti-discrimination laws should apply. Collective action is required to resolve questions involving rights and duties, norms and standards. There must be learning, cooperation, interdependence, and mutual empowerment along both sides of the GCC equation.

REFERENCES

Note: Please refer to the original article in this volume for references not here included.

Atkinson, A. B. (1970), "On the Measurement of Inequality," *Journal of Economic Theory*, 2: 244-63.

Cohen, G. A. (1995), *Self-Ownership, Freedom, and Equality*. Cambridge, UK: Cambridge University Press.

Dugger, William M. (1998), "Against Inequality," *Journal of Economic Issues*, 32:2 (June), 286-104.

Dworkin, R. (1981), "What is Equality? Part I: Equality of Welfare," and "What is Equality? Part 2: Equality of Resources," *Philosophy and Public Affairs*, 10: 185-246; 283-345.

Frankfurt, H. (1987), "Equality as a Moral Ideal," *Ethics*, 98: 21-43.

Handy, Charles (2003), "Helicoptering Up," *Harvard Business Review*, August, 80-93.

Matsuyama, Kitminori (2002), "The Rise of Mass Consumption Societies," *The Journal of Political Economy*, 110:5 (October), 1035-72.

Neale, Walter C. (1991), "Who Saves? The Rich, the Penniless, and Everyone Else," *Journal of Economic Issues*, 25 (December), 1160-66.

Okun, A. M. (1975), *Equality and Efficiency: The Big Tradeoff*, Washington, DC: The Brookings Institute.

Sen, Amartya (1992), *Inequality Reexamined*, Oxford, UK: Oxford University Press.

Sen, Amartya (1997), "From income inequality to economic inequality," *Southern Economic Journal*, October, 64: 2, 384-401.

Index

T - #0475 - 101024 - C0 - 212/152/19 - PB - 9780789026064 - Gloss Lamination